"Engaging, persuasive, and thought provoking. Morton discusses the potential role and consequences of geoengineering and puts forward his own carefully considered views on the subject. *The Planet Remade* is a tour de force of wide-ranging scholarship as well as a soundly argued polemic."

—John Shepherd, University of Southampton

"Morton accessibly describes the potential and risks of geoengineering and puts them in the context of climate change and other large-scale interventions that humans have had on the earth system or might seek to have in the future."

—Tim Kruger, University of Oxford

THE PLANET REMADE

Oliver Morton

The Planet Remade

How Geoengineering
Could Change the World

PRINCETON UNIVERSITY PRESS
Princeton and Oxford

Published in the United States and Canada in 2016 by Princeton University Press, 41 William Street, Princeton, New Jersey 08540
press.princeton.edu

First published in Great Britain by Granta Books, 2015

Extract from 'The Love Song of J. Alfred Prufrock' taken from *Collected Poems* Estate of T.S. Eliot and reprinted by permission of Faber and Faber Ltd.

'Democracy' Words and Music by Leonard Cohen © 1992, Reproduced by permission of Sony/ATV Songs LLC/Sony/ATV Music Publishing UK Ltd, London W1F 9LD

'Handlebars' Words and Music by Andrew Guerrero, Jamie Laurie, Jesse Walker, Kenneth Ortiz, Mackenzie Roberts and Stephen Brackett © 2008, Reproduced by permission of Flobots Music Publishing/Sony/ATV Sounds LLC/Sony/ATV Music Publishing UK Ltd, London W1F 9LD

The right of Oliver Morton to be identified as the author of this work has been asserted by him in accordance with the Copyright, Designs and Patents Act 1988

Library of Congress Control Number 2015946728
ISBN 978-0-691-14825-0

Typeset by Avon DataSet, Bidford on Avon, Warwickshire

Printed on recycled paper

Printed in the United States of America

1 3 5 7 9 10 8 6 4 2

For Joe Morton,
Dominic Harrison,
Eva Herle Schaffer
and Jack Bacon

– My millennials

Contents

Introduction: Two Questions   1

    Climate Risks and Responsibilities   5

    The Second Fossil-Fuel Century   8

    Altering the Earthsystem   22

    Deliberate Planets, Imagined Worlds   26

Part One: Energies

1   The Top of the World   35

    Discovering the Stratosphere   38

    Fallout   43

    The Ozone Layer   47

    The Veilmakers   54

2   A Planet Called Weather   57

    The Worldfalls   62

    The Trenberth Diagram and Climate Science   66

    Steam Engines and Spaceship Earth   71

3   Pinatubo   83

    Volcanoes and Climate   86

    Predictions and Surprises   93

4   Dimming the Noontime Sun   100

    Rough Magic   107

    Promethean Science   112

5   Coming to Think This Way   124
     Martians and Moral Equivalents   129
     The Day Before Yesterday   135
     The Rise of Carbon Dioxide Politics   139

6   Moving the Goalposts   148
     From Plan B to Breathing Space   156
     Expanding the Boundaries   165

Part Two: Substances

7   Nitrogen   175
     The Making of the Population Bomb   184
     Defusing the Population Bomb   189
     Far from Fixed   195
     How to Spot a Geoengineer   201

8   Carbon Past, Carbon Present   209
     The Anthropocene   219
     The Greening Planet   229

9   Carbon Present, Carbon Future   243
     Ocean Anaemia   251
     Cultivating One's Garden   259

10 Sulphur and Soggy Mirrors   268
     Global Cooling   274
     Cloudships   283
     Bright Patchwork Planet   288
     What the Thunder Didn't Say   298

Part Three: Possibilities

11 The Ends of the World   305
    Control and Catastrophe   312
    Doom and Denial   317
    The Traditions of Titans   323
    A Tale of Two Cliques   332
    After Such Knowledge   338

12 The Deliberate Planet   344
    The Concert   347
    Small Effects, and Bad Ones   359
    And Straight on 'til Morning   369
    Envoi   375

Acknowledgements   379

References, Notes and Further Reading   383

Bibliography   393

Index   415

Introduction

Two Questions

Let us go then, you and I,
When the evening is spread out against the sky
Like a patient etherized upon a table

T. S. Eliot 'The Love Song of
J. Alfred Prufrock' (1915)

In March of 2012, in a large-windowed conference hall on the snowy campus of the University of Calgary, I heard two simple questions. The man asking them was trying to help his audience get the most out of their day by giving them a clear understanding of where they, and others, stood when it came to action on climate change. To that end he asked them:

Do you believe the risks of climate change merit serious action aimed at lessening them?

Do you think that reducing an industrial economy's carbon-dioxide emissions to near zero is very hard?

Although this book is about more than climate change, it is because of the challenges of climate change that I have written it. And the two questions which were posed that morning by Robert Socolow, a physicist from Princeton University, seem to me a particularly good way of defining your position on the subject. So take a moment to answer them, if you would. The book's not going anywhere without you, and I think it will be a better read if you position yourself with respect to those questions before getting stuck in.

Here's a bit of context.

There is no serious doubt that the atmosphere's greenhouse effect is a key determinant of the Earth's temperature. Nor is there any serious doubt that carbon dioxide is a greenhouse gas, or that humans have been adding to the level of carbon dioxide in the atmosphere for the past few centuries by burning fossil fuels. In 1750, before the industrial revolution, the carbon-dioxide level was 280 parts per million. In 1950, when the great global boom of the second half of the twentieth century was taking off, it was about 310 parts per million. Today it is 400 parts per million. The bulk of that change has been due to the burning of fossil fuels. If you disbelieve any of those statements, you have been misled. I am not going to take the time to try and disabuse you in this book, and you should read on in expectation of frustration.

There is, however, a lot of room for doubt about the level of climate change the planet will see over the next decades and centuries. The best current estimate is that if fossil-fuel use continues on anything like current trends, the Earth is likely to end up at least two degrees Celsius warmer than it was before the industrial revolution, and possibly quite a lot warmer still. Change by one degree or two over a century or so may sound minimal, but it would be unprecedented in human history. Models of what happens to the climate in worlds in which fossil-fuel use is unconstrained point to severe, even cataclysmic, consequences in the form of damage to agriculture, greater harm done by extreme weather, the loss of biodiversity and sea-level rise over timescales of decades to centuries.

That said, different models provide different possible climates at any given carbon-dioxide level – some are more sensitive to the gas than others, in the language of modellers – and it is possible that the models on which warnings about climate change have mostly been based are, for some reason, skewed towards an unrealistically high sensitivity. It is also possible that humans and their natural world will be able to adapt to changed climates more easily and cheaply, and with less suffering, than most people concerned about climate change now believe. Thus it is possible that, even though carbon dioxide is unarguably a greenhouse gas, and a lot of it is being added to the atmosphere, climate change due to human action will not in the end be a planet-changingly big deal.

The question, though, is not about the possibility of benign outcomes. It is about your willingness to do something about the risks of bad or even catastrophic ones. A catastrophe does not have to be certain for steps to avoid it to be worth taking.

Now here's some context for the second question. The International Energy Agency, which compiles such statistics for governments, says that when the industrial nations committed themselves to cutting their carbon-dioxide emissions at the Kyoto climate-change conference in 1997, 80 per cent of the world's energy demand was met with fossil fuels. Renewable energy sources furnished just 13 per cent of the energy used; 10 of those 13 percentage points represented energy from biomass, including the wood burned on fires and in stoves by more than a billion people without other options. Wind, solar and hydropower provided just three percentage points.

In 2012, after 15 years of post-Kyoto political action on climate, wind, solar and hydro still provided 3 per cent of the world's energy needs; fossil fuels provided 81 per cent. Industrial carbon-dioxide emissions in 2013 were more than half as high again as they were at the time of Kyoto.

So how do you answer the two questions?

I answer them Yes and Yes. Yes, the risks posed by climate change are serious enough to warrant large-scale action. And Yes, moving from a fossil-fuel economy to one that hardly uses fossil fuels at all will be very hard.

To judge by what they say, and what policies they support, most people in favour of action on climate change are in the Yes/No camp: they want to act on the risks; they don't think that getting off fossil fuels is a terribly hard problem. Their way forward is to argue ever more strongly for emissions reductions; they believe these would be quite easily achieved were it not for a lack of political leadership willing to take on the vested interests of emitters.

Most of those against action on climate are in the No/Yes camp: they don't think climate is very much of a worry; but they do think that getting off fossil fuels is difficult, even impossible. Their leaders tend to focus on the weaknesses they see in the science and politics underlying the case for action on emissions, and on the drawbacks of renewable-energy systems.

Neither of these approaches works for people like me in the Yes/Yes camp. Yes/Yes people need different responses: responses which seek to lessen the risks of climate change without impractically rapid cuts in fossil-fuel use; or responses which seek to change society so deeply that such reductions become feasible. I think that deliberate modification of the climate – climate geoengineering – could offer a response of the first sort. It is to outline the promise and attendant perils of that idea, and to appreciate its antecedents and its implications, that I have written this book.

But that is not all I want to do. This is a book not just about a particular set of ways in which the world could be changed – it is about a world already changed in all sorts of ways that are not spoken of as clearly as they should be. It is a world in which the impact of the human is far greater than it used to be: a world in which the global economy has become something akin to a force of nature, in which the legacies of past generations and the aspirations of generations to come dwarf the impacts of hurricanes and volcanoes. Some people reject or denounce the implications of this change; others blithely accept them in a way that underplays their magnitude. I think those implications need to be opened up, inspected from different angles, interrogated, analysed, appreciated. Only then will it be possible to make the necessary judgements and choices.

Thinking about geoengineering is a worthwhile end in itself. But it is also an exercise in building up the imaginative capacity needed to take on board these deep changes the world is going through, and which it will continue to go through whether or not anyone ever actually attempts to re-engineer the climate. The planet has been remade, is being remade, will be remade. This book is an attempt to help people imagine the challenge those changes will bring.

Climate Risks and Responsibilities

Before going any further, though, let me justify my Yes/Yes. If you are already a Yes on either count, and impatient to boot, please feel free to skip ahead a section or two, as appropriate. If you are a No on one or the other count, let me see if I can bring you round, or at least clarify our disagreement.

First, the Yes as to risks from climate change that merit serious action. Some economic analysis suggests that there could be net benefits to the first degree or so of climate change, thanks to increased agricultural productivity in temperate zones, increased rainfall in some dry areas, less harm done by the cold at high latitudes and various other factors. If climate change due to carbon-dioxide emissions is a net benefit to the world, some people argue, countries should take no concerted action against it.

But in no such projections is everyone better off. And those who end up worse off are for the most part the people who start off poorer and who are responsible for fewer carbon-dioxide emissions, such as farmers and the urban poor in developing countries. I think that if carbon emissions do harm to those unfortunates, then people in rich countries and rich people in poor countries – the two groups, each about a billion strong, whose ways of life account for an overwhelming share of emissions – have a duty to act. That duty persists even if the emissions are somehow helping other people, such as rich-world farmers or old people at risk of death in cold winters. In most peoples' moral systems, you do not get a free pass to do harm to one set of people just because you are doing good to another set.

I understand that cutting emissions is by no means the easiest way to help the people whom those emissions put at risk. Easier immigration to rich countries; well-implemented development aid; political reform that puts more weight on the livelihoods of the poor; a more open trade system: all these policies offer more immediate solutions. But helping in those ways, while admirable and a good idea regardless of what else is done, would not fully excuse the climate harm. Just as it is wrong to help some people with one hand while hurting others with the other, you can't knock people down with impunity just because you are willing to slip them some cash first and pick them up off the floor afterwards.

Not everyone will accept this reasoning, and some who do accept it will, as a matter of pragmatism, feel justified in settling for net good even when it involves harm to some vulnerable people. But there are other reasons for believing that climate change merits a serious response. Climate change may be neither as big a problem nor as poorly tolerated as most of those who study it think; but even were this the case, there would remain a fair chance that it will be pretty bad, and a smaller chance that it will be very bad indeed. Let's say there is a 50 per cent chance of net harm, and a 5 per cent chance of this harm being very severe. To me, those odds justify serious action. A 5 per cent chance is one in twenty: pretty close to the odds that, on throwing a pair of dice, you will get either a double six or snake eyes. Not likely, but a long way from unheard of. When I was told on reasonable authority that my risks of a cardiac event in the next fifteen years or so were about 6 per cent, I resolved to make some changes in the way I lived my life. A little later I actually managed to act on those resolutions.

A straightforward reading of the latest assessment by the scientists of the Intergovernmental Panel on Climate Change (IPCC) would suggest that the risks are higher than those I just gave; many scientists and almost all environmental activists would put them much higher. But if you think, as I do, that figures as low as 50 per cent and 5 per cent justify action, it doesn't really matter for the purposes of this discussion if the figures are actually higher. Provided that threats to the world at large move you, you have already bought into the case for finding a way to act.

If you require more specific threats – threats to yourself, your loved ones, your descendants and theirs – things are not so clear-cut. I will not pretend that climate-change risks are all that high for reasonably well-to-do people in developed countries in the next half-century or so, and I imagine that describes most of my readers. My choice to worry about the more general threat is, I recognise, a choice. You may choose differently.

So those are my reasons for a Yes to the first question. What about those of you who answered No? Some, as mentioned, may find threats to humanity at large insufficiently motivating. Among the others, there seem to me to be two ways you might have come to your No. You may think that my one-in-twenty chance of catastrophe is small enough to live with. Alternatively, you may think that a 5 per cent chance would indeed justify serious attempts at risk reduction, but that the odds are actually longer than that – that the chance of catastrophic harm is, say, one in a hundred.

In the first case, the one where you are less risk-averse than I, there's probably not much I can do to change your mind. Please read on, though. I hope that readers do not have to agree with all the premises of this book to find its ideas stimulating and its effect on their imagination rewarding. I also suspect that some of those ideas are going to sound disturbingly risky to many readers. It will be nice to have a few people who laugh in the face of danger come along for the ride.

In the second case, the one where you think the risks are less than one in a hundred, I think you are displaying an indefensible level of certainty about how the climate works and how much carbon dioxide will be emitted over the next century. I can see that it would be nice to feel that level of certainty. But I just don't see how you can if you've looked at the issue seriously. Given the uncertainties involved, to be sure that a climate disaster is that unlikely shows a self-confidence so well developed as to be indistinguishable from folly.

The Second Fossil-Fuel Century

What about the other Yes? Why should moving off fossil fuels be so difficult? The answer lies in the scale of the problem and the speed of the change required, and – fair warning – it will take me rather longer to run through than the first Yes did.

The 30 billion tonnes of carbon dioxide emitted in 2013 came from burning three trillion cubic metres of gas over the year; from burning almost three billion barrels of oil in each of its months; from burning a bit less than 300 tonnes of coal in each of its seconds. The infrastructure needed for all that burning was almost as complex as it was essential.

To stabilize the climate by means of emissions reduction means replacing the whole lot.

The world has made huge investments in the facilities that extract fossil fuels from the ground and burn them – mines, oil wells, power stations, hundreds of thousands of ships and aircraft, a billion motor vehicles. Leaving aside the political lobbying power that such investment can command, there would be a limit to how quickly that much kit could be replaced even if there were perfect substitute technologies to hand that simply needed scaling up. If the world had the capacity to deliver one of the largest nuclear power plants ever built once a week, week in and week out, it would take 20 years to replace the current stock of coal-fired plants (at present, the world builds about three or four nuclear power plants a year, and retires old ones almost as quickly). To replace those coal plants with solar panels at the rate such panels were installed in 2013 would take about a century and a half. That is all before starting on replacing the gas and the oil, the cars, the furnaces and the ships.

And the challenge of decarbonization is not just a matter of replacing today's extraordinary planet-spanning energy infrastructure; you have to replace the yet larger system it is quickly growing into. The twentieth century began with a world population of 1.6 billion, none of whom enjoyed the energy-intense affluence of the citizens of today's modern industrialized states. Today's emissions are for the most part a result of the fact that two billion people now lead such lives.

But there are five billion more people in the poorer countries not leading such lives. About a quarter of those people lead lives illuminated only by sun, moon and fire, with no reliable access to modern energy supplies of any sort. They deserve better. All of those people should be able to lead the lives that the affluent two billion lead today, with access to the industrial and agricultural goods and services that copious energy makes possible. And so should their children and grandchildren.

The world's population is expected to grow from seven billion today to more or less ten billion by 2100. By that time the number of people enjoying rich-world energy privileges should also reach ten billion. So the challenge is to achieve for an extra eight billion people in the twenty-first century what was achieved for two billion in the twentieth century. Meeting that challenge implies a lot more energy use. It may be that a prosperous person of 2100 ends up using no more energy than a European did in 2000. That would be unprecedented, but it is perhaps not implausible. Energy use has, after all, plateaued in some places. Even under that assumption, though, the twenty-first century has to see either the continuation of a world divided into haves and have-nots or a massive expansion of the energy systems of developing countries.

It was the prospect of such an imminent expansion that made the grand agreement on climate change which much of the rich world was seeking at the Copenhagen climate summit of 2009 unattainable. The proposal that shaped those rancorous debates was that world carbon-dioxide emissions should halve by 2050, with developed countries going further and reducing their emissions by 80 per cent or more. The idea was that such deep reductions by rich countries would give developing countries a chance to go on using fossil fuels for a bit longer – a bit of room for growth. In shaping the proposal this way, its architects were recognizing the fact that, at the moment, developing countries are powering their growth with fossil fuels. Given the energy systems they have today, other technologies could not provide them with comparable amounts of power at the scale and cost they need.

The problem with the Copenhagen proposal was that while populations in the developed world are mostly stable, many of those in the currently developing world are still growing. At the time of Copenhagen, there were about six billion people in developing countries. In 2050 there will be more than eight billion in those countries. Take this into account and the Copen-hagen deal offered no real room for growth; the extra fossil-fuel use envisaged in the developing world was only enough to balance its countries' increasing populations. In terms of carbon-dioxide emissions per person, the developing world would have had to stick very close to its then current levels, which were less than half of Europe's, less than a third of America's. Unsurprisingly, the deal did not get made. New fossil-fuel-burning capacity has been added round the world since Copenhagen at an even higher rate than it was being added before.

Though emissions from developing countries are unlikely to diminish any time soon, many rich nations remain committed to the reductions of 80 per cent that they spoke of at Copenhagen. Indeed, Britain has gone so far as to enshrine them in law. Such targets imply that emissions will be reduced by 4 per cent every year for 40 years. For comparison, when France converted its electricity infrastructure almost entirely to nuclear power between 1970 and 1995 it managed a reduction in emissions of just one per cent a year. Britain's 'dash for gas', a large-scale shift away from coal that followed the liberalization of the electricity market in the 1980s, reduced emissions by the same rate in the 1990s. America's dramatic shift to greater natural-gas use in response to the shale-gas revolution of the past decade has led to emissions reductions on a rather smaller scale. There are very few precedents for 4 per cent year-on-year emission reductions that don't also involve an economic collapse – and they can only last for so long.

This fits with the lessons that Arnulf Grübler, an academic at the International Institute for Applied Systems Analysis outside Vienna, has drawn from decades spent studying the history of energy systems, and in particular the 'energy transitions' in which one energy technology displaces another; the steam engine replacing the draft animal and the waterwheel, for example. One general principle, he says, is that energy transitions have been slow – they take about a century.

Things are different now, say the mainstream environmentalists and the environmentally conscious politicians in the Yes/No camp. Previous energy transitions were for the most part realized with no overarching plan. This one will be deliberate. And there has already been a renewables revolution on which the transition can be built. In the past couple of decades wind and solar power have been deployed on an industrial scale for the first time. There are now single installations with a capacity of between 100 megawatts and a gigawatt – facilities similar in size to advanced gas-turbine plants at the bottom end and nuclear plants at the top end.

(A note here on power and energy: power is the rate at which energy is made available or used over time; it is measured in watts, and multiples of watts. A human body burns up the energy in food at a rate of about 100 watts. The 1.5-litre engine in a compact car like a Toyota Corolla generates power from gasoline at about 50,000 watts, or 50 kilowatts (50kW). A really big wind turbine turns the energy of the wind into electricity at a rate of 5,000,000 watts, or five megawatts (5MW). A big power station typically runs at a billion watts or so – a gigawatt. The energy use of a major economy like America's, Europe's or China's is a thousand times larger still: something like a trillion watts, which is a terawatt (1TW). Energy is what you get if you multiply power by time. Use a kilowatt of power for an hour, and you have used a kilowatt hour.)

The attraction of renewables goes beyond drastically reduced greenhouse-gas emissions. Burning fossil fuels produces a wide range of 'aerosols' – tiny particles floating in the atmosphere (aerosol spray cans are so called because they turn their contents into such particles). Millions of lives are shortened each year because of the harm these aerosols do when inhaled; power plants that burn coal are particularly grievous offenders. Chemical contaminants created by generators and engines – nitrogen oxides and ozone – also do a lot of damage, both to people and to crops. And the supply of fossil fuels can fluctuate wildly, either because of changes in the market or because of politics. The fuel costs for some renewables, on the other hand, are fixed and very low – wind, sunshine and the tendency of water to flow downhill come for free, and the plants grown to burn as biomass can often be furnished pretty cheaply, too.

It is a fine list of benefits. But there is a second lesson from Grübler's studies of past energy transitions to be confronted. They have, in the main, been driven not by the availability of new ways of providing energy, but by new ways of using it: transitions are pulled by demand, not pushed by supply. Electricity and internal combustion engines were adopted because they allowed people to do things they hadn't done before, and people demanded those new energy services in ever-greater numbers. The requisite fuel supplies, generating technologies and distribution systems raced to catch up. There is simply no precedent for a wholesale change that doesn't offer users appealing new possibilities in terms of the way they use the energy — for a change that is pushed through rather than pulled along. And as far as the end user is concerned, renewable electricity is just another form of electricity — it offers no advantage as a means of powering things, even if generating the electricity that way has various charms. Its benefits are felt at the level of the system, not at the level of the individual buyer. That means a renewable-energy transition will need significant pushing.

As with Grübler's observations about the time transitions take, this points merely to decarbonization being unprecedented, not impossible. But the best example in recent history of an energy transition that governments tried to push through, rather than simply letting users pull, is not very encouraging. Governments in various countries pushed quite hard for a transition to nuclear power in the 1960s and 1970s. In many of them, though, the technology's early growth subsequently stalled.

This was in part an economic matter. In America, in particular, early promises of cheap and plentiful power failed to pay off as companies saw the costs of nuclear plants go through the roof at the same time as the growth in demand they had expected failed to materialize. But on top of this, the 1970s saw a catastrophic turn in the way people thought of and talked about the future; nuclear power played a role in this turn, and suffered from the consequences of it. I think it is worth looking at that process in a little detail, not just because of what it says about energy transitions, but because it throws light on our main themes. As will become apparent at various times in the course of this book, little else can hold a candle to the energies of the nucleus when it comes to imagining the impacts of world-changing technology.

In its early years nuclear power benefited from a carefully crafted position as the epitome of the scientific progress taking the world forward into a better future. Although there was a persistent undercurrent of cold-war anxiety, there was a general enthusiasm for the future that nuclear power was held to offer. Initial concern about nuclear power imagined it posing an insidious and ubiquitous threat, contaminating the world through its very existence in rather the same way that nuclear fallout did. It was neither a plausible concern nor one that gained much traction. Many environmentalists focused instead on the technology's environmental benefits; in terms of cleaning up the air people breathed nuclear plants promised to be a great step forward from coal plants.

In the 1970s, though, the original fear of nuclear business-as-usual was at first reinforced, and then displaced, by a fear of nuclear accidents. The notion of the meltdown focused nuclear anxiety on specific events, and relied on increasingly widespread concerns about the military-industrial complex and the technocratic hubris of governments. At the same time, it maintained the underlying sense of an invisible, intangible and global threat that makes all concerns about radiation so unsettling and prone to irrational exaggeration. The double vision in which specific accidents were also global threats reflected the sheer scope of the effects imagined: a ball of radioactive slag produced in a meltdown passing right through the Earth (the 'China Syndrome'); a meltdown poisoning its surroundings for geological periods of time. On top of these fears about what would happen if the nuclear-bomb-like energy of a reactor got out of control were worries about the levels of control that organs of state security might impose on the public to stop any saboteurs seeking to bring about disaster. The power needed to keep the genie bottled up was as worrying, to some, as the power of the genie unleashed – maybe more.

The new nuclear fear was not the only factor behind the stall in the transition to nuclear energy. In America, as noted, the technology proved to be far more expensive than its proponents had hoped, in part because of the rushed way in which it was rolled out. By the time of the Three Mile Island accident of 1979, which did a great deal to cement anti-nuclear fears in the American imagination, no new nuclear power plants had been ordered in America for more than three years. But the new fears added to nuclear power's woes, and indeed its costs, by making permission to build plants harder to gain; the current mixture of expense and public disquiet goes a long way to explaining why most nations with nuclear-power programmes get less than 20 per cent of their electricity from them and have expanded them little since the 1980s. (The great exception is France, which has a culture of technocratic planning, a trust of engineers on the part of both the public and the policy-making elite, a history of valuing its energy self-sufficiency, governments consistently happy to take a direct role in running the energy system and low fossil-fuel reserves. Nuclear reactors generate almost 80 per cent of its electricity.)

There are now environmentalists who would like to shock the stalled nuclear transition back to life as a way of fighting climate change. They argue that nuclear power's obvious problems – the risk of accidents, the production of radioactive waste and the facilitation of bomb building – are nothing like as bad as they have been painted, and pale compared to the damage climate change could do, and they are mostly right.

Only one nuclear accident – that at Chernobyl, in 1986 – has led to significant loss of life. Current assessments of the Fukushima meltdown suggest that there will be no discernible deaths as a result. Compare that with more than a million who die with coal-ruined lungs every year. New nuclear reactors built to the standards demanded by experienced government regulators with the power to have their decisions respected will be significantly safer than older designs. Long-term storage of waste has been politically mismanaged in many countries but is neither a particularly pressing problem – safe interim storage solutions that can last for decades, even a century, are tried and tested – nor a fundamentally intractable one. Though many civilian nuclear-power programmes have been linked to weapons development, those links have often proved breakable: neither Argentina nor Brazil is currently pursuing a bomb; South Africa gave its bombs up. Proliferation is a grave risk, but doing without nuclear electricity would not lead to a proliferation-free world. North Korea and Israel have produced nuclear weapons with no civilian power programme at all.

Despite all this, there seems little likelihood either that the green movement will pivot to nuclear power en masse or that the number of reactors will grow substantially. They remain more expensive watt-for-watt than fossil-fuel plants, most hydro-electric dams and some wind installations, and they only come in large sizes, which means you have to buy a billion watts or so at a time at costs of tens of billions of dollars. While smaller reactors would alleviate some of that problem, their development is difficult – nuclear power is, given the items involved and the regulations that surround them, a hard area in which to innovate. And nuclear energy enjoys none of the demand-pull that was crucial to earlier energy transitions; for a domestic or industrial user, nuclear electricity is no better than any other sort.*

---

* This is true for civilian power; in the military, however, nuclear electricity does have a key advantage, in that it can be used to power submarines which would otherwise have to take on air through a snorkel so that they could burn diesel. It was this small but crucial niche that led to nuclear reactors capable of generating electricity being developed in the first place.

Many of those pressing for a nuclear renaissance accept some or most of this. But they persevere despite knowing that their chances are slim. In part this is the willful blindness of the enthusiast – people who fully take on board how hard it is to change things often do not try. There is also a dislike of the irrational and a worry over seeing it dominate policy; the unrealistically exaggerated fear of radiation that drives mainstream green attitudes regarding nuclear power baffles and offends such people. They look askance at Germany's decision, after the Fukushima disaster, to listen to its long-standing anti-nuclear movement and close down a fleet of reactors that was producing safe, copious, clean and (because their capital costs were paid off) cheap electricity. Being pro-nuclear serves them as a badge of their willingness to break from green orthodoxy and to look at the world as it is. But most importantly, they have a deep-seated belief that renewables cannot on their own produce the sort of decarbonization that reductions in climate risk require, especially when the needs of the as-yet-undeveloped world are fully taken into account.

Renewables have constraints that go beyond their current costs. Wind and solar energy are intermittent. They become unavailable in ways both easily predictable – there is no solar energy at night – and less so. Sometimes the wind will fail to blow over quite large areas for days or weeks at a time. This need not be as much of a problem in the future as it would have been in the past; information technology will make it easier for 'smart grids', smart appliances and, indeed, smart people to cope with such fluctuations by managing demand; consumers will probably consent to such management if it lowers bills. But intermittency still drives up the costs and complexity of power supply if you want to get most or all of your electricity from renewables and you don't have access to a great deal of hydro-electric capacity – a largely zero-carbon source that can be ramped up or down very quickly to balance out the intermittencies of other supplies.

Wind, solar and biomass also all take up a lot of space. A wind farm on a moderately well-chosen site produces one or two watts for every square metre of land in its footprint. A square kilometre thus delivers about a megawatt, and so it takes a hundred square kilometres to deliver the 100MW you might expect from a gas turbine that fits into a plant not much larger than a warehouse. Crops grown to burn as biomass are an even more dilute source of energy – harvesting them provides at best a watt or so per square metre in temperate climes, though in the tropics things can be a bit better. Solar power does better, but still takes up quite a lot of land.

What does this mean for an industrialized country like the United Kingdom? The rate at which British people use electricity is about 40GW. To generate this at one watt per square metre would mean devoting 20 per cent of the area of the country to renewable energy (to be fair, you can graze sheep on the same land. But still . . .). If you want to deliver ever more of the nation's energy in the form of electricity (for example, if you want to replace fossil-fuel-powered cars with electric ones, or gas-fired domestic heating with electric heat pumps), you will need to be generating more electricity, and thus using even more land. Efficiency can help – efficiency can always help – but not enough to transform the situation. You also have to consider that, to some extent, more efficient energy systems encourage people to use more energy.

Even though they are intermittent and profligate in their use of space, renewables have a role to play; but the presence of fossil fuels will squeeze that role for as long as the fuels are allowed and affordable. Quite a lot of work on renewables in the 1970s and 1980s was premised on the idea that this would cease to be the case – that fossil fuels would become sparse and unaffordable. Similar ideas were prevalent in the mid-2000s when Europe framed an energy and climate strategy that gave pride of place to renewables: fossil-fuel prices, it was claimed, would in the long run both rise and become more volatile. At the time of the Copenhagen summit it was argued that if Europe were to become a renewables superpower it could avoid those costs and get ahead of the rest of the world.

But fossil fuels have become cheaper, not more expensive, and look likely to stay quite affordable for rich countries for decades to come. Attempts to make fossil-fuel prices higher through carbon taxes and similar schemes could, in principle, change this, forcing a lot of investment into alternatives. But in most places they have not attracted enough political support to stick, and it is not clear that they can. They would stand a better chance in a world that coordinated its actions internationally; when people talk about the low costs of a transition to renewables, they are imagining it taking place in such an optimal world. But that world has not yet been achieved.

Instead of increasing the costs of fossil-fuel generation through a carbon price or tax, governments have preferred to subsidize renewables. One problem with this is that it doesn't encourage people to stop using fossil fuels in existing plants; it just rewards people for building alternatives. Another problem is that the more renewables get built, the pricier the subsidies get. Germany's current *Energiewende*, a national policy which aims to cut carbon-dioxide emissions from the power sector drastically while at the same time retiring all of the country's nuclear plants, seems set to find out how far such subsidy approaches can go; they have cost well over 100 billion euros to date. In general, few economies have wind, solar or biomass supplying much more than 20 per cent of their electricity market (the same sort of level, possibly coincidentally, that has been achieved in the other push-not-pull attempt at an energy transition — that of nuclear power). That's a large enough fraction to transfer a significant amount of money to the builders and buyers of wind turbines and solar. But it is not enough to change the course of the planet's climate.

This reflects a general issue with environmental policies;

they are often aimed at pleasing voters and lobbies with green interests more than they are geared to achieving the stated environmental ends. People see wind turbines being built in prodigious numbers, and see solar cells on roofs, and think they are looking at a solution. In fact, in part because of the low energy-density of renewable energy, these impressive – and, to some, infuriating – sights are achieving very little in terms of providing enough power for a world of ten billion reasonably well-off people.

Renewable efforts do not have to be paltry. The *Energiewende* is no small thing. And it is possible that Germany – a rich, technologically potent country which is rather good at sustaining national consensus – may be able to convert itself almost entirely to renewable energy. That said, in 2013 Jane Long, formerly the associate director for energy and the environment at Lawrence Livermore National Laboratory in California, chaired a study of the prospects for emissions reduction in her state – also rich and technologically potent, though less adept when it comes to political consensus. Long's study found that by encouraging energy efficiency, completely replacing fossil fuels as a source of electricity and greatly reducing their use in industry and transportation, California might cut its emissions by 60 per cent over 40 years. With a justifiable pride in her home state's record on innovation and commitment to environmentalism, Long says that if California can't do better than that, no one else can. And compared with that scenario, which made use of nuclear power as well as copious renewables, Germany is aiming for a higher target with one arm tied behind its back.

But the challenge of decarbonization would not be met just by a few environmentally conscious economies cutting their emissions by 60 per cent, or 80 per cent, even if they could; all the big emitters need to get on paths that take them to 100 per cent if the level of greenhouse gas in the atmosphere is to be stabilized in this century, with the electricity sector carbon free a lot earlier. Just half of them doing so does not cut it – and perversely, by reducing demand, it might well reduce the cost of fossil fuels to the other half.

It was the fact that all the big emitters need to be involved that drove the hoopla over Copenhagen. But the belief that the world can in some way come together to agree to do as a whole what its large economies are not obviously willing to do individually is illusory. There is a value in international negotiations: they can help shore up a sense of purpose; they can provide something by way of sticks and carrots. But an international agreement will not lead any government to follow climate policies that are clearly not supported at home for reasons of ideology, cost, or any other factor. And an international agreement on climate questions is also peculiarly hard to come by.

There are many reasons why this is so, most of them linked to the fact that the people who do most of the emitting do not face most of the risks from climate change, and also to the fear that if some act but not all, then those who do not act will get as much of the benefit as those who do. Most crucial of all, though, is the problem that whatever benefits there are and to whomever they accrue, they will not be felt for decades. The climate depends on the cumulative carbon emissions – on the total stock that has been added to the atmosphere over centuries, not on the rate at which they are added at any particular time. Any plausible cuts in carbon-dioxide emissions made today would have more or less no effect until the mid-century. By that time the costs of inaction might be horribly plain – but there will be no time machine with which to come back and set the necessary cuts in motion on the basis of that future knowledge. As Hans Joachim Schellnhuber, an influential expert on climate-change impacts, puts it, 'climate change is too slow a problem to solve in time'.

The costs of action and the lack of an international mechanism do not mean there will be no decarbonization in the decades to come; but I suspect it will be more like that seen in China than on the scale imagined in Germany or California. China is building more renewable capacity than any other nation, and is ramping up a big nuclear programme, too. It is also enacting ambitious energy-efficiency measures. But this is only slowly reducing the proportion of its energy it gets from fossil-fuel use. Its current plans have its carbon-dioxide emissions continuing to rise until 2030 – at which point 80 per cent of Chinese energy will still be coming from fossil fuels. How quickly it might fall after that is anyone's guess.

As China, so the world. An investment in non-fossil-fuel sources of energy great enough simply to keep up with increasing demand is a huge commitment. An investment big enough to displace fossil fuels entirely does not look to be remotely on the cards under current conditions. Between 1750 and 2000 humans released half a trillion tonnes of the carbon that the Earth had stored up in fossil fuels. It is very hard to believe that, over the coming century, they will not release a trillion tonnes more.

Altering the Earthsystem

Yes/Yes is compelling to me, and I hope it now looks pretty reasonable to you. But as Rob Socolow pointed out in that big-windowed Calgary hall, it is a minority view.

Those who oppose climate action sit firmly in the No/Yes camp. For some of them, the Yes drives the No. If you understand that, Yes, action on fossil fuels is hard – hard technically and hard on people who get hit with higher electricity and fuel bills, as well as hard on people who have investments in oil, gas and coal – concluding that No, it isn't necessary is quite convenient. People in the No/Yes camp have a fair bit of motivation to search for reasons to doubt the science behind calls for action, a search with which organized lobbies have been happy to help.

The Yes/No camp saves its doubt for people who point out the impracticality of an energy transition on the scale required to make a big change in the risks. Politicians who accept the need for climate action insist that it will be relatively painless, maybe even an enjoyable improvement, bringing jobs and growth. The greens who accept that sharp reductions in fossil-fuel use will indeed have costs often imagine them falling predominantly on big businesses. Some also argue that the costs are, in a way, illusory. While less affordable energy and consequent drops in consumption look like a 'cost' to economists, some greens see the latter, in particular, as a benefit.

The rich world contains quite a few people who have found that they can lead happier lives with less stuff, and the same might be true of many more, if we could only see our way to making that choice (I say 'we' here because I accept that I may be among those who, because of the ingrained mindset of consumerism, are failing to follow a course of action which might make them happier). But as Pat Mooney of the Canadian environmental group ETC (it stands for Erosion, Technology and Concentration, and is pronounced 'et cetera') pointed out to me a few years ago, people who see some evidence that choosing a path of lowered consumption would make them happier and yet do not choose to act on that evidence are very unlikely to make the same choice for the benefit of others. And it is also unlikely that they will acquiesce in being forced down the path of happiness unchosen. 'Lead the life I want you to lead or the planet gets it' is not only an unattractive position, it is an ineffective one.

The world's political leaders are resolutely Yes/No or No/Yes, and most of the public seems OK with that. But the people Socolow was addressing in Calgary had pretty much declared for Yes/Yes simply by turning up. If they had not both taken responding to the risks of climate change seriously and believed fossil fuels were hard to get rid of they would not have bothered to attend a meeting on innovative chemical-engineering techniques for pulling carbon dioxide out of the air, a technology known in climate-change circles as 'direct-air capture'. Used at a scale large enough to have an appreciable effect of the atmospheric carbon-dioxide level, direct-air capture would be a form of geoengineering – which I define as the deliberate modification of the earthsystem on a global scale.

Where did that word 'earthsystem' come from? Why not just say modifying the Earth? Because I think people can sometimes be helped to see things afresh by expanding the language that they use, and this is one of those times. The Earth sounds and feels like a thing; when I say it – or rather, when you read it – in a context such as this, you like as not see a blue-and-white globe floating in space. When I say earthsystem instead, you probably see nothing at all. But it is my hope that, over time and through usage, you will come to feel something.

What I hope you come to feel, like wind on the skin, or the tremor in the ground from rushing water nearby, or lightning sensed through shut eyelids, is something dynamic. The earthsystem's essence lies not just in rock and water, or in air and plants, but in all the Earth's interplay of energy and matter. It lies in the flows of energy which drive the cycles of carbon, water, nitrogen and life's other essentials that roll ceaselessly round the planet and down through the eons. And it also lies in the way that those movements, and the transformations which they bring, shape the course of the flows of energy that drive them. These flows and cycles are the defining processes of the earthsystem.

The forces and feedbacks between these flows of energy and matter make the earthsystem subject to changes, just as other complex systems are. The system's processes can shift from one state to another quite quickly. Some of the biggest transitions that geologists see recorded in the rocks of the Earth, transitions which they use to distinguish one period of the planet's history from another, can be understood as shifts in the way the earthsystem works, shifts from one state to another.

Today human technology is driving change on a similar scale to that which has punctuated the story of the planet's past. What, at the beginning of the age of science, Francis Bacon dubbed 'human empire' – the expansion of power and knowledge, barely distinguishable from each other, out into the world – has become something like a force of nature. Dams are changing the flow of rivers, engineering works are altering the processes of erosion, agriculture is redefining the global cycling of nutrients and the patterns and pace of extinction – which is to say, evolution. And humankind's greenhouse gases are changing the rate at which energy flows through the planet's living systems. These effects on the earthsystem are now so marked that some scientists favour the idea of defining the current period of human empire as a new phase in the planet's history: they call it the Anthropocene.

And this is another reason for wanting to talk of the Earth as a system. Because the way that humans are imposing the Anthropocene on the planet is not just as a set of seven billion primates but as a system – the system that makes up the world in which we live, a world of social and economic and political and emotional forces just as real as the physical forces of the earthsystem, and quite as subject to feedback and changes of state. If you see the environment as just a thing – a planet, a lump of rock and water and air – then this socially mediated world, a world always, to a human, coloured by subjective experience, will perforce seem separate from the environment, as different in kind as a performance is from a stage. If you see the human world's context as the earthsystem, though, then the two systems begin to intermingle: the boundaries that have previously kept the social from the natural begin to blur. Neither has a monopoly on action. Both have impacts on the other. Economic feedbacks and climate feedbacks, political forces and radiative forcings, do not exist in different frames of reference. They all provide the settings for each other. It is through the earthsystem and its processes that the human world changes the geophysical facts of the planet; it is to changed states of the earthsystem that the human world finds itself responding.

When the change that humans bring to this new Anthropocene state of the earthsystem is deliberate, I see it as geoengineering; in this book, that term will cover any deliberate technological intervention in the earthsystem on a global scale, not just those aimed at countering, or ameliorating, the changes that people are making to the climate without deliberation. The notion of deliberation matters; to the earthsystem, a change made in passing may be no different to one made on purpose, but in the human world there is a difference between the changes you make and those that you plan, between having an effect on the future that you can foresee and having an effect that you intend. The extinction of the dodo is one thing; that of smallpox is another.

The effects of piling more and more carbon dioxide into the atmosphere can be foreseen, though not in as much detail as one might like. But they are not exactly intended. If the effects were to be reduced by machines capable of sucking carbon dioxide out of the atmosphere, like those under discussion in Calgary that morning, the resultant change would be intended, even if not all its impacts could necessarily be foreseen. Thus large-scale direct-air capture of carbon dioxide would be a way of giving the Earth a climate other than the one it would expect, given the amount of carbon dioxide that human activities have emitted. And that is what climate geoengineering aspires to more generally. Climate geoengineering can be pursued in very different ways, but the aim is always to decouple the climate from humanity's cumulative emissions of carbon dioxide. It is to unshackle, if only to a very limited extent, the future from the past.

Deliberate Planets, Imagined Worlds

If direct-air carbon-dioxide capture of some sort could be implemented safely on an arbitrarily large scale it is hard to imagine that it wouldn't be. Sucking carbon dioxide out of the atmosphere as fast as it was pumped in would seem to more or less solve the climate problem, as long as somewhere could be found to put the carbon dioxide thus sucked out. Maybe it could be stored in reservoirs underground; maybe it could be turned into solid carbonate rock; maybe it could be turned back into hydrocarbon fuel, so that such fuels would never run out.

Unfortunately, at the moment direct-air capture cannot be implemented on a remotely large enough scale because there is no proven technology for taking carbon dioxide out of the air that is practically or economically up to the job. And if your goal is to pull carbon dioxide out of the atmosphere at anything like the rate at which it is currently being pumped in, it's a very good bet that no such technology will ever exist. Some of the reasons why this is the case – as well as the promise direct-air capture might still offer while never meeting such an all-encompassing goal – will be discussed later. What I want to bring to your attention here is not the detail of the technology, but something about the people trying to make it real.

There were four groups actively working on the idea at the Calgary conference, and three of them had something striking in common. They were all fronted by charismatic North American physicists, the sort of people who impress and inspire students by showing the near-inexhaustible ability of physics to provide answers, and by encouraging them to ask questions to which the answers are truly interesting. They are the sort of men who make knowledge – both theirs and, once you learn from them, yours – feel like power. Men of human empire.

They are also the sort of men who can attract the interest and admiration of wealthier and more powerful men. All three of the physicists whose work was under discussion in Calgary were professors at prestigious universities, but their air-capture work has been mostly done under the aegis of companies they started for this purpose with the help of investments by rich sponsors. Klaus Lackner, of Columbia University, the first person to make a splash working on direct-air capture, was able to take his ideas about 'artificial trees' forward thanks to the backing of the late Gary Comer, the man who founded Lands' End clothing; Peter Eisenberger, also of Columbia, who dreams of using solar power to turn the carbon dioxide captured from the air back into fuel, attracted Edgar Bronfman of Seagram. David Keith, then at the University of Calgary and now at Harvard, landed the biggest fish of all. The main investor in his direct-air-capture company, Carbon Engineering, is Bill Gates.

To some, this will seem proof enough that climate geo-engineering is a pernicious capitalist plot. Further proof, if needed, might be adduced from Richard Branson's enthusiasm for such schemes (he has set up a prize to reward progress in pulling carbon-dioxide from the atmosphere). When Branson said, in 2009, 'if we could come up with a geoengineering answer ... then Copenhagen wouldn't be necessary. We could carry on flying our planes and driving our cars' you could all but hear green hackles rise at such get-out-of-jail-free sentiments from an owner of airlines. But if it is all a plutocrats' plot, it is a very poorly contrived one. The Calgary meeting had been arranged by Keith to discuss the cost of the schemes in question. A report by a panel of the American Physical Society, chaired by Socolow, who has a long background in energy and climate studies, had derived costs for capturing carbon dioxide from the air of around $600 a tonne, possibly much more. That was far higher than the figures Eisenberger and Lackner had floated, and more than double the less ambitious figures Keith's company talked about. The meeting was intended to thrash out the differences.

It managed to get some way towards that goal; members of Socolow's panel admitted that their estimate might be slightly high, though they didn't think that would make any difference to the technology's feasibility. Keith still thinks that they are wrong, and that it is conceivable that his direct-air-capture technology might turn a profit in places where there is an industrial need for more carbon dioxide than can easily be brought in on trucks and where, in addition, the government has set a significant price on carbon. But he is talking tens of thousands of tonnes, not tens of billions. No one could have come away from Calgary thinking that direct-air capture was anywhere close to being a viable tool for large-scale climate geoengineering. Anyone investing in direct-air capture as part of a plot to take over the climate is making a mistake.

If, then, direct-air capture doesn't really matter, why think about it? Because it may matter, in time – and because knowing that helps put the present into context and lets you imagine the future more fully.

In the Yes/No and No/Yes camps, the details of the future don't much matter. The fear of bad outcomes motivates both climate activists and their foes, but the precise details don't matter. Both sides see themselves as averting a future that they don't like more than creating one that they do.

The Yes/Yes position requires a richer imagination – one that allows that the future may be quite different. It is in the Yes/Yes world that you will find people who are open about the need they see to fundamentally change the world economy so that it no longer demands or delivers constant growth – an option today's liberal democracies, even green-looking ones like Germany, scarcely countenance – or to return large numbers of people to a relationship with nature centred on the land. It is in the Yes/Yes world that you will find people who think, with regret but with clear eyes, that more or less all that can be done about the risks of climate change is to equip the afflicted with the means to ride those risks out, and that even then adaptation may be beyond the reach of billions. It is in the Yes/Yes world that you will find the most plaintive Cassandras, convinced that catastrophe is now inevitable.

And it is in the Yes/Yes world that you will find people imagining a planet where the earthsystem is manipulated in such a way that climate and carbon emissions are no longer so tightly bound. There is much to criticize in such thinking. It can be horribly simplistic. It can feed on, and give rise to, ideas about 'the control of nature' that are neither plausible nor palatable. It can be used to justify inaction. But I believe it can also open up doors, doors both practical and utopian. I think there may be ways in which climate geoengineering could really reduce harm. I also think that imagining geoengineered worlds that might be good to live in, in which people could be safer and happier than they would otherwise be, is worth doing. A utopia does not need to be attainable – indeed, by definition it cannot be. But that is not a reason to reject utopian thought. It is part of the reason for embracing it.

The possibilities of utopian imagination, though, are undercut, even betrayed, if the group doing the imagining is too small. That is currently the case, I think, for geoengineering. Listen to the discussion of the topic going on today and you will hear natural scientists who are cautiously curious about the ideas but have no real interest in trying to make them practical; you will hear social scientists and philosophers interested in providing critiques of the modes of thinking that shape the discourse; you will hear environmentalists who see in it, or project on to it, everything they dislike about centralized action, about capitalism, about mechanistic world views; you will hear the fantasies of the rich and powerful and the fears of the frightened and doctrinaire. It is too small a set of voices.

The way a society imagines its future matters. And who gets to do the imagining matters. The purpose of this book is to spread the tools with which to imagine a re-engineered earth-system a little more broadly. In doing so, it looks at the scientific possibilities under discussion. It also looks at the history of that discussion, at the beliefs people have held about the proper relationship between climate and humanity, at the political contexts that have grown up around those beliefs. I fear that may make it sound like the driest sort of imagining. I hope it will prove not to be. If nothing else, I think there is a particular appreciation of the wonder of the earthsystem that can be gained only by imagining how it could be changed.

The ultimate challenge is not just to picture what an earthsystem subject to some level of deliberate design might be like. It is to picture a world in which you would feel happy about such a design being realized. It is about finding happiness and exercising compassion on a planetary scale – a project that will have to be as political as it is scientific or technological. The goal is to help you imagine a world attractive enough that many would welcome it, but robust and provisional enough that its creation does not require everyone to agree on every aspect of it; a world that requires neither uniformity of outlook nor the suppression of dissent, but offers ways for justice and sympathy to spread out through the human world and into the earthsystem beyond.

PART ONE

Energies

1

The Top of the World

*One might say that immensity is a philosophical
category of daydream. Daydream undoubtedly feeds
on all kinds of sights, but through a sort of natural
inclination, it contemplates grandeur. And this
contemplation produces an attitude that is so special,
an inner state so unlike any other, that the daydream
transports the dreamer outside the immediate world to
a world that has the mark of infinity*

Gaston Bachelard,
The Poetics of Space (1958)

The sun is shining, but the sky above is Bible black. It takes on
colour only lower down, first deep violet, then, just above the
encircling horizon, a band of blue and white. The descending,
brightening sweep of colour gives a swelling curve to the sky.

Within that encompassing blue-white band, the bright-below
Earth, too, is curved. It bends away in every direction towards its
blue-lipped rim.

The only straight line in this whole vast, round, empty world
is the wing.

You are 22 kilometres up, well inside the stratosphere – a
realm which, although it is about as peripheral as a part of the
Earth can be, plays a central role in the story to come. If climate
geoengineering ever takes place, there is a good chance that it

will take place up here, in the Earth's attic. If it does not take place, it may well be for fear of the damage it could do to this bright-lit void.

Even if it were not a crucial locale for geoengineering schemes, though, the stratosphere would still have much to recommend it as the starting point for a book about the environment, its protection and its politics. Its short history – it was discovered only in 1902 – weaves together threads of scientific exploration, military ambition and environmental concern. Beyond that, though, in its liminal way the stratosphere seems to me a perfect setting in which to begin a book which looks at the way the earthsystem works and ways it might work differently, a book about the boundaries between physical planets and imagined worlds.

You are an inhabitant of the Earth's surface who has, in all likelihood, seen more of that surface than your ancestors would have dreamed possible; you are probably the sort of person who can imagine crossing an ocean for a holiday: and so you think that you know the Earth. But less than a day's walk vertically above you, your planet offers an environment beyond your ken, a realm without local features or breathable air, a windy but oddly weatherless stack of atmospheric layers sliding around from equator to pole without storms or clouds. The rules that govern the workings of the lower atmosphere are turned on their head up here, and common-sense ideas about the world you have picked up on its surface hold no sway. In understanding the world below, science can feel like an optional extra. Here it is indispensable.

The stratosphere is closed – a volume of about 15 billion cubic kilometres with well-defined boundaries at its base and at its top. At the same time as being finite, though, it is all-encompassing – no bit of the world below lacks a stratosphere above, no trip beyond the world can avoid passing through it. In this, it is a realm not simply described by science, but oddly akin to science itself: limited but all-encompassing. Like the stratosphere, science is, in its way, alien to everyone; it is at the same time, and by the same measure, common to all, sheltering the just and unjust alike. It provides a viewpoint from which the world is bigger and stranger than it seems from the surface. The world thus revealed is more abstract, too, and there is no denying that something is lost in that. Yet a sort of universality is gained, and a liberating rootlessness.

I prize that rootlessness. I also worry about it. That is why, in our thinking, I would not have our scientifically informed imaginations waltz around this vast curving ballroom completely unconstrained, like dancing giants of the mind. That is why I insist, as you look out to the blue-white-bright horizon, that you also see the wing, straight and true and joined to your point of view. Because there must be a wing. With the exception of the very occasional balloon, it is only with wings that people rise this high into the stratosphere. And I would not ask you to picture this abstract not-quite-place, this featureless more-than-place, without also having you acknowledge the means by which people come to see such things.

There are stratospheres around other planets. Mars has one; so does Jupiter, and Saturn; Saturn's moon Titan is in the club, too. Spacecraft have measured their heights and their temperatures and sent profiles of them back to Earth – just as orbiters closer to hand have done for the Earth's own stratosphere. But that Earthly stratosphere is not just known from the outside, as the others are, in the planetary way; up here, on the wing, you see it from within, in the way that worlds are seen. There have been no births in this part of the world, though there have been deaths, and few have spent much time here. But what those few achieved here has had human impact. What could be seen from up here helped to determine the early course of the cold war. The damage that might be done to this thinnest of airs did much to define the evolving global environmental consciousness of the

1970s and 1980s. Nor do you have to enter this high realm to partake, a little, of its splendour. Everyone who has ever treasured the quality of light just after sunset, where the blue scattering of the stratosphere comes into its own, has felt their world touched by this planetary periphery. Like all stratospheres, that of the Earth is tied to what lies beneath it by gravity and radiative-transfer mechanisms and atmospheric dynamics. But there are also ties of history, politics and wonder.

A paramount expression of those ties is the wing, and that which the wing entails; the people who worked the moulding and riveting, who designed the cross section, who defined the mission, who built the company that built the aircraft, who elected the politicians who contracted for the aircraft to be built. Without all of them, you couldn't see all of this. Even in a daydream of almost empty immensity, I insist on this rule: you can't imagine the end without imagining the means. And the means are human.

The wing tip edges up. The harsh sun arcs gently across the sky.

Discovering the Stratosphere

The wing is shaped as it is, long, strong-shouldered but thin, because that is what it takes to lift you this high. You are as far above an everyday airliner as that airliner would be above the ground. The surface far below stretches in vast ambit. The horizon would be almost 600 kilometres away were it not obscured by haze and cloud. Those impediments, too, are far below you. No normal cloud can come close to this height, nor could any mountain; the planet's crust would buckle under the weight of a peak that tall. If there were a mountain range built on top of summer thunderclouds, their spreading anvil tops forming its buried roots, you would still pass over its peaks with ease.

You are not just far above the land and sea and cloud. You are above most of the air, as well. Nine-tenths of the atmosphere is below you. What is left is too thin to breathe, too thin to hold warmth, too thin to brighten the night-black sky. The everyday sky of lower places is blue because of the way the molecules of the atmosphere break up sunlight. The longer wavelengths, the reds and greens and yellows, pass through the air relatively unscathed; but the blue, short-waved and flighty, is diffused. While the rest of the light takes a direct route from sun to surface, some of the scattered blue spreads itself across the vault of the sky. Up at 22 kilometres, though, the sunlight has not seen enough molecules for much scattering to take place. That is why the sky above is black and the untwinkling stars shine steadfast and true. You are only just this side of the edge of space.

That is not the only difference. In the lower atmosphere, the air gets colder as you rise higher. For centuries scientists had believed that this was a constant trend – that the air got thinner and colder until it petered out altogether in the vacuum of space. It was because of this deep belief that Léon Teisserenc de Bort, who discovered and named the stratosphere at the turn of the twentieth century, was so surprised when the instruments he was sending up into the sky on balloons seemed to show that, above a certain height, the air got no cooler. Sometimes it even warmed – surely there was some sort of mistake? But after balloons by their dozens had told him the same thing, he decided that there wasn't.

The discovery de Bort announced in 1902 came to be seen as 'the greatest ever made in meteorology', in the words of an English scientist a generation later, because the way the atmosphere changes temperature with height is fundamental to how it behaves. Warm air is buoyant, and rises. In the warmed-from-below air near the surface this buoyancy keeps everything in a state of flux; warm air endlessly rises into the colder air above, stirring things up and causing a great deal of weather. But in the realm to which de Bort had sent his balloons, this cooling trend not only stopped, it was reversed. Warmer air sat on top of cooler air, leaving no scope for buoyancy to create instability; circulation in the stratosphere, it was to turn out, was side to side, not up and down. Layering was not just possible, it was inevitable. Hence de Bort's name for the new realm he had discovered: the stratosphere is so called because *stratos* means layer. The lower atmosphere, in contrast, he dubbed the troposphere, from *tropos*, to turn or stir.

At the end of the nineteenth century dividing the Earth into concentric spheres with Greek prefixes was becoming popular, a terminological expression of a deeper shift in the way people thought and talked about the planet. The first scientists to take the study of the Earth as their particular domain had been geologists, and they had come to the Earth through rocks, rocks that could be admired in landscapes, collected for museums and mined for money. Their Earth was primarily a history, because it was history, they came to understand, that explained which rocks were to be found where. They sliced up the history of the world ever more finely in space and time, all the time arguing over the sequence and nature of the events for which they thought they saw evidence.

The physicists who turned their imaginations to the planet as a whole (and to other planets too) towards the end of the nineteenth century took a different approach. Only by emphasizing the whole over the parts and the idealized over the specific could the numerical approaches they prized be brought to bear on their new subject. They found dividing the Earth into simplified spheres much more congenial than dividing its history into hard-to-perceive periods. So under the atmosphere (a term which had, as it happened, first been applied to imagined gases around the moon, and only later used to describe the air around the Earth) there was a lithosphere – the stiff rocky shell of the Earth's surface – and a hydrosphere – the oceans. By the early twentieth century all sorts of specialists wanted spheres of their own; glaciologists termed the icy parts of the world the cryosphere, soil scientists took as their subject the pedosphere. Seismologists discovered new spheres within the Earth, atmospheric physicists found new ones in the sky; they eventually stacked three ever-more-tenuous shells – the mesosphere, the thermosphere and the exosphere – on top of the stratosphere.

Those higher realms, though, quickly become otherworldly. They cannot be reached with wings and only barely with balloons; they were not explored before the age of rockets. The stratosphere is the highest realm humans have visited, rather than simply passed through. The first of them to do so, 30 years after its discovery, peered out of tiny windows in metal gondolas hung beneath vast balloons. They rose up in part for adventure, in part for glory, in part for science. There were strange radiations at the top of the world: hard ultraviolet light not seen in the lower atmosphere; the newly discovered 'cosmic rays' held by some to be the birth pangs of new matter. Their more fanciful chroniclers saw the stratonauts pushed up against the boundaries of humankind's narrow reality; Gerald Heard, who wrote science commentaries for the BBC in the 1930s, talked of them being poised to feel the 'untamed energy of the outer universe', of coming close to 'that ocean of energy in which all the suns and raging stars are but mist and spray'.* This sense of being on the edge of immensity is at the heart of the experience of the sublime, a response to the power and scope of nature which, in the words of Edmund Burke, 'fills the mind with grand ideas, and turns the soul in upon itself'. The stratosphere, then and now, offered the sublime in heady drafts.

After the Second World War, flights to the stratosphere became much more common, and wing-borne to boot. They became less about what lay beyond, and more about what sat below; they

---

* Heard was devoted to the breaking of boundaries, participating in his friend Aldous Huxley's experiments with LSD later in his life, as a polymathic Californian.

also became much more predominantly American. In partially taming some droplets of Heard's ocean of energy, the Manhattan Project changed the way strategists thought about power on a planetary scale. They found in the stratosphere a frontier-free high ground from which the warriors in charge of the new nuclear arsenals could look down; America's first great expression of global power, the B-52 bomber, was accordingly named the Stratofortress. The nuclear age also realigned the interests of the military and those of scientists. Geophysicists, aware that, unlike other physicists, they were largely bereft of laboratories, had come to think of the upper atmosphere as something akin to a replacement, a natural laboratory, and the military, impressed by what physicists had put into the bomb bays of its Stratofortresses, proved happy to offer them better access to it, with rockets and new sorts of aircraft. It also offered them new opportunities to experiment in it.

If there is an emblem for this view of the world, it is the Lockheed U-2, created to serve America's national security, still flying more than 50 years later in the name of science. A remarkable reconnaissance aircraft conceived, built and flown in utter secrecy, it took its pilots far higher than a B-52 could, up to the heights where this chapter began, heights beyond the reach, at least at first, of any interceptors or anti-aircraft missiles. There, the U-2 pilots carried out one of the most important missions of the cold war, attempting to count and pinpoint the USSR's nuclear weapons. From an impervious height, the U-2's elite pilots used cameras finer than any made before to gaze in near omniscience on a world that seemed all but limitless.

And yet, as they did so, the pilots themselves were subject to the tightest of limits. They could see across whole countries; but they could not scratch their own noses, their faces trapped within the helmets of pressure suits no pilots before them had needed. Their wings spanned 40 metres; but they could not stretch their arms or legs, or even reach all the controls in their cockpit without makeshift tools.

Their flight was made possible by the work of thousands of people down below, work carried out at the behest of governments that represented millions more. Yet in the early days of the U-2, the pilots were as alone as it is possible to be, flying thousands of kilometres across enemy territory in radio silence for hour after hour, navigating by the diamond-steady stars (their cockpits had sextants built in). And for all the unimpeded weather-free emptiness of their realm, their flight was constrained within the tightest of aerodynamic envelopes. In air that thin, the difference between flying too fast – and thus being struck down by turbulence – and flying too slow – and thus stalling – was about 20 kilometres an hour. Bank too steeply, and one wingtip would start to tremble as the other stopped providing any lift at all. The pilots had a name for this constraint. They called it the coffin corner.

And yet, for all the boundaries pressing in on them, they could and did appreciate the grandeur of their situation. Even in the coffin corner they could feel at ease among immensities, and joyful in their flight.

Fallout
The U-2 was not just equipped to see. It could sniff as well, detecting some of the aerosols of the upper air *in situ* and also bringing down samples for further study. This was so it could measure a new addition to the stratosphere: nuclear fallout.

Between 1945 and 1963, the United States and the USSR tested hundreds of nuclear weapons, the most disturbing experiments ever carried out by warrior-physicists. Like the radioactive tracers then starting to be used to take pictures of blood vessels, the fallout created and lifted up in these blasts allowed previously invisible processes to be studied. In the mid-1950s Roger Revelle, an oceanographer who was director of the Scripps Institution of Oceanography, used it to look into the absorption of carbon dioxide by the ocean. His studies convinced him, and later his colleagues, that industrial carbon-dioxide emissions could and would change the climate. In tune with the geophysicists' way of thinking about the Earth as a laboratory, he referred to this anthropogenic change as a 'large-scale geophysical experiment of a kind that could not have happened in the past nor be reproduced in the future'.*

The U-2s were sent up to sample the bombs' radioactive fallout before it fell, thus providing scientists with insights into both the weapons being tested and the circulation of the stratosphere. Some pilots preferred these missions to the photo-reconnaissance work. For reconnaissance, the aim was to fly as precisely as possible along paths laid down in advance by the intelligence agencies. In sampling missions, on the other hand, the pilots, guided only by the clicking of their Geiger counters, were given the initiative, seeking out the thickest parts of the invisible radioactive slicks, gently climbing and swooping as they looked for the thin layer where the passing fallout was concentrated, flying both with and against the invisible grain of the stratosphere.

Spread around the world by the stratosphere, fallout from nuclear tests eventually settled on soils, in shallow seas, on glaciers and icecaps. With the right instruments, that layer of fallout will be discernible tens of thousands of years from now. It is not necessarily the longest lasting of the marks humankind has left on the face of the Earth, but it is one of the most widely spread and, thanks to the fact that atmospheric testing ended less than twenty years after the first nuclear explosion, one that is very tightly constrained in time. If the Anthropocene ever becomes an

---

* A nice coincidence: Revelle's co-author on that paper, Hans Suess, was the grandson of Eduard Suess, the Austrian geoscientist who first introduced the notion of cutting the Earth up into a lithosphere, a hydrosphere and so on in the 1870s.

officially sanctioned geological term – although it was conceived more as a rhetorical prod, the meticulous boundary setters of the geological profession are now making moves towards giving it a strict official definition – it will need, as all geological periods do, a stratigraphic marker, a boundary layer before which the Earth had not entered the Anthropocene, and after which it had. The sediments enriched in fallout during the 1950s are often offered up as a candidate.*

The global spread of fallout was of interest not just to geologists of the future; it mattered to anti-nuclear campaigners in the there and then. In the late 1950s and early 1960s, fallout from nuclear tests became the first truly global environmental issue to engender widespread protest and political debate.

America's Atomic Energy Commission (AEC), responsible for making and testing the nation's bombs, held that risks to human health from the fallout produced in nuclear tests were very small at any great distance from the site of the test. One of its arguments, expounded by chemist and AEC commissioner Willard Libby, was that for fallout to travel long distances, it would have to get into the stratosphere, which he saw as empty and basically static. Fallout that got up there, this argument went, would stay there for a decade or so, which was time enough for it to spread evenly around the world; it would come back down so slowly, and in such a diluted form, that it would pose no serious problems.

Others held a more dynamic view of the stratosphere and a more cautious view of fallout. A programme of U-2 flights that started in 1957 helped to swing the argument their way (though the results were classified, so not all the parties to the discussion could make use of them). To begin with, they pointed out that the fallout was not, for the most part, getting into the 20-kilometre-plus reaches of the stratosphere where this chapter began, a realm sometimes known as the overworld. It was mostly

* We will look at what I believe to be a better one in Chapter Eight.

in the lower stratosphere, and it stayed there typically only a couple of years, rather than the ten or more that Libby had been claiming.

Nor did it come down evenly. In the 1940s and 1950s, atmospheric scientists began to appreciate the ways in which the behaviour of the tropopause – the boundary between the troposphere and the stratosphere, defined as the height at which the air stops cooling with altitude – changes from place to place. At the equator, the tropopause sits at 20 kilometres or so, at the base of the overworld and out of reach of any aircraft less capable than a U-2. As you head towards one or other of the poles, though, it gets lower; at high latitudes it drops down to about 12 kilometres, meaning that everyday airliners can get up into the stratosphere. This gradient is tied to the fact that the stratosphere, as well as being party to winds from west to east and east to west, also circulates from the equator to the poles. Air gets in from the troposphere at the equator and flows downhill, as it were, towards the poles. Fallout-enriched air lifted into the lower stratosphere by the rising mushroom cloud of a nuclear test would be carried along with the flow.

That poleward flow, though, is not even. The constant stirrings of the troposphere, with their rising and falling air masses, mean that the tropopause is uneven; as American meteorologist Oliver Wulf put it, the stratosphere flows across a 'stony river bed', bumpy both in time and space. And the flow is twisted, too. Because the Earth is spinning, the current of air moving north or south from the equator gets turned to the east or west, depending on the hemisphere, as it goes, and this torsion stresses the tropopause. The stress is expressed in winds of great speed that meander round the planet like rushing rivers of air at the top of the troposphere – the jet streams. In the zones where these jet streams occur, air from the stratosphere mixes more freely with that from the troposphere than it does elsewhere. This meant that fallout not only came down from the lower stratosphere quicker than Libby and the AEC thought; it also did so preferentially in particular parts of North America, Europe, Russia and China – that is, in the belt of land above which the northern jet stream winds its way.

The new idea of the stratosphere as a selective delivery mechanism for fallout, rather than a slow-draining reservoir for it, was ably disseminated by Barry Commoner, a biologist who would go on to become one of the leading environmental activists of the 1960s and 1970s. In an influential book published in 1970, *The Closing Circle*, Commoner formulated four rules of ecology, the first two of which were 'Everything is connected' and 'Everything has to go somewhere', with the explanatory corollary, 'There is no away into which things can be thrown'. The fallout's passage through the stratosphere exemplified both of these rules. The stratosphere was what connected weapon tests in the Pacific with radioactive strontium levels in the bones of children in America. It was not an away into which things could be thrown. It was a component of the earthsystem with its own complex behaviour, linked to what went on below not just by the dynamics of the climate but by the turning winds of politics.

The Ozone Layer

The stratosphere had a supporting role in the first great global environmental issue – that of fallout. It sat out the second such issue – the global spread of organic pesticides such as DDT. But it took a leading role in the third – that of damage to the ozone layer.

Scientists had known since the second half of the nineteenth century that the lack of short-wavelength ultraviolet radiation in sunlight at the surface of the Earth was down to the fact that it was absorbed by ozone in the atmosphere, and some had suspected that that ozone was concentrated at high altitudes. Soon after de Bort discovered the stratosphere, Alfred Wegener, more famous for coming up with the theory of continental drift, suggested that this newly discovered sphere might be the ozone's home, and that turned out to be the case.

The ozone was not just contained by the stratosphere – it was what gave the stratosphere its upside-down character. Absorbing ultraviolet radiation makes the ozone hot. Because the ultraviolet is stronger at height, and because the air higher up is less dense and thus more easily warmed, this absorption heats the top of the stratosphere more than the bottom.

But the ozone does not just sit there maintaining the stratosphere's stabilizing hotter-at-the-top structure. It is constantly being created and destroyed. Ultraviolet radiation breaks down the two-atom molecules of which common-or-garden oxygen consists into individual atoms, which are highly reactive. Some of these atoms react with other two-atom oxygen molecules to make three-atom oxygen molecules – ozone. Some of them react with existing ozone molecules to recreate the more stable two-atom molecules. To make matters more complicated, ultraviolet light breaks down ozone molecules, too – releasing single atoms which go on to take part in further frolics.

These interlinked reactions were first explained by a British physicist, Sydney Chapman, in 1930. He showed how the fact that all of these reactions go on at different rates, some dependent on the temperature, some on the amount of ultraviolet, affects the amount of ozone found at a particular time or season in a given part of the stratosphere; the balance between the processes that create ozone and those that destroy it changes according to the circumstances.

Chapman's spectacularly impressive work established the idea of an 'ozone layer'. It is not, in fact, a well-defined layer. The ozone is distributed through much of the stratosphere, and even at the altitude where it is most concentrated – up in the overworld, at about 25 kilometres – it makes up no more than a four-hundred-thousandth part of the very thin air. If it were a well-defined layer, though, it would be a remarkably thin one – at room temperature and at sea level it would be just 2.5 mm thick – and that fascinated people. Whether talking to each other in academic texts or to the general public in popular articles, scientists speaking of the ozone layer in the 1930s and 1940s never passed up an opportunity to mention this thinness. In an age of concentric geophysical spheres, the very idea that the Earth could have a shell so thin, and that humans could measure it so precisely, was a source of wonder.

It was also seen as evidence of a certain cosmic providence. In 1933 a report from the Associated Press on work by Charles Abbot, astronomer and president of the Smithsonian Institution, made the point in the *New York Times* under an excited stack of headlines: '1/8 inch of ozone alone saves life – Smithsonian Reports on Study of Dr. Abbot of Gas Wall 40 Miles From Earth – sun's death rays barred'. But such a dramatic take on the issue was unusual, and to some unwelcome. *The Times*'s science editor, Waldemar Kaempffert, used his next weekly column to chide his colleagues on the news pages for the inexplicable prominence they had given the report. Many aspects of the atmosphere, Kaempffert pointed out, were necessary for life – the ozone layer was nothing special. When Abbot himself wrote about the subject for the general public, in *The Sun and the Welfare of Man*, the impossibility of life without an ozone layer merited a paragraph or so. In *Exploring the Upper Atmosphere* by Dorothy Fisk, a delightfully eclectic British author of the same vintage, it gets noted in a similarly scant way.

Indeed, Fisk and Abbot both spend considerably more time celebrating the fact that the ozone layer is not thicker – since if it were, humans would be deprived of the longer, softer ultraviolet wavelengths, too. This radiation had recently been shown to be the component of sunlight that caused human skin to generate vitamin D; with less ultraviolet there would be more rickets. In the 1930s, rickets and vitamins were seen as much more interesting than some abstract notion of the habitability of the planet. The ozone layer, Ms Fisk pointed out, was beyond any human influence; vitamins, on the other hand, could be destroyed by over-cooking on the gas cookers spreading into more and more kitchens. That was the sort of threat on which people should concentrate.

At no point in any of this did anyone use the word that today seems almost synonymous with the ozone layer – 'fragile'. The layer was thin, yes – but there was no reason not to think it robust and not to expect it to be permanent. Chapman speculated about the possibility of making holes in it on a temporary basis for the benefit of astronomers, thinking they would learn a lot if they could see the universe as revealed by its ultraviolet emissions; but it was hardly a very serious proposition, and in time the possibility of simply putting the ultraviolet telescopes on to satellites came to seem far more practical. In the 1950s a few scientists interested in the very early history of life gave thought to the fact that, before photosynthesis endowed the atmosphere with significant amounts of oxygen, there would have been no ozone and thus a lot more ultraviolet. But if they saw the vestiges of the ozone layer's beginnings, they saw no prospect of its end.

In the 1960s this began to change. New data showed that Chapman's brilliant chemical analysis could not be the whole story; given the strength of the sun's ultraviolet rays (which could now be measured directly by satellites), more ozone had to be being produced in the stratosphere than Chapman's chemistry predicted. Because Chapman's explanation of the stable level of ozone depended on a balance between the rate of its production and the rate of its destruction, a higher level of production meant there had to be extra processes of destruction, too. None of the gases present in the stratosphere in bulk seemed to be responsible, so the culprit must be a trace gas, one present at lower concentrations even than the ozone itself.

This, more than anything, opened the door to worries that the ozone layer might, indeed, be fragile. Harold Johnston, one of those who spent the 1960s looking for the trace gases responsible for destroying the ozone, later said that he thought, before the fact, that people were 'inclined to assume [that the atmosphere was] impervious to human intervention' simply because there was so much of it. There were, after all, more than a million tonnes of atmosphere per person back then (a bit less than a million now, thanks to population growth). There were hundreds of tonnes of carbon dioxide for each person, and even the ozone layer, so often singled out for its thinness, added up to more than a tonne each. Such amounts seemed to put the earthsystem beyond touch; it simply seemed too big to budge.

The trace gases with which Johnson came to be concerned, though — nitrogen oxides, which scientists abbreviate as NOx — were able to punch above their weight. Each molecule of NOx could shift the stratosphere's chemical balance in such a way as to remove many molecules of ozone. So just a few kilograms per person could make a difference — a small enough amount, Johnson noted, that the inadvertent production of NOx by human industry might drive a global change. There is a general lesson there: humans have found two ways of making a difference to the workings of the earthsystem. One requires large, species-wide effort, such as millennia spent devotedly farming, or a century's concentrated effort devoted to the burning of fossil fuels. The other requires finding a small thing that makes a big difference, as NOx can in the stratosphere — something that offers leverage. The same will apply, as we shall see, when you want to make such differences deliberately. Finding a powerful lever is the key to moving the earthsystem.

By the 1960s there were plans in America, the USSR, Britain and France to develop new airliners that would fly higher in order to fly faster; in doing so they would inject NOx from their exhaust gases into the ozone layer proper. (Then, as now, many airliners passed through the stratosphere at high latitudes, where the tropopause is low, but such traffic is not in itself a worry when it came to ozone, which is almost all found at higher altitudes). Rockets, meanwhile, were starting to punch their way right through the stratosphere, leaving NOx and more in their wake – and if the Earth was really on the brink of a 'space age', there were going to be a lot more rockets. From the mid-1960s onwards the possibility that rockets and jets might damage the ozone layer on a global scale was being looked at by scientists, by military planners and by concerned environmentalists.

In 1970 America decided not to go ahead with the development of supersonic airliners. It was a crucial moment in environmental history – the first time that, in part because of public concerns about the environment, a futuristic technology of the sort that had long enlivened the covers of popular science and science fiction magazines was deliberately foregone. Fears about the health of the stratosphere played only a part in this – by the time there was plausible scientific evidence of the damaging effects of the proposed airliners' exhaust, the programme was already doomed by its cost and by a more immediate, if less global, environmental concern about sonic booms. But continuing research that built on what Johnston and Paul Crutzen – a wily and astute atmospheric chemist who, a couple of decades later, introduced the term 'Anthropocene' into science's vocabulary – had discovered about NOx showed that a fleet of 500 supersonic airliners would reduce stratospheric ozone levels by 10–20 per cent worldwide. The idea that the ozone layer was fragile, and that human activity could significantly affect a fundamental part of the earthsystem, was firmly established.

In the end, though, it was not through the nitrogen oxides produced by jet exhausts that humans began to do serious damage to the ozone layer. It was through the chlorofluorocarbons

(CFCs) made to cool fridges and squirt hairsprays and for all sorts of industrial uses, all as far removed from the ozone layer as Ms Fisk's vitamin-destroying gas cookers had been.* They were very long lasting and very chemically inert – until they rose to the stratosphere, where ultraviolet light tore off their chlorine atoms in a way that made them powerful catalysts for ozone destruction. These reactions gave them incredible leverage: a single CFC molecule could destroy 100,000 ozone molecules. Worldwide production of a few hundred grammes per person per year was enough to have a real impact on the earthsystem. In the 1970s CFCs started to be seen by many as a potential environmental disaster.

In the years after a hole in the ozone layer was discovered over Antarctica in 1985, a NASA U-2 devoted to civilian research was dispatched to investigate, flying some of the longest and most dangerous missions in the aircraft's history. Jim Anderson, one of the atmospheric chemists who led the work, today speaks movingly of the burden of responsibility that came with knowing that men were risking their lives to take the instruments he and his colleagues had made into the Antarctic skies. The data they brought back allowed Anderson and his colleagues to show that the hole had been created by CFCs. While the flights were under way, diplomats were negotiating a treaty – the Montreal protocol – to eliminate the production of CFCs. It has since become perhaps the single most successful global environmental agreement.

When it was discovered at the beginning of the twentieth century, the stratosphere was a realm of pure science, described only by numbers, visited only by instruments. A lifetime later, its peculiarities had been mapped quite literally to the far ends of

---

* It is worth noting that the CFCs made the refrigerators sitting next to those gas ranges in well-appointed kitchens much safer; ammonia had previously been used as a refrigerant, and it sometimes leaked, killing people. CFCs were saving lives as well as costs.

the Earth by brave pilots, its lower reaches traversed by hundreds of thousands of airline passengers, its integrity challenged by the products of warfare and industry and its protection undertaken by the governments of the world.

The Veilmakers

What more change could another lifetime bring? Come and see. Back up at 22 kilometres, imagine an aircraft similar in shape to the U-2 but bulkier, with stronger shoulders for its long, taut wings. Imagine it not sampling the air, but adding to it – a mist of liquid trailing out behind it all but invisibly, spread through the layers of stable air a few tonnes at a time, eventually forming a layer of aerosols finer than dust. Look to windward and, far off, another aircraft does the same. They are part of a small fleet that flies up here all the time, neither to study nor to spy, but to add to a spreading veil that brightens, very slightly, the haze of the horizon off to one side and lightens, very slightly, the sky as seen from below – a layer that diminishes, just a bit, the amount of sunlight that makes it through the stratosphere to the Earth's surface.

In this big round world these new wings have come to draw a line.

The most widely argued-over form of climate geoengineering is one that would use such a veil to make the world cooler than it would otherwise be. There is much about such an idea to fear. The stories of the wings, and the pilots they carry – the stories that tie the stratosphere into the future world below it – may be stories of ill-informed folly, as those of the SST's swept back wings would have been, or of empire. The pilots of the veilmakers, if the aircraft are designed to need pilots, may be in military uniform, doing with discipline a job that one country and one country alone has decided upon, seeking to tailor the environment to that nation's interests while paying no heed to the interests of any others their actions may harm, either then or further into the future. They may fly with countermeasures to protect them from attack. Their geoengineering may, in itself, constitute an attack on others.

Then again, the pilots might simply be skilled functionaries – professional employees like those who take aircraft from A to B for Federal Express or easyJet in their shirtsleeves. They may be contracted to some international organization that has reasonable standing, and offers convincing reasons, for what it is doing. They may see their task as a distasteful necessity; they may, though, take joy in it, as well as pride. They may be enthusiasts smiling and chatting with tourists who have paid to come along, they may embellish their veilmakers with names from history and fiction – *Beagle* and *Challenger*, *Pequod* and *Surprise*, *Eagle* and *Discovery*, *Tannhauser Gate* and *The Ends of Invention* and *But the Sky, My Lady! The Sky!* Or maybe they wear not flight suits or shirtsleeves but vestments. Maybe they are adepts called to be part of an order that protects the people of the Earth from the extremes of climate.

And whether they are priests or warriors, fanboys or functionaries, they may be misguided.

The rest of this part of the book looks at science's current understanding of how energy flows through the earthsystem, at how veils of fine particles in the stratosphere can provide leverage to alter that flow, and at what effects leaning on such levers could have. There will be details and abstractions, histories and theories, computer modelling and moral philosophy.

Through it all, though, spare the occasional thought for the stratosphere itself, for a play of scattered light on the edge of space, for a boundary, a protection, a vulnerability.

Think of the people up there. Think of the wings that lift them and the stories that build the wings. But, for all that I cherish the idea of looking out at those wings, remember that looking out this way is a privilege. Most people will not look out at the wings; they will look up at them, far, far above, untouchable as angels. Like most of those who look up at wings of all sorts today, they may have little say in where they fly, or at whose command. Some may look with wonder; others may look with unease, with fear, with anger.

Who built those wings? What did they build them for? To what other uses might they be put?

What insignia catches the sun as the aircraft banks on to a new course?

2

A Planet Called Weather

*The notion of the global environment, far from marking
humanity's reintegration into the world, signals the
culmination of a process of separation.*

Tim Ingold, 'Globes and Spheres:
The Topology of Environmentalism' (1993)

Six years after Tony LeVier, an American test pilot, took off in the
first U-2, Yuri Gagarin became the first man to go right through
the stratosphere and come out the other side. The missiles the
U-2 had been built to peer down on had grown powerful
enough to throw people or weapons, according to taste, around
the planet or into orbit. Superpower rivalry required that the
prowess of these missiles be proved competitively, and so it was.
Less than a decade after Gagarin's flight Americans were thrown
as far as the moon, and the conflict that threatened to end the
world enabled a new appreciation of it.

This chapter is about that new appreciation, and the changes
that it brought with it. It is about how the earthsystem works
– something which needs to be understood before you think
too deeply about how to change it – and how humans think
about it. It is thus both about climate science and about the way
people think about their environment.

Before the space age, the environment had always been, by
definition, the thing one was in. The environment was immediate;

57

when it needed fixing it stung the eyes, it dirtied the washing that hung on the line, it sickened the children who drank from the tap. The conservation of the distant wild also had a place on the periphery of politics, but the central environmental concern was maintaining the quality of people's surroundings. Clean air acts were not passed for the benefit of the air, but for the benefit of those who breathed it.

In the 1960s, though, environmentalism expanded its remit to the distant and the difficult to detect. The new threats of fallout and pesticides were perceivable only with expert knowledge; they also operated on the scale of the whole world. Interconnected food webs could take DDT from cornfields in Iowa to the blubber of Antarctic seals just as the stratosphere could deliver fallout thrown up at Bikini Atoll into the tooth enamel of German infants. Seeing these threats and dealing with them required a scientific outlook that went beyond anyone's immediate surroundings and encompassed the entire globe. It was much the same sort of outlook that the worldwide struggle of the cold war was embracing for its own purposes – one that stood outside the everyday to see the global whole more clearly.

This new way of seeing strengthened environmentalism. It also changed it. Where once damage to the environment had been obvious, now it was subtle; the senses were not to be trusted in the face of these invisible threats which preserved the appearance of wholesomeness. Pollution was no longer an obvious stain on the world but a hidden sickness within it, one that could hollow it out and turn it into a toxic replica of itself. This uneasiness was underlined by the new role of cancer in environmental thought. Cancer was what first made fallout scary, and soon after that it did the same for pesticides too – Rachel Carson drew the analogy right at the beginning of *Silent Spring*, the 1962 book which made pesticide residues a global environmental issue. Eating away from within, already there before you knew it was killing you, cancer offered a bodily correlate to the feeling of a pervasive wrongness about the world. It matched fears of uncontrolled growth in the consumer economy with the horror of uncontrolled growth within.

As the world revealed by science, rather than the world as experienced by its inhabitants, became ever more important to environmental action, the planet revealed by such science became a subject of environmental concern in its own right. The Earth itself needed to be protected. This might seem surprising. The planet is a vast thing, after all. It offers each of its human inhabitants not just a million tonnes of atmosphere, but a billion tonnes of ocean and a trillion trillion tonnes of lithosphere. Could such a behemoth not take care of itself?

But perhaps the Earth's size, like the purity of its air, was just another illusion. Perhaps the Earth was, contrary to one's untutored perceptions, small.

The sense that the Earth was getting smaller was hardly a novelty. It had been a recurrent theme in European thought more or less ever since Europeans first started to annex far-off bits of the planet for their own ends. By the twentieth century it had become hard to find anyone mentioning the size of the planet for any reason other than to remark on the fact that it was diminishing. The insight was ceaselessly presented as surprising, but it really wasn't. First the world had been tied together by steamers and cables and timetables, then by radio and aircraft and general staffs. By the 1960s, mobilization for a world war – and, indeed, the fighting of one, should it come to that – was a matter not of months but of minutes. In the 1963 speech that led to the atmospheric test ban in 1963, thus laying to rest the first global environmental issue, John F. Kennedy made the diminished globe the key fact for discussion of the human future: 'In the final analysis, our most basic common link is that we all inhabit this small planet.'

The Apollo programme that Kennedy had started two years previously provided what would come to be the definitive iconographic evidence for this idea. Everyone said the world was small: the Apollo astronauts had the experiential and photographic proof. From the U-2, the Earth looked utterly vast – as big as the sky. From the moon it looked no bigger than an ashtray at the other end of a bar. 'You can hide the Earth behind your thumb,' reported Jim Lovell, pilot of *Apollo 8* and commander of *Apollo 13*. 'Everything that you've ever known, your loved ones, your business, the problems of the Earth itself – all behind your thumb.'

The Earth now had a face, and a persona. It was an object for celebration – at the Earth Days that began in America in 1970s – and protection. It was remarkable, unique, small – fragile. Lowell's striking *Apollo 8* image of the disk of the Earth blotted out by a thumb brought to mind a compelling, if subliminal, comparison to another disk and digit – the nuclear 'button' covered by a finger about to send thousands of lesser rockets on flatter, deadlier trajectories. The cold war and the environmental movement were not linked only by a new interest in monitoring and understanding the world as a whole. They were linked by their understanding that its smallness could be equated with fragility.

Fragility might seem simply to be a corollary of small size, but it isn't; try stepping on a ball bearing. Or think about the way people talked about the ozone layer in the 1930s. That it was strikingly thin was always part of what made it interesting. The idea that it was fragile never even cropped up. The fact that the rich, bright-lit, fascinatingly filigreed swirls and patterns of the earthsystem, as seen from 400,000 kilometres away, looked fragile to the people seeing them – rather than, say, seeming to churn with potential and promise – reflected how those people looked.

It is not in any way wrong to recognize the fragility of the human world. But to transfer that sense of fragility to the planet, as environmental rhetoric has done in the decades that followed the Apollo programme, is off-putting, at least to me. To talk of 'saving the planet' rather than of preserving and enhancing the boons the planet offers to the people who live on it divorces environmental rhetoric from the moral causes that most concern me. You might see this as quibbling on my part: you might think that when people talk of 'saving the planet' they actually just mean something like 'preserving the environment' or 'protecting nature', and to some extent they surely do. But planet-speak does, subtly and pervasively, change the focus.

The overlapping and conflicting ways in which one can think of nature as both outside and inside, something both beyond the social and, as human nature, at the heart of the individual have been the stuff of art, philosophy, argument and quiet contemplation for centuries. There is no such accumulated nuance around planet-speak. By seeking to raise the stakes – it's about the whole planet, dammit – planet-speak weakens the ties between the natural and the human. It channels concern away from your life or the lives of others, towards the well-being of an abstract geophysical entity. Like a spaceship, it takes environmental concern out of the world.

Concern for the planet itself is also, as a matter of blunt fact, misplaced. Unlike humans and their societies, unlike many of the landscapes and creatures that people love and depend upon, the planet is not fragile in any way. The Earth has persisted as a habitable planet for four billion years, moving its inhabitants in and out of galactic dust clouds, surviving asteroid impacts the energy of which dwarfed that now locked up in nuclear arsenals, riding out the shockwaves of nearby supernovae and adapting itself to the challenge of a sun which has, over that time, grown more than 40 per cent brighter. In terms of its experience, the

Earth is huge – its history stretches back as far as a third of the lifespan of the universe. In terms of the hazards it has faced over that time it looks anything but fragile.*

The other thing that was emphasized in commentary on the Apollo pictures was the Earth's seeming isolation. Just as seeing the Earth as small and fragile makes it seem a more appropriate subject for care, so does seeing it as isolated, a vision that deprives it of support and implicitly increases its value. But the Earth's isolation, which seems so complete when you see it displayed against the velvet background of space like a pearl on a velvet bolster, is in one crucial way illusory. Materially, it is true, no other worlds push up against that which humans inhabit. Very little matter arrives from outer space – a drizzle of dust, some shooting stars, those discommodious but happily uncommon asteroids. And even less leaves. There's a constant whisper of gas whipped away from the exosphere by the solar wind, and in the past few decades there have been hundreds of spacecraft (though many of those in low orbits will, eventually, return). Their military, scientific, economic and cultural value has been immeasurable. Their tonnage has been trivial.

Think about energy instead of matter, though, and things look very different. Energy flows to and from the planet at a prodigious rate. An appreciation of this flow of cosmic energy is fundamental to understanding the earthsystem, the science of climate change and the potential of climate geoengineering.

The Worldfalls

The energy arrives in the form of 170,000 trillion watts of sunlight. A bit less than a third of that torrent splashes straight off back into space. Scattering by the molecules of the

---

* Perhaps it does not matter to you that the Earth has been around for an appreciable fraction of the age of the universe; but if its comparatively great age does not matter, why should its cosmically negligible size? Judge the Earth for the wonder it is.

atmosphere is responsible for some of that backlash – part of the same scattering which, seen from below, makes the sky blue. Scattering by aerosols plays a role, too, and about 7 per cent bounces straight off the surface. But most of the sunlight reflected from the planet comes off the clouds.

In terms of telling your planet from your world, it is interesting to note that this is not how things were imagined before the space age. In the first half of the twentieth century, artists working for science fiction magazines, *National Geographic*, astronomical observatories and other bodies with the foresight to care about the matter assumed that what people would see from space would be the aspects of the planet that mattered in the human world – coastlines, cities, the features learned in geography lessons. Their Earths look like classroom globes, with any clouds sparse and carefully arranged so as not to obscure the important stuff that lies beneath. In photographs of the planet seen from distant space, though, it is meteorology, not geography, which dominates the view. In low orbit you will see bright deserts, dark seas and lush forests, but as you recede towards the moon, ever-changing clouds impose themselves over everything, a broken band around the equator, a white train of choppy waves in temperate climes, typhoon knots over the oceans. Eternally romantic about the sea, the science fiction author Arthur C. Clarke used to quip that if it had been discovered from space, the Earth would instead have been called Ocean, given that most of its surface is covered by its waters. But the seas reflect very little light. Going just on what you can see from out by the moon, the best name for the Earth would be Cloud – or Weather.

In total, the clouds, the aerosols and the surface reflect back 30 per cent of the sunlight. The amount left over remains remarkable: 120,000 terawatts is almost 10,000 times the power that flows through the planet's industrial civilization – all the world's reactors, turbines, cars, furnaces, boilers, generators and cooking fires provide about 15 terawatts. Yet so smooth is the

120,000-terawatt flow that you hardly notice it; it just feels like the way the world is.

Here's a way to try and get to grips with such energies. In the nineteenth century Horseshoe Falls, the most familiar, forceful and dramatic cataract at Niagara, was perhaps the most celebrated image of nature's sublime power in America. Picture it in full spate; feel its spray and its roaring and its endlessness. Now increase the height of the falls by a factor of 20; you have a kilometre of leaping water, higher than Angel Falls in Venezuela. And now increase the flow by a factor of ten. That's 300 tonnes of water falling over each metre of the lip of the falls every second, compared with the 30 tonnes at Niagara.

And then widen the falls. Stretch them until they span a continent, with billions of tonnes of water falling over them every second. Go on widening them until they encompass the equator itself: an unending kilometre-high wall of water 40,000 kilometres long, cutting the world in half with its incessant thunder, deafening leviathan in the abyss. That is what 120,000 terawatts looks like. That is what drives the world in which you live.

To make those Worldfalls real, all the sunlight the Earth absorbs would have to be employed by fearsome engines devoted to pumping water from the ocean at the bottom of the falls back to the ocean at the top. In the real world a great deal of the sunlight does indeed power what, to physicists, count as engines – but they are far more diverse, in size and effect. The grandest does what the fearsome engines of the Worldfalls do – it lifts water. Evaporating water from the surface so that it can condense into clouds and fall back elsewhere uses about a third of the energy that the planet absorbs as heat. Lesser engines drive the winds and ocean currents, from the trade winds and the great ocean gyres to the sea breeze of a summer's evening, all in aggregate taking heat from the warm equator to the chilly poles – the engines of the Earth are forever smearing out contrasts in temperature. And tiniest, but most vital: molecular engines in the earthsystem's plants, algae and photosynthetic bacteria use about a thousandth of that sunlight to move hydrogen ions through membranes. These are the engines of photosynthesis, and it is through their work that a tiny sliver of sunlight is turned into chemical energy. That is the work that fuels all the living things you have ever seen.

All these engines, like all engines everywhere, share a common feature. While they redistribute energy, they do not destroy it. Engines turn some energy from a form in which it can be used into one in which it cannot – that is what an engineer means by 'work' – but the energy is still there at the end. It is just spread around the environment as 'waste heat'. Some of the 120,000 terawatts of sunlight that hits the Earth does some work; some does not. It all ends up as heat, regardless.

Except this is not really an ending up. The laws of thermodynamics insist that everything that has a temperature has to emit radiation. Things at the Earth's temperature emit it at infrared wavelengths. And this is the ultimate fate of all that 120,000 terawatts; it heads off back to the cosmos as infrared radiation just as fast as it comes in as sunlight. No energy is lost, because energy cannot be created or destroyed. It just comes and goes. But it can do so in a roundabout way. This is the great thing about being a planet with an atmosphere.

The average temperature of the Earth's surface is about 14°C. If you were to look back at the Earth from the moon in the infrared, though, and work out the planet's temperature from what you were seeing, you would calculate the temperature as −15°C, more or less. This is because the atmosphere absorbs some of the outgoing infrared. The gases that are particularly good at this absorbing – water vapour, carbon dioxide and methane – are the ones thought of as greenhouse gases.

As well as absorbing infrared, every bit of the atmosphere also emits some, because the laws of thermodynamics insist that everything that has a temperature has to emit radiation appropriate to that temperature. This emission takes place in every direction – up, down and sideways. So some of the infrared goes back down towards the surface, and some goes up. Some of this upwelling infrared then gets absorbed by higher layers, and so on, until you get to a layer where an appreciable fraction of the infrared can make it out into space unmolested. At this, the point of departure for the infrared going back out into the cosmos, the air has become very cold: it is down to around −15°C. But there's a world of warmth below.

The Trenberth Diagram and Climate Science

You can see the net effect of all this in the Trenberth diagram on the next page, a way of visualizing the flow of energy through the earthsystem that has been worked on over decades by Kevin Trenberth, an atmospheric scientist from New Zealand now based at the National Centre for Atmospheric Research (NCAR) in Boulder, Colorado, with the help of various colleagues. It is not as pretty a picture of the planet as those taken from the Apollo spacecraft, but it has charms both pragmatic and didactic. Where the iconic views of the Earth from space leave the flow of energy invisible, the Trenberth diagram shows nothing else.

In the diagram, the width of the arrows is proportional to the flow of energy that they represent, with the flow measured in terms of the number of watts per square metre of surface. Thus, on the left-hand side, the arrow of incoming sunlight gets thinner as energy is stripped away by the reflecting clouds or the absorbent atmosphere. The key to why the Earth's surface is warm can be seen in the big, thick arrows on the right-hand side. The surface is getting much more energy in the form of infrared from the sky above than it is in the form of direct sunlight. Similarly, the lower atmosphere is getting more heat from the ground below than from the sun above. The flows back and forth within the system are much larger than the net flow in and out. The net imbalance between the in and the out

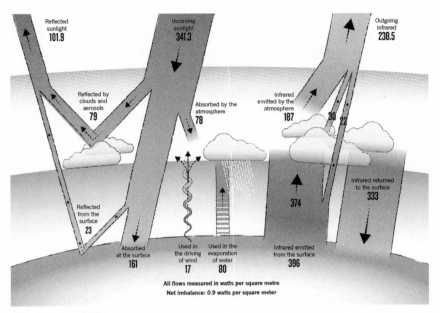

The Trenberth Diagram

– the underlying cause of the climate problem – is too small to be seen.

Not all the energy absorbed by the surface and then sent back to the atmosphere travels in the form of infrared radiation. Some warms the air directly, driving convection, the warming-from-below that is responsible for much of the troposphere's churning. Then there is the energy which, when absorbed, causes water to evaporate – the energy that powers the engines of the water cycle, lifting 18 billion tonnes of water into the sky every second. When that water vapour condenses back into liquid in the form of cloud droplets – which it all does – the energy that was used to evaporate it is released back into the atmosphere, warming it up. Added up around the whole world, the rate at which heat is pumped into the troposphere by water condensing is equivalent to what you would get if you let off one Hiroshima bomb every second. That's why the world has a lot of weather.

Looking at the Trenberth diagram, the flows it represents seem, if not simple, then basic. So they should. They are the fundamental facts about the way the earthsystem works. But this does not means their scale is obvious, or easy to measure, or known beyond dispute. In this, the diagram stands for the state of the art of pretty much all of climate science. Its overall shape is pretty well established. But the understanding of some of its details is in some ways insufficient, and their quantification approximate.

This is because measuring things on a global scale is very hard, and the details of how to do it can be the subject of disagreement – all the more so in an arena where, as in climate science, results may be politicized. So, for example, satellite measurements of the atmosphere's temperature made using infrared instruments have been the subject of heated debate between various groups over past decades, with a great deal turning on the satellites' exact orientation, the timing of their orbits, and their precise altitude. Ground-based measurements hold other problems. The average temperature of the planet's surface, would seem the simplest of all climatic measurements to obtain. But it is an unholy mess to calculate, relying as it does on observations made at thousands of locations with differing equipment and degrees of diligence. This work is done not by huge, well-financed teams working under a unified and clear-sighted command, but piecemeal in institutions that have sort of chosen, sort of inherited the burden. There has been a cottage industry of bloggers and others intent on suggesting that much of this work is poorly done, and there are undoubtedly some mistakes. But good-faith examinations of the data show that warming is real, and most agree pretty closely on the rate: a global average of 0.16°C per decade since the 1970s.

One of the reasons to trust these results is that linked aspects of the earthsystem can sometimes be measured in very different ways. The rising trend in global temperature can be believed in with confidence because measurements of ocean-surface temperature show it just as clearly as measurements of rising air temperature over the continents. The satellites show it too. If different measurement systems with different biases and different instrumental oddities give comparable results then it becomes much easier to believe them.

The models which climate scientists use to examine the earthsystem's workings also have their problems. Like the spy plane and the satellite, the computer was a creation of the cold war that provided a new way of seeing the earthsystem's workings. From the very earliest days of electronic computing, one of the goals that inspired computer makers was the idea of modelling the atmosphere in such a way as to predict the weather, and, later, the climate. Now as then, computer models cut the earthsystem up into chunks, building the ocean and the atmosphere up as a succession of layers, dividing each layer up into a grid of cells more or less along lines of latitude and longitude (though at the poles it gets trickier). The state of each cell at a given point in time is calculated on the basis of the radiation flowing through it, the clouds it contains and what was going in the neighbouring cells just beforehand using fundamental physics. The modern ones, though, cut the world into far more cells than their predecessors did, and can also cut time into shorter steps. The British Met Office's HadGEM3 climate models can offer their users up to 85 layers of atmosphere and 75 layers of ocean and cells measuring less than one degree of latitude and longitude on a side; this requires a computer capable of hundreds of 'teraflops' – which is to say, one that can make almost a million billion calculations a second.

For all this mathematical mojo, the models are far from perfect – Brian Hoskins, one of Britain's foremost climate scientists, cheerfully acknowledges that they are 'pretty lousy' in many respects, though improving. Getting them to work is something of a dark art; changes meant to improve things can make them worse. And though the simulations they come up with look like the real climate in many ways – they have the same sort of regional climates, in terms of temperature and rainfall; they have the trade winds in the right places; their atmospheres divide into a troposphere and stratosphere realistically, and so on – in others they don't really match up. El Niño events – shifts in the world's climate associated with transfers of heat back and forth across the tropical Pacific – are caught reasonably realistically in some, less so in others. South Asian and African monsoons are often not rendered particularly realistically. El Niño events and monsoons are not little details. They matter to the livelihoods of billions.

When the models do not capture the specifics of these phenomena, they certainly can't be relied on to predict how such things will change in a world that is warmed by higher greenhouse-gas concentrations. But to think that such prediction is a realistic near-term aim for climate science is a misunderstanding. Climate models are much more about understanding possibilities and investigating processes than they are about making predictions. They are useful for saying what sort of things will likely happen as greenhouse-gas levels rise. They are not suited to identifying specific events likely at specified times.

This is in part because there is a lot of background variability in the climate, some of it random, some due to earthsystem processes not yet understood. Take the fact that less surface warming was observed in the 2000s than in the 1990s – a deficit seen by some who argue against action to reduce the risks of climate change as evidence that the models erred in their predictions and should not be trusted. One likely reason for the slow down seems to be that more heat is being stored in the deep oceans. The rate at which heat gets into the deep oceans naturally may vary over time in a way that models do not capture; it may be being changed by some under-appreciated consequence of earlier warming. Trying to work out which of those is the case is a much more interesting undertaking than simply pointing to the temperature discrepancy – because understanding such processes will improve estimates of how much warming can be expected in decades to come.

This bears out what the eminent statistician George Box used to say: 'All models are wrong . . . but some models are useful'; climate modellers like the line a lot, although they can be wary of using it in public. Because their work is of great importance – there is no better guide to what the earthsystem will look like as greenhouse-gas levels rise – it is very tempting to misconstrue it. Those urging a strong response to the risks of climate change will point to results from particular models that show terrible outcomes after very high emissions and imply that they are certain to come about without immediate action, rather than the far edge of a range of possibilities. Those who favour doing nothing will highlight discrepancies and inconsistencies. Those who truly want to understand the earthsystem, and its responses, will see mismatches between one dire result and another less perturbing one, or between the output of models and data from the real world, as hints at processes that need to be understood better. They will accept and respect the whole thing as a vital, tentative work in progress, one both alarming and, intellectually at least, exciting.

Steam Engines and Spaceship Earth

On top of the effects increased levels of greenhouse gases are having on the Trenberth diagram, thickening that arrow from sky to surface, other human activities are also at play in the earthsystem – so many of them that more or less every arrow except the first shaft of incoming sunlight is now under some appreciable level of human influence. This is as strong an argument as any for thinking it reasonable to talk about the Earth having entered the Anthropocene.

The amount of sunlight that a planet, or part of a planet, reflects back into space defines the property known as albedo. As we have seen, the Earth's albedo is dominated by its clouds, but the surface matters too, and so do the changes that humans make to that surface. A landscape covered with trees is less reflective than one covered with grass and cereals; this lower albedo means the amount of heat the landscape absorbs from the sun is higher. By cultivating lots of formerly forested areas, humans have made a small change to the global albedo; smaller change, though locally important, has been made where they have filled cities with tarmac and glass.

The amount of heat exported to the atmosphere in the form of water vapour also has a hefty human component. Again, the balance of trees versus crops matters, as water can return to the atmosphere through the leaves of trees at a different rate from that at which it passes through crops; so does the question of whether, and how, those crops are irrigated. When it comes to winds the human effect is much less pronounced, but changes to the surface have their effects. More effects could be on the way – build them big enough and wind farms could slow down winds on a regional or even a continental scale.

Then there's the atmosphere. It is carbon dioxide which accounts for most of the extra long-term greenhouse effect due to human activity, increasing the infrared flux at the planet's surface by just under two watts per square metre. But emissions of methane, nitrous oxide, CFCs and the ozone-friendly HCFCs that have replaced CFCs also play a role, adding another watt per square metre, all told. Such additions are called 'radiative forcings': see them as changes in the amount of radiation that force the earthsystem to accommodate a wider, or narrower, arrow.

By warming the surface, all these greenhouse gases encourage evaporation, thus increasing the amount of water vapour in the atmosphere. Water vapour is itself a greenhouse gas, so this addition of water vapour amplifies the warming due to the greenhouse gases, a phenomenon known as water-vapour feedback.

Water vapour is also the raw material for clouds, which are

the great complicating factor of the climate just as they are the defining feature of the planet's face. The lack of certainty over how sensitive the world's temperature is to greenhouse gases – doubling the carbon-dioxide level could warm the planet by less than 2°C or by almost 4°C – can in large part be seen as an argument about the arrows pointing to, from and through the clouds in the Trenberth diagram. Which will get thicker: the arrows headed down from the cloud to the surface? Or the arrow representing sunlight bouncing from the clouds back into space? If they both thicken, which will thicken more? And what will those changes, in turn, mean for other arrows, constrained as they are always to balance each other?

Things get more complex still when you remember that greenhouse gases are not the only things humans put into the atmosphere. They also put in aerosols of various sorts. Soot particles absorb sunlight and emit infrared, thus warming the bits of the atmosphere they are found in. Other particles act more like mirrors, reflecting and scattering sunlight and decreasing warming. Still others do a bit of both. And all of them affect the way that those pesky, confounding clouds behave.

None of these changes and forcings looks huge on a Trenberth diagram – a slight thickening here, a slight thinning there. But they do not have to. Changes that are small in proportion to the overall flows can still have big effects. The changes between present-day conditions and those of an ice age don't look that big – a bit less carbon dioxide in the atmosphere weakens the greenhouse effect, a bit more ice on the ground increases the albedo. But those little changes give you a world in which there is a kilometre of ice sitting on top of Montreal and Manchester, in which there are paths that let people walk from Asia to America, in which plant life on the continents gets reduced by a third to a half. You don't have to throw a machine as big as this far out of whack to have a serious effect.

Is it a machine? Trenberth's diagram certainly makes it look like one. Representing the size of an energy flow with the width of an arrow, as Trenberth does, is a visual convention instituted for the comparison of steam engines by an Anglo-Irish engineer in the late nineteenth century, Matthew Sankey.* That an engineer's approach to representing steam engines should end up applied to all sorts of other systems – and, in the hands of climate scientists, to the flows of energy that animate the earthsystem itself – should not come as a surprise. Thermodynamics, the science of energy, heat, work and power, began as the science of steam engines; as it developed over the nineteenth century, it became clear that it had to be a study of those engines and the environment in which they operate – that is to say, a science of systems and flows.

Ever since the ideas of thermodynamics and its systems approach became available people seeking to reform and reimagine other sciences have been borrowing from them to try and understand flows as various as those of money through the economy or libido through the unconscious. In the 1930s the English botanist Arthur Tansley was drawing a specific parallel with the physicists' approach when he suggested that the way organisms lived with each other in a given environment should be thought of as an 'ecosystem' defined not just by its inhabitants but also by the flows of chemicals and energy that united them.** In his 1926 book *The Biosphere* the Russian geochemist Vladimir Vernadsky described a system of great engines using the flow of

---

* Sankey worked in the British Army's mapping operation, the Ordnance Survey; when he needed a steam engine to power the printing process he decided to map out the flows of energy through the contenders to decide which would best suit the job. In his original diagrams every inch of an arrow's width represented a power of 100,000 British thermal units per minute.

** Tansley's embrace of the systems approach was also related to the course of training analysis he undertook with Sigmund Freud. His fascination with Freudian psychology and its picture of the human mind shaped by a system of pressures and repressions reinforces the wide currency this way of thinking was achieving.

the sun's cosmic energy to reshape the planet's surface, blurring the lines between physics, chemistry, geology and biology as he did so. Alfred Lotka, an American thinking along similar lines at the same time, added evolution to the mix, interpreting natural selection as a force that ceaselessly improved the efficiency with which energy flowed through the living world. In the 1970s – unaware, like most at the time, of Vernadsky and Lotka – the British scientist James Lovelock argued, again using thermo-dynamic arguments, that life was playing a crucial part in driving the great cycles of the Earth and remaking its environment in ways that seemed to be to its benefit – the Gaia hypothesis.

The notion that nature is made up of systems that shape and condition each other took time to make its way into the mainstream. But it is now common to speak of 'Earth system science'. Various new ways of seeing the world provided by the cold war – radioactive tracers, computer models, the sight of the Earth from space – helped the process along.

So did another development in the same military–industrial milieu – a new interest in ways of sealing people off from the world. Nuclear submarines had to stay sealed up for weeks, or months. Space capsules had to be self-contained microworlds in which everything not recycled was used up. Fallout shelters, too, needed to be closed off and self-reliant. A new branch of ecological science called 'cabin ecology' was born out of the need to make such systems work. The name had a double resonance: the sealed cabin of the submarine or spacecraft; the self-reliant cabin of frontiers gone by. Sankey diagrams were sketched out for moon bases and missions to Mars in which all materials were recycled, in which all flows were balanced, in which everything was sealed off.

In the late 1950s Dave Keeling, a young researcher working for Roger Revelle at the Scripps Institution of Oceanography, applied the physical tools of cabin ecology to the Earth as a whole. He adapted devices that had been developed to measure carbon-dioxide levels in submarines (and thus keep their air breathable) into instruments to measure the carbon-dioxide level of the atmosphere. The increase he found over the years and decades that followed became perhaps the single most important piece of evidence for anthropogenic global warming. Others made the systems-thinking link from capsule to planet through imagination, rather than hardware. They pictured their homeworld as a sealed 'Spaceship Earth' equivalent to the capsules from which it was seen. The visionary thinker R. Buckminster Fuller probably invented the idea and certainly promulgated it. 'Spaceship Earth' became a rallying call, a way of expressing humanity's common interest, a way of understanding the global environment as a system of sustenance. Computer models of natural and industrial flows of energy and material were employed to take inventory on its cargo and determine its 'carrying capacity' – a term originally used for the number of people who could be put on to a steamship, but since given new meaning as the amount of life an ecosystem could support.

The metaphor of Spaceship Earth is powerful. It is also troubling. Ships are built for a purpose and come with a certain model of discipline attached – especially among cultures which have used them as tools of empire. Neither of these things applies to the Earth.

When you see them hanging unconnected among the stars, it is easy to mistake a space station for a moon, a planet for a spaceship. But they do differ. Though the Earth might yet be given purpose – purposes are something with which humans populate their world – it did not come with one built in, as a ship does. The workings of the earthsystem are not preordained. They are just what they have turned out to be. In this they are like the workings of Tansley's ecosystems, which emerge through the interactions between their inhabitants and the environment. This process of emergence gives ecosystems system-wide properties – push them in a particular way and they will respond, sometimes in a predictable way. But those properties were not preset. Indeed, Tansley's purpose in coming up with the idea of the ecosystem was precisely to have a way of talking about nature that did not require everything to have a preordained place and all relationships to conform to set norms. One of his reasons was that such views, being promoted in Oxford in the 1930s by followers of the former (and future) South African prime minister Jan Smuts and his new philosophy of 'holism', encouraged people to expect nature to be hierarchical, and to go from that to expecting different human races to have a proper hierarchy, too, with everyone in their 'natural' place.

The metaphor of the spaceship plays up holism and hides the contingency of the earthsystem. It implies that there is a single ship-shape way that the Earth should run. It encourages the notion that there is a fixed limit to the Earth's carrying capacity, just as there is a fixed complement for a vessel – an argument that has been used to justify brutal ideas about population control. And it can be used to divide humans into officers, crew and supercargo. The people developing the metaphor in the 1960s tended to cast themselves as the officers, and some took a dim view of goings-on below decks. 'The crew of spacecraft Earth' was 'in virtual mutiny to the order of the universe', Edgar Mitchell, the lunar module pilot of *Apollo 14*, wrote in the 1970s. Mutiny has penalties, and the Earth would enact them – unless its deputized authorities in the officer class got there first. The influential environmentalist Garrett Hardin used talk of a troubled Spaceship Earth as a way to advocate the suspension of humanistic moral values in favour of 'lifeboat ethics' derived from the harsher parts of naval law and practice.

Ultimately, the weakness and unpleasantness of the Spaceship Earth metaphor stems from the same problem as that which I see encouraged by the use of Apollo images as environmental icons: it leads people to divorce the physical planet from the human world. Such a divorce does not have to sound quite as ghastly as it did when Hardin talked of it. The idea that there are 'planetary boundaries' which should limit a whole range of specific human influences on the earthsystem so as to keep it close to the pre-industrial status quo, an idea which has been taken up by an influential group of earthsystem scientists in the past few years, seems to embody a reasonable prudence. But the sense remains that the planet imposes limits on human behaviour that have a force more akin to a law to be obeyed than to practical constraint that might be altered.

This is very hard to square with the reshaping of both the human world and the earthsystem necessitated by the previously unimaginable two-century transition from two billion people to ten billion, and from near universal peasantry to affluence, security and personal autonomy. This is a transition that can only change the environment – and it has already done so in ways that go well beyond those notional planetary boundaries. The question is not how to 'save the planet' as it was, but how the planet can be remade in a way that works while respecting the rights of the people living on it, and the value that they place on it. It is a task that calls for imagination and compromise much more than for naval discipline. It is a task of homemaking, not ship handling.

The relationship of people to planet is not that of a crew to a vessel, or of a parasite to a host, or of subjects to a law, or indeed of a cancer to a body. The fingers-in-every-arrow idea of the Anthropocene is that human enterprise is now part of most of the earthsystem's flows and cycles, and that more and more of the earthsystem is dependent on the political and economic systems of its human components; in the earthsystem, world and planet are increasingly indistinguishable. This is, as it happens, close to the vision that Alfred Lotka offered in his *Elements of Physical Biology*, first published (and largely ignored) in 1925. Showing singular foresight, he not only anatomized the earthsystem's great chemical cycles – its flows of carbon, nitrogen and the like – he in each case noted the human contribution to them, and the striking recent growth of that contribution. As the historian Sharon Kingsland has written, his 'way of thinking about evolution . . . was meant to demonstrate the unity of man and nature, to show that human activity was intimately tied in with the operation of the vast world engine'.

Seeing the world this way underlines a profound irony. The affluent, mostly urban life achieved by some and aspired to by more is often experienced as a sundering from nature. The space opened up by that sundering is well suited to planet-saving sloganeering that seeks to make of the environment a non-human victim. But the infrastructure that this affluent life requires, in terms of flows of energy and resources appreciable even on the Worldfalls scale of the earthsytem, leaves humans in the aggregate more and more 'intimately tied in with the operation of the vast world engine' even if living this new, affluent life takes them away from the experience of being in nature.

This intimacy underlines the importance of looking at the earthsystem not from the outside, but from within. The arrows of the Trenberth diagram do not sit still on the page like a planet seen from airless distance. They reach out of it – and they reach through you. They warm your face as the sun, they buffet your limbs as the wind, they soak your skin as the rain. They fire your metabolism with sunlight, and as the sweat evaporates from your brow they moisten the sky.

For an image of how you, industrial humanity and the earthsystem come together, look not for a planet or a diagram. Take inspiration from greatest visual artist of the industrial revolution, J. M. W. Turner. Part of what made the revolution's steam engine so world-changing was that it changed the world around it not just mechanically, economically and objectively; it did so subjectively, too. People started to see the world as not merely containing these strange new devices, but as being one of them. It happened in the sciences, through thermodynamics, but it happened more widely, too. When the critic John Ruskin – always fascinated by atmospheric phenomena – wrote in the 1830s of a 'vast machine' that might predict the weather, a machine not that different in spirit from the teraflopping computers now running climate models at the Met Office and elsewhere, it was in part because of a new sense that the weather was, itself, a vast machine.* In the 1970s the French philosopher Michel Serres wrote of this thermodynamic worldview as the key to understanding Turner. Turner not only painted the engines which were changing the Victorian world in an elemental way – he painted as if he were inside a world-changing engine of his own, a plenum of cyclic movement, of power, of energy, of light.

Picture the famous (possibly apocryphal) occasion on which Turner had himself lashed to the mast of a ship tossed in a storm before painting the magnificent *Snow Storm: Steam-Boat off a Harbour's Mouth*. A frail body is exposed to savage power; an extraordinary spirit then represents that exposure to others. The earthsystem is huge and powerful and all around; the power to understand it and appreciate it is within. Turner did not tame the waves that lashed him. But through his art he made them more than mere reality. To say that the earthsystem is a machine need not be the same as saying that it is an artefact, or that it is manageable, or that it is separate from the nature that sits inside humans. It is not to deny it as an experience, or as a source of the sublime. Instead, it can add to all of that another layer of understanding – a new understanding of the planet, of the world and of the human place in them.

★

---

* In fact, as made real today, the 'vast machine', as Paul Edwards points out in his excellent book of the same name, is not just the computers, but also the immensely complex set of instruments, social practices and people that feed them data and interpret their results, an insight that makes the inextricability of the humans in the system even clearer.

Not being a spaceship, the earthsystem was not set up to work a certain way. As humans build themselves into its workings, it will adapt and change, as Tansley understood that ecosystems do when a new species is added. The problem is that that change may not be to the human advantage. The challenge to the expanding human world of the twenty-first century is broadly taken as one of minimizing those changes, perhaps by altering the flows in the earthsystem that humans feel they control – flows of capital, flows of resources, flows of political power.

The challenge of geoengineering is to imagine changes beyond that human realm, but to its benefit. To do so means finding ways to slim or thicken arrows in the Trenberth diagram and similar representations of the flow of matter and energy through the earthsystem. With a human economy running at 15 terawatts in a 120,000-terawatt earthsystem – with all humanity's industrial energy amounting to just a five-kilometre-wide cataract in the equator-spanning Worldfalls – this might seem absurd. But there are levers one can use.

Archimedes is said to have said that, given a lever long enough and a place to stand, he could move the Earth. But he gave no indication of what he imagined would serve as the lever's unmoving fulcrum, and at the heart of this chapter has been the message that there is no such place to find. You cannot stand outside the earthsystem, any more than you can stand outside the world. You may luxuriate in solitude at the stratospheric edge of space, but you only do so thanks to the wings that thousands in your world came together to make. Look back from beyond the moon with Jim Lovell, and you do so from a capsule ecology that started off as a little bit of the earthsystem and was separated from it for the most worldly of reasons – rivalry between superpowers.

Geoengineering is about finding levers with which to move the earthsystem. It is also about finding worldly fulcrums that can take the pressure of those levers, and guide their force with precision – that stop the lever from slipping and doing damage.

Such a fulcrum would not be a physical place. It would be an institution, a shared goal, a new understanding of nature, an obligation of solidarity, a political achievement – all of those and more. It would be something that builds on the world-spanning imaginations of global environmentalists and cold warriors but goes beyond them – or perhaps goes deeper – in order to provide a position from which people can think fairly and decently about deliberate changes to the earthsystem.

Such a fulcrum, such a seat of balance, may never be built. If it is, though, it will be because of a need for, and understanding of, the specific lever that is to rest on it. The obvious way to find such a lever is to look at what happens when the flows change naturally for seemingly insignificant reasons. One potentially helpful change of this sort has been observed, and many of its repercussions mapped out. Scientists know of a way to cool the earthsystem because, not that long ago, they saw such a lever in use.

3

Pinatubo

A volcano may be considered as a cannon of immense size.

Oliver Goldsmith, *A History of the Earth and Animated Nature* (1840)

In July 1990 an earthquake of magnitude 7.8 – the most powerful earthquake recorded on the planet that year – rocked the Philippines, damaging and displacing not just buildings, roads and people but also the deep roots of Mount Pinatubo. Pinatubo, a volcano near the middle of the Philippines' largest island, Luzon, had been dormant ever since Europeans first came to the archipelago. Its lower slopes were rich farmland. Its subterranean guts contained about 100 cubic kilometres of molten rock.

The following April local people alerted geologists to eructations of steam and noxious gas issuing from various parts of the mountain. Seismometers picked up small but frequent earthquakes far below. The mountain's plumbing had sprung a leak during the earthquake; fresh, hot basaltic magma was flowing up into the cooler, stodgier magma that had been there before, heating and mobilizing it. Geologists from the Philippines and America assessed the risks and alerted local and national government to the need to plan an evacuation. By the beginning of June, a dome of rock was growing in the crater at the mountain's peak as the first lava reached the surface. Plumes of ash began to shoot into the sky.

The eruptions got larger until, over nine hours on June 15th, the mountain disgorged so much lava that it hollowed itself out, its peak collapsing into the void within to form a crater two and a half kilometres deep. The pillar of erupting lava reached 35 kilometres into the sky, its head spreading out like a mushroom cloud 400 kilometres across. In the pillar's core, tens of millions of tonnes of gas that had been held at high pressure in the magma chamber drove billions of tonnes of rock into the sky. And almost as fast as the pillar's core rose its flanks collapsed, their hot ash tumbling back to the surface with crushing power. Vast flows of debris spilled down the mountain's side. Five years later valleys on the lower slopes remained smothered by ash layers 200 metres thick, their insides still as hot as molten lead. Elsewhere the ash became mud – in an almost biblical coincidence, tropical storm Yunya hit the islands during the eruption, its eye passing 75 kilometres northeast of Pinatubo, its rains lashing the mountain slopes. Some valleys that were entirely filled with ash the day after the eruption were so thoroughly scoured out by the torrential rain that within a week they were deeper than they had been to begin with – the fastest erosion geologists have ever witnessed.

It was the mud that killed most of the hundreds who died. The tens of thousands who didn't die, despite living where they did, owed their lives to the well-executed evacuation. Within a few weeks, most of the people whose world had been catastrophically disrupted by the planet's fury were embarking on the hard work of putting their lives and communities back together.

But up in the stratosphere the mountain's cloud of ash and dust spread out around the world. So did the gas that had propelled it there, including more than 20 million tonnes of sulphur dioxide. In the stratosphere the sulphur dioxide was slowly oxidized, a chemical reaction which produced sulphate ions; these charged particles quickly combined with water vapour. There is very little water vapour in the stratosphere – the upper levels of the troposphere are so cold that almost all the water which rises into them falls back down as ice. But desiccated as the stratosphere is, it still contains some water vapour, and sulphate ions can turn this vapour into tiny droplets of sulphuric acid. The sulphur from Pinatubo thus created a tenuous aerosol mist that spread around the world.

Aerosols all tend to fall through the air in which they find themselves. But they do not all do so at the same rate. The sulphate droplets were smaller than the ash the volcano had thrown up at the same time, and so while the ash stayed in the stratosphere for just a few months, the sulphate aerosol lasted a couple of years. And because they were so tiny, the droplets had a remarkably high combined surface area, considering the relatively small amount of water and sulphur it had taken to make them. Imagine you have a tonne of water. If it is in a single spherical drop it will have a surface area of a bit less than five square metres. Divide it into drops a millionth of a metre across, and the surface area grows to six square kilometres. Make the drops a fifth smaller still and you get 30 square kilometres, give or take. The total surface area of the stratospheric aerosol created by Pinatubo is hard to calculate, because the exact size distribution of the particles isn't well known. But it was probably similar to the surface area of a large desert – definitely bigger than the Mojave, likely smaller than the Sahara. And like a desert, the tiny particles contributed to the planet's albedo by reflecting sunlight back out into space.

Most of the sunlight that encountered this new desert in the sky was able to get through it undeflected. There was still a lot of space between droplets (even the mighty Gobi covers less than 2 per cent of the earth's surface). Of the light that hit the droplets, most suffered only a glancing blow; it was deflected, but it still got through. A small fraction of the incoming sunlight, though, bounced back the way it had come. So some of the sunlight that would normally have hit the planet's surface didn't, and the world began to cool.

Volcanoes and Climate

Tambora, a volcano in what is now Indonesia, erupted in 1815 with a force as great as that of any known eruption during the past two thousand years. It created a stratospheric veil twice as thick as Pinatubo's, with cooling effects felt around the world. The Indian monsoon failed when the subcontinent's surface did not get hot enough to pull moist air in from the surrounding seas. There was famine in southern China. In Europe the following summer it was not only cold, there were storms unlike any seen before. Country roads became choked with displaced, starving peasants; glaciers descended from their high redoubts. In New England there was frost in July; the year was looked back on as 'eighteen-hundred-and-froze-to-death'. But no one linked this global climate emergency to the volcano; there was simply no way for the science of the time to do so.

By the time of the 1883 eruption of Krakatau – smaller than Tambora, larger than Pinatubo, also in what is now Indonesia – Victorian men of science had the means with which to perceive the global effects of the blast. Barometers around the world shuddered as the sound of the blast, too low for humans to hear, rolled over them. From Australia to the Americas, recently developed instruments that measured the colour and intensity of sunlight saw changes in both; at night, observers recorded strange new auras around the moon. The sunsets were spectacular, with the stratospheric aerosols reflecting the red light of the dropping sun back down from the stratosphere like a carmine wash across the sky (the effect was captured in the background of Edvard Munch's *The Scream*, the first version of which was painted from memory a few years later). A decade before the invention of seismology, the Victorian world had experienced the first globally detectible geophysical event to be understood for what it was.

The scientific establishment was fascinated. But it spared no thought for the climate. Like most people, scientists see the things they are interested in at the time, not the things that will matter to them later. To people looking back from a world where climate change is a central scientific and political issue it seems perverse that when the Royal Society, Britain's national academy of science, gathered data on the effects of Krakatau from scientists around the world, it paid no attention to what had been recorded by the Empire's more-than-ample stock of thermometers. But to Victorians, the world's average temperature was not an issue. Climate was not a global concern; zones and regions had climates, not the planet as a whole.

At least, not on a year-to-year basis. In the years before Krakatau, though, a growing number of geologists, and some of the physicists seeking to take the whole Earth as their subject of study, had become convinced that there had been climate changes in the past which were coordinated across the planet: the ice ages. Evidence that glaciers in the Alps had once stretched much further than they did in the nineteenth century had been supplemented by signs that much of northern Europe and northeast America had been at one time covered by ice. The way the ice had scraped the landscape beneath it showed that it had come and gone repeatedly.

How such cycles of climatic change could come about would remain a fascinating problem for a century; until anthropogenic global warming took its place, the origin of the ice ages was the great enduring question for those seeking to understand the climate. Given the sheer drama of mountain ranges made of ice crawling over continents, of shorelines shifted by hundreds of kilometres, of strange woolly wildlife walking across un-recognizable landscapes – and of all this happening at a time recent enough that humans had already mastered the arts of fire and the spear – how could it not be?

So though the cooling that Krakatau prompted was not recognized at the time, the mere fact that the eruption had had global effects was enough for some of the scientists interested in ice ages to start paying attention to volcanoes. One camp saw them as possible instigators of the ice ages, their ash blocking out the sun and cooling the planet in a glacier-friendly way. Another saw them having the reverse effect. In the 1860s John Tyndall, an English physicist, had shown that the greenhouse warming provided by carbon dioxide explained why the surface of the earth was warmer than that of the moon. Since volcanoes produced prodigious amounts of carbon dioxide, perhaps they could warm the planet as a whole.

The great Swedish chemist Svante Arrhenius – one of those late-nineteenth-century scientists who thought straightforward principles of physics and chemistry could be applied to the planet as a whole – calculated how much carbon dioxide it would require to lift the planet out of an ice age, including in his calculations the different warming expected at different latitudes and the amplifying feedback due to water vapour. With just pencil and paper, though with a great deal of the latter, he worked out that a doubling of the carbon-dioxide level could lead to 5°C of global warming. If ice age carbon-dioxide levels had been low, he concluded, volcanoes could have brought those ice ages to an end. He also noted that industrial burning of fossil fuels might warm the planet yet further – but since he thought they would continue to be burned at early-twentieth-century rates, and because like most Swedes he was not averse to the idea of a bit of warming, he didn't worry about the matter.

For most of the twentieth century, though, the link between volcanoes and climate was a fairly speculative one. The realization that the climate emergency of the late 1810s could be linked to Tambora strengthened the arguments of those who thought volcanoes could cool the climate. But a dearth of suitably large eruptions deprived them of any confirming data. When, in the 1970s, the link between volcanoes and climate came into its own, it was not thanks to an eruption; instead, it was as a result of the insights made possible by comparing the earth to its siblings across the solar system.

In November 1971 *Mariner 9* became the first spacecraft to orbit another planet; its close-up pictures of the surface of Mars were eagerly anticipated. Unfortunately the surface was nowhere to be seen. Mars was wrapped in a dust storm that hid all its features from view. For the geologists eager to puzzle out the planet's surface features the storm's slow subsidence, revealing first mountain tops, then high plains, was a frustrating tease.

For *Mariner 9*'s more geophysically inclined scientists, though, the dust offered the sort of thing they loved: a global experiment. *Mariner 9*'s infrared spectrometer could see through the thick veil and measure the temperature of the surface below. And as the veil thinned, the spectrometer saw the surface warm. The Martian dust was acting in the way it had been suggested the dust from earthly volcanoes did – scattering light at the top of the atmosphere so as to cool the surface below. Nothing so large scale and clear-cut had ever been directly observed in the thicker, messier, more turbulent atmosphere of the earth. In the words of Brian Toon, one of the scientists involved, the Martian dust storm and its aftermath constituted 'the only global climate change whose cause is known that man has ever scientifically observed'.

Seeing Mars being cooled by stratospheric dust got Toon, his friend James Pollack and their mentor Carl Sagan interested in the effects volcanoes might have on the Earth. Their studies of a third planet showed them that dust was unlikely to be all, or even most, of the story. The upper atmosphere of Venus reflects a remarkable amount of sunlight, and data from another Mariner mission had shown that this reflection was being done by sulphuric acid particles. If there was a way to get sulphur into the Earth's upper atmosphere, some similar shininess might result.

And a few years before an Australian researcher, S. C. Mossop, had shown that there was indeed a way of getting sulphur into the stratosphere. The 1963 eruption of Agung, a volcano on Bali, had been unremarkable in size – a lot smaller than Pinatubo – but propitious in timing. Unlike all the planet's previous eruptions, it took place at a time when people were able to visit the stratosphere and take a military-environmental interest in how stuff moved around in it. When a U-2 flew through the plume downwind of Agung the visible residue left on the plane's windshield proved 'painfully acid to the tongue', the plainly pragmatic Mossop reported in *Nature*, and was thus, he assumed, at least in part a sulphuric-acid aerosol. ('More sophisticated tests are in preparation,' he added.)

Sagan, Pollack, Toon and various colleagues put together the cooling they had seen on Mars, the bright clouds of Venus and the sulphate particles found in the Earth's stratosphere after Agung's eruption in a 1976 paper which argued that volcanoes did cool the earth, and that they did so with stratospheric sulphates, not with the dust that previous western exponents of the volcano–climate link had concentrated on. James Hansen, a NASA scientist who focused on the Earth's climate, used the ideas and data from the work by Pollack and Toon to model the climatic effects of Agung and compare them with the climate records for 1963. There was a broad enough agreement not to rule the effect out, but the state of the art in the 1970s allowed only the crudest modelling and the data were not great either, so the case was hardly proved. The next major eruption was a Mexican volcano called El Chichón, which went off in 1982.[*]

---

[*] The eruption of Mount St Helen in 1980 was not really in the big league. It is well known mostly because it happened in America and was thoroughly studied as a result.

Again, studies of the climate effect were inconclusive, in large part because a growing El Niño event swamped any cooling effect from the volcano.

When Pinatubo went off with a much bigger bang Jim Hansen's moment had come. He had at his disposal better models and much more powerful computers than when he first looked at volcanoes in the 1970s. And he was also playing for higher stakes. Hansen's modelling work had convinced him that human emissions of carbon dioxide were driving the world's temperature up. Testimony he had given to Congress in 1988 – a year in which much of America suffered a peculiarly hot summer, mostly due to another El Niño – had been crucial to lifting the idea of global warming out of the realm of scientific debate and into that of politics. But he knew how crude the computer models which produced his extrapolations were. The fact that they more or less captured the contemporary climate was encouraging. But given the amount of fiddling they required, the fact that they matched might have come about at least in part because of unconscious bias by their creators – they might have been 'tuned' to the world as it is.

What the models needed in order to show their mettle was an experiment: a sudden, distinct change to the earthsystem. If the models predicted the outcome of such an experiment with some accuracy, then the credence to be put in them would increase. Unlike Agung and El Chichón, Pinatubo looked like offering a big enough bang to rise above the background hum of the Worldfalls' engines. If the models Hansen and his colleagues had developed could get Pinatubo right, they would show that, despite their imperfections, the models really did capture the way the climate worked; if they got the effects of stratospheric aerosols right, then what they said about the effects of carbon dioxide would be more believable.

The models predicted that the stratosphere would warm, mostly because the sulphate aerosols would absorb infrared radiation from above and below. This was not in any way counterintuitive – stratospheric warming had been seen after Agung and, for that matter, above *Mariner 9*'s dust-wrapped Mars. But the models got the magnitude right: two degrees of warming, roughly twice what had been measured after El Chichón. The amount of light scattered back into space in the models agreed with the measurements made by satellites. And so, more impressively, did the drop in temperature of the lower atmosphere. By summer 1992, a year after the eruption, when Hansen's model said the radiative forcing due to sunlight – the big incoming arrow on the Trenberth diagram – would be down by three watts per square metre, the surface was about 0.5°C cooler, on average, than would have been expected without Pinatubo, which was very close to what the model predicted. If it had not been for yet another confounding El Niño it would have been a few tenths of a degree cooler still, pretty much dead on the prediction. Hansen's model was pretty good on both the magnitude and the timing; the cooling effect tapered off over the following three years, in line with its predictions.

Not all the models that tackled the question did equally well. And they all had a slight tendency to overestimate the cooling. But the models were getting things broadly right. What was more, they seemed to be getting them right for the right reasons. For example, when scientists removed the water-vapour feedback from their models, the models failed to reproduce the observed cooling; when they kept the water-vapour feedback in, the models matched the observed temperature shift well. Doubts about whether climate models captured the water-vapour feedback correctly were the basis for significant criticism of their results in the early and mid-1990s. Ten years later, people so inclined could still find plenty in the models to criticise. But the water-feedback issue was largely put to rest, in part because of the opportunity provided by the natural experiment of Pinatubo to test things.

Predictions and Surprises

Pinatubo showed that computer models of the climate had a modicum of predictive power and that volcanoes really did have a cooling effect. It also revealed that a layer of reflective aerosols in the stratosphere had a range of other global repercussions, many unexpected, all of relevance to people who might be tempted to try and replicate such cooling deliberately.

First, something expected. The prodigious surface area of the stratospheric sulphate aerosols did not just scatter sunlight. It also provided a place for chemical reactions to happen. Some atmospheric chemistry is much more keen to get its reactions on when one of the molecules involved is stuck to a surface, and some of the reactions by which CFC-derived chlorine destroys the ozone have this predilection. It is because the polar stratosphere has clouds in it for part of the year, whereas the rest of the stratosphere is more or less permanently cloud-free, that ozone depletion due to chlorine has mostly been seen at the poles. The surfaces of the polar-stratospheric-cloud droplets provide places on which the ozone-depleting chemistry can take place.

The post-Pinatubo sulphate aerosol did not form clouds, but it supplied a lot of new surface area for ozone-depleting chemistry, and not just at the poles. In 1992 the amount of ozone contained in the stratosphere as a whole dropped lower than at any other time on record. A decade's worth of thinning was seen in a year. In 1992, a year after the eruption, the ozone hole over Antarctica let in more ultraviolet than ever before. In 1993 and 1994, while Pinatubo's aerosols were still at work, things got even worse. The hole has never been as deep again since.

That is quite a spectacular effect for an eruption that, though big in contemporary terms, was a long way from being the largest of the past few centuries, let alone millennia. Tambora provided the stratosphere with a lot more sulphur. The Taubo eruption around 26,000 years ago was bigger still; the Toba eruption

74,000 years ago dwarfs them all. Why didn't these eruptions tear the ozone layer to shreds? The answer is that Pinatubo happened after humans had put CFCs into the stratosphere, and before there was chlorine up there, the aerosol surfaces would have had no ozone-destroying chemical reactions to catalyze. Indeed, by interfering in the complex stratospheric chemistry of NOx, they might have led to short-term increases in ozone levels.

Other expected effects had unexpected consequences. The aerosols were warmed by the sun and by infrared coming up from the surface; the infrared they emitted as a result warmed the stratosphere around them. This warming was greatest in the tropics. So in the lower stratosphere the difference between the temperature at the equator and that at the poles – the driving force for so many of the earthsystem's engines – increased. The greater flow of heat away from the equator strengthened the circumpolar jet streams.

When the northern circumpolar jet is strengthened this way, its effects are felt in the circulation of the troposphere below, where it tends to lock the weather into a stable, self-reinforcing pattern. Imagine a cloverleaf centred on the pole. In the three leaves – one over the North Pacific and Alaska, one over Greenland, and one over European Russia down to the Black Sea – the winter is particularly cold. In the three spaces between the leaves – Canada and the United States, Western Europe, and northeastern Asia – it is unusually mild. The idea that such a cloverleaf pattern followed eruptions had been discussed for a while among a few experts, notably in Russia, where they take their winters seriously.

The regional winter-warming pattern underlines the fact that no climate shift, whether due to volcanoes or greenhouse gases – to nature or people – is uniform. There are always variations in time and space. The general cooling after volcanic eruptions brings local and seasonal warmings. The general warming due to greenhouse gases sees regional exceptions and unexpected changes in pace. The world has been getting warmer decade by decade since the 1960s, but there are plenty of years cooler than the year before, and plenty of places where it has not got any warmer at all. A significant number of surface-temperature records show a downward trend over that time, just as some places have warmed at twice the planet's average rate.

The patchwork nature of change also shows up in the way ecosystems respond. Consider Pinatubo's effects on two living things that loom large in the environmental imagination – coral reefs and polar bears. In the Red Sea, Pinatubo, through its cooling effect, killed coral reefs. When the surface of the Red Sea cools, it becomes easier for the waters to circulate between the shallows and the depths (seas, like the stratosphere, tend to be warmer at the top than at the bottom, and are thus often stably stratified). The upwelling of deep water brings with it nutrients that produce blooms of algae and photosynthetic plankton, and these blooms kill off the corals. When this happened in the winter after Pinatubo, researchers went back over past records and found similar coral die-offs following earlier eruptions.

In Hudson Bay, although the post-Pinatubo winter was warmer than usual thanks to the jet-stream effects, the spring was cold, thanks to the aerosol cooling. The cold spring meant long-lived ice. That meant very good feeding for that year's polar bear cubs, which as a result grew up unusually big and strong. For years to come, rangers watching the bears could pick out the 1992 cohort by their size. They called them the Pinatubo bears.

Regional warming amid a general cooling was not Pinatubo's only surprise, nor its most consequential. As mentioned above, the atmosphere's water-vapour feedback played a role in the post-Pinatubo cooling. The cooler planet saw less evaporation, which meant less water vapour in the atmosphere. That meant less greenhouse effect. It also meant less rain. Various teams of scientists have pieced together histories of how much it rains and snows the world over, some compiling data from satellites, some from rain gauges. According to a paper by Kevin Trenberth and his colleague Aiguo Dai, there is one thing on which all these datasets agree: 1992 saw less precipitation over the continents than any other year in those records. A larger proportion of the planet's surface suffered drought conditions in 1992 than at any other time in the 55-year instrumental record that they looked at. In some places there were food shortages. As yet, no one has studied agricultural productivity statistics to see whether the dryness had any systematic effect on global food supply or security; there were regional food shortages in some places, but so there are in many years.

One reason why this drying year might not have seen bad harvests was that Pinatubo had a final surprise: in its aftermath the plants of the less-moist-than-usual Earth put on a distinct growth spurt. The evidence for this comes from Dave Keeling's records of the atmosphere's carbon-dioxide level, begun in the late 1950s. From year to year the level always goes up. But within each year the level oscillates to the rhythm of the seasons. In the northern hemisphere's autumn and winter carbon dioxide goes up more steeply; in spring and summer it comes down. This is because in spring and summer the northern hemisphere's plants (which vastly outnumber those of the southern hemisphere) are growing fast enough to outstrip, for a season, the steady outpouring of carbon dioxide from fossil fuels. In autumn and winter things droop and decay, and a fair chunk of the carbon dioxide taken up earlier in the year returns to the atmosphere, boosting the fossil-fuel effect. The two processes are not mutually exclusive – there is photosynthesis in the winter and decay in the summer – but one dominates for one half of the year, and the other for the other.

The fall in carbon dioxide during the northern spring of 1992, the year after Pinatubo, was much steeper and deeper than usual. Overall, the net increase in the atmosphere's carbon-dioxide level in 1992 was the lowest since 1963, when industrial emissions were less than half those of 1992.

In part, maybe in large part, this was because of the post-Pinatubo cooling. The rate at which soil bacteria and fungi turn the dead bodies of plants into carbon dioxide through respiration depends on the temperature; warmer is riper is more productive. Because the sulphate aerosols in the stratosphere made 1992 anomalously cool, soil–respiration levels would have been lower than normal; more carbon would have stayed locked in the soil and the level in the atmosphere would have reflected that.

But many earthsystem scientists believe that there was something else at play, too: the effect of diffuse light. Some of the light that hits the stratospheric aerosol particles is, as we have seen, reflected out into space – a process called back scattering. But some is 'forward scattered' – dealt a glancing blow that leaves it still heading for the surface, but at an angle to its original path.

Considered together, the two forms of scattering make the sun a bit dimmer and the rest of the sky a little brighter. So a given spot on a leaf will receive less sunlight directly from the direction of the sun than it would otherwise do, and a little more light from other directions. And this seems to suit a lot of leaves rather well. Direct sunlight can be overpowering; it burns up the molecular engines of photosynthesis, forcing the leaf to put time and effort into their repair. Indirect light is good, because it gets to parts of the leaf that would otherwise be in shadow. So diffuse lighting can be preferable even if the total amount of light is a little less.

The Pinatubo aerosol diffused the light received by all the plants in the world; there is debate about the effect this diffusion had, but there is some evidence that, on balance, the plants approved. A study carried out at Harvard Forest, a bit of deciduous woodland in New England that researchers have been studying for a century or so, estimated that at noon on clear days in 1992 this diffuse-light effect increased photosynthesis by 23 per cent.* A year later it was still 8 per cent up. The degree to which that is representative of the global response is open to question, but many scientists believe that the diffuse light seen after Pinatubo contributed to the record rate at which plants took up anthropogenic carbon dioxide that year, despite lower rainfall than would have normally been expected. The Keeling record suggests that plants responded similarly, though less dramatically, to the eruptions of El Chichón and Agung.

Other users of the sun would not have been as well served. Power stations which generate solar electricity by focusing sunlight with mirrors in order to drive a turbine – as opposed to the much more familiar and widespread solar panels, which turn light directly into electricity – are sensitive to how much light comes directly from the disc of the sun and how much from the sky as a whole. There were no such plants operating in 1992, but a prototype in America's Mojave Desert saw a distinct drop in its power output after the eruption of El Chichón in 1982. A Pinatubo-like eruption today would decrease the output from such plants, which would not be a big deal in terms of world energy (the installed capacity of this sort of 'concentrated solar power' is roughly a gigawatt), but would matter to the owners of the plants – some of whom imagine that, in the future, there will be a lot more of this sort of solar power around.

Just over a century after Krakatau's effects on temperature were not even seen as worthy of study, Pinatubo showed the world that volcanoes could indeed cool the climate (though they are not the answer to the great enduring question of the ice ages). Armed with all manner of new instruments, from lasers that scanned the sky from below to satellites that looked down through it from above, and boasting networks for bringing together observations from around the world that were even better than those of Queen Victoria's Empire, scientists showed

---

* Such calculations were only possible because the forest had been studied for so long – long records are among the climate researcher's truest friends.

not just the fact of the cooling, but also its mechanism. And they did all that in the context of increasing alarm over global warming both within academia and beyond it – the year that Pinatubo cooled – 1992 – was the year that the United Nations Framework Convention on Climate Change (UNFCCC) was signed at the Rio Earth Summit. It's hardly surprising that some people wondered whether Pinatubo could be more than an experiment – whether it could, instead, be a prototype.

4

Dimming the Noontime Sun

Those things we call artifacts are not apart from nature.
They have no dispensation to ignore or violate natural
law. At the same time they are adapted to human goals
and purposes. They are what they are in order to satisfy
our desire . . .

Herbert A. Simon, 'The Sciences
of the Artificial' (1996)

A volcanic eruption does not have to be as powerful as Pinatubo to inspire awe and terror. Matt Watson, a British volcanologist who also works on climate geoengineering, experienced a heady rush of the sublime during the 1995 eruption of Soufrière Hills on the island of Montserrat. It was the eruption of the year, rather than the eruption of the century, but the sight of the sky turned solid and rushing towards him remains both vivid and almost incomprehensible – 'It feels like falling off the Earth.'

The idea that human ingenuity can deliver the same impact as Vulcan's wrath seems implausible, even impious. But it is an everyday truth of the Anthropocene. Humans outstrip nature all the time in ways that just a few centuries ago – at the time, say, of Tambora – would have shocked all those romantic spirits looking to high mountains and wild weather for the thrill of the sublime. They bridge the deepest chasms, light the darkest night, dam the greatest rivers; they surpass almost every limit you would imagine for mammals not even two metres tall with brains smaller than melons. The sense of an awe-inspiring Other against which humans measure themselves is now as readily provided by their own works as it is by those of nature; there is now, in a phrase of the historian David Nye, a technological sublime, an overwhelming of the imagination, as he puts it, that 'undermines all notions of limitation, instead presupposing the ability to innovate continually and to transform the world'.

This usurping of nature can be spectacular – Nye points to huge engineering projects, atomic weapons, rockets to the moon. But it does not have to be. Some of the tasks of titans can be accomplished with comparative ease. Creating a veil in the stratosphere looks like one of those tasks.

The manner in which Pinatubo delivered 20 million tonnes of sulphur dioxide to the stratosphere was, after all, monstrously inefficient. The gas rose because it was hot and buoyant; keeping it hot and buoyant when, left to itself, it would quickly have cooled as it rose was the business of the billions of tonnes of hot rock thrown up into the air with it. A system designed to lift the gas alone could do it with a lot less palaver. In principle, a tonne of anything can be lifted to the overworld with about 70 kilowatt-hours of energy, which is not a great deal – less, in fact, than you get from burning ten litres of gasoline. If you spread the process out over a year, you could lift 20 million tonnes into the stratosphere with a constant output of about 160MW – the sort of power you can get from a power station running a couple of modern gas turbines.

In practice, mechanisms for lifting gas are not quite that efficient. But they are still quite manageable, in energy terms. The obvious way to do it would be to lift the gas in aircraft, an option that was explored quite thoroughly in a 2010 study by Aurora Flight Sciences, a company that builds, among other things, stratospheric research aircraft. The study was commissioned by David Keith, the researcher whom we met in the introduction at the University of Calgary, arguing about how much it might cost his company, Carbon Engineering, to suck carbon dioxide out of the atmosphere. Professors don't often have the wherewithal to commission studies on subjects they are interested in from commercial aircraft makers. But Keith, probably the single most influential researcher in the climate geoengineering field, is not a typical professor. Among other things, he and his friend Ken Caldeira, an earthsystem scientist at the Carnegie Institute, have for most of the past decade held the purse strings of a fund created by Bill Gates and some of his former employees to pay for research on climate and climate engineering. (This is separate from Gates's investment in Carbon Engineering, though obviously there's some common cause in terms of Gates's respect for Keith's abilities.) Money from this fund has paid for a fair amount of the research into climate geoengineering that has gone on since 2007, including the Aurora report.*

The Aurora analysis concluded that, at a pinch, you could deliver a million tonnes of something with a density about the same as that of water to some parts of the stratosphere with various technologies already up and flying – such as a fleet of 14 Boeing 747-400 jumbo jets. Setting up the system would cost a bit less than $1 billion, and a year's operating costs would be another billion or so.

It is worth noting that a number of people think just such an airliner scheme is already going on, except on a much grander scale. There is a network of observers, activists and fellow travellers around the world devoted to the belief that the condensation trails left by some commercial and other aircraft are not clouds fed by water vapour left in the wake of the aircrafts' engines, as all authorities insist, but instead noxious substances deliberately sprayed into the atmosphere as part of a covert government programme: not contrails, but 'chemtrails'. There is

---

* Among other things, the fund has supported a number of meetings I have attended; it has not paid me any honoraria.

no real consensus on what the purpose of these chemtrails might be – mind control and compulsory sterilization are sometimes suggested, as are weather modification and even the deliberate encouragement of global warming – but whatever it is it is assumed to be nefarious. For evidence, chemtrailers point to photographs showing aircraft trails in a profusion or pattern for which they can see no explanation other than a covert spraying programme: trails showing that the planes are looping round on themselves, rather than following a path from A to B; trails that follow one plane but not another, despite the two seeming to be in exactly the same part of the sky (a purportedly telltale phenomenon known as an OWOW – 'One with, one without'); trails with different appearances coming from the wings of a single plane. They pay detailed attention to pictures that show small nozzles and vents on the wings and engine nacelles of airliners, and to discontinuities and discrepancies in the data available on websites that track civilian aircraft. They are beset by a sense of the wrongness of the world, a wrongness of which the uncanny signs in the sky are evidence.

There are clear similarities between the beliefs of chemtrailers and those of many other conspiracy theorists. I know of explanations for many of the phenomena that they attribute to covert spraying which I find far more persuasive, and when they describe things for which I know of no alternative explanation – chemtrailers have become a fixture of public discussions of geoengineering in the past few years, and are normally eager to talk – my response is always to think the effect infinitely more likely to be explained by a quirk of aircraft engineering or atmospheric physics of which I am unaware than by a huge conspiracy of remarkably close-lipped pilots, engineers and others.

Chemtrailers come to these geoengineering events in part to confront speakers they suspect of being part of the conspiracy, in part to get their message across to the audience and in part because they want to see how the things the scientists are saying can be used to further the chemtrailers' own understanding. They are in my experience mostly civil; they do not impute evil motives to everyone in the discussion, though they are puzzled at the scientists' inability or unwillingness to see what is surely obvious. Are the scientists liars or are they dupes? Is their treatment of geoengineering as a future possibility rather than a current undertaking a double bluff? Are there multiple programmes, wheels within wheels? Sometimes things get nasty. Keith has received death threats from people animated by such beliefs, and many researchers – including some who are themselves very sceptical about the merits of climate-geoengineering schemes – get abuse by email or Twitter. That said, the chemtrailers I have met have seemed mostly peaceable, if often frustrated and sometimes angry. Their worry that people over whom they have no control are changing the skies above them is a potent one, even if their evidence is utterly unconvincing.

To return to the practicalities of real (which is to say, purely notional) stratospheric geoengineering, as assessed by the Aurora report: spraying out aerosols behind jumbo jets would be far from ideal. With a ceiling of 14 kilometres (45,000 feet), 747s can't get above the tropopause in the tropics, where it is at its highest. They would only be able to put aerosols into the stratosphere in temperate and high latitudes, and then only into its lowest reaches. That would limit the aerosol's ability to stay up for a long time and spread around the world.

A high-performance fighter such as an American F-15 can get above the tropical tropopause. But it would not be able to loiter there, and such aircraft are poorly suited to spraying out gas. Because their payloads are much smaller you would need almost ten times as many fighters as you would jumbos, and a far larger ground operation; fighters take five times more maintenance per hour of flight than passenger jets do. If you can get past the various regulatory and political problems that putting together an air force of that size might involve, you would be looking at all-in costs of $10 billion or more a year.

A better solution, the report found, would be to design a new aircraft from scratch. To fly between 20 and 25 kilometres you'd want something a bit like a fattened U-2, with its long wings pointing almost straight out from the fuselage and engines modified to operate in air too thin for a normal jet. Using rules of thumb derived from previous aircraft projects, Aurora estimated that developing such an aircraft and buying a few dozen would cost about $7 billon, and operation costs could be kept nicely tight – all in, a bill of about $2 billion a year. Scale up to five million tonnes a year and some of the assumptions get altered – you might want a bigger aeroplane, you might want to look at propellers instead of jets – but the cost per tonne doesn't change much.

These are the aircraft I was thinking about when picturing wings in the stratosphere in Chapter One. Climbing from a handful of bases around the world, they would fly in long loops for an hour or so at a time, spraying out a few tens of kilogrammes of aerosol for every kilometre of flight to add imperceptibly to the blue-white light-scattering haze circling their horizon. Workaday aircraft with whimsical names going up to the top of the world and back down five times a day.

Aeroplanes are not the only option. Other engineers who have looked at the problem favour vast balloons holding aloft 30-kilometre hosepipes attached to pumping stations on the surface. In energy terms, such a system might be the cheapest imaginable. Aircraft have to spend energy lifting themselves, their payload and a fair bit of their fuel up into the stratosphere before discharging their payloads; a balloon and hose can be kept aloft for as long as the hydrogen or helium that provides their buoyancy holds out. And there are no high accelerations to endure, so the structure can be exceedingly light. That said, the balloons would have to be very large indeed – larger even than the 100-metre-diameter behemoths that NASA uses to lift specialized telescopes into the stratosphere, successors to the pioneering cosmic-ray-watching flights of the 1930s. They would have to deal with powerful winds. And the combination of forces twisting the pipes and pressure from the fluids being pumped up inside them could pose big problems. While their advocates think such hurdles can be overcome, the balloon is not an easy option.

Others might prefer what are called 'hybrid airships' – currently experimental aircraft that gain some lift from thrust and aerodynamics, some from buoyancy. The fact that such hybrids have a passing resemblance to the bulbous-yet-streamlined *Thunderbird 2*, a staple of my television-fed imagination when a child, makes me hope that they might be on to something, but no detailed studies have been done. There has been discussion of using guns with shells full of pressurized gas, which is another way of keeping almost all the system on the ground, but which would tend to have higher costs in thrown-away shells.

I suspect that winged aircraft are the best bet simply because the world has a lot of experience building and flying them. But how the veilmaking might be carried out is a secondary detail. What matters is that it can be done with pretty straightforward developments of contemporary engineering. And that seems beyond dispute; given an initial investment of around $10 billion and an organization no bigger than a pretty small airline, you could be lifting a few million tonnes of whatever you fancy into the stratosphere within a decade.

Compared with the cost of addressing climate risks through mitigation, $10 billion is peanuts – less than the cost of a single American nuclear power station (they are cheaper in other parts of the world, though possibly also less reliable). Germany's ambitious renewable-energy plans may end up with capital costs almost a hundred times higher. This is what led Scott Barrett, a professor of international relations at Columbia University, to give one of his papers on the subject the title 'The Incredible Economics of Geoengineering'.

That title can mislead. David Keith points out that monitoring the thickness and development of a stratospheric veil with the accuracy you would want could cost a good bit more than just making one. Others might argue that the costs of geoengineering should always include compensation for people harmed – which, with some schemes, under some conditions, might end up a significant bill. Alternative forms of climate geoengineering could be much more expensive than veilmaking – and might still be preferable to a veil on grounds other than cost. And, as we shall see, any such effort worth supporting would need to be integrated into a larger set of climate policies and actions far costlier than its geoengineering component. There is a lot more to economics than up-front costs. But the low level of those up-front costs – and what that says about the idea's feasibility – is still striking.

Rough Magic

Not only do volcanoes waste a lot of energy getting sulphur into the stratosphere, the sulphur they put there does not form aerosol particles that are optimized for cooling. The distribution of particle sizes in the veil produced by Pinatubo is not that well understood (quite a few climate scientists are eagerly awaiting the next such eruption, hoping to capture the details much better than was possible in the 1990s). But it seems pretty certain that they were larger than is ideal for cooling.

This is because of the way the sulphate particles were produced. Imagine millions of tonnes of sulphur dioxide injected into a clean aerosol-free stratosphere. Over time the sulphur dioxide oxidizes into sulphate ions that encourage aerosol particles to form. The first ions created by the oxidation of the gas will produce small particles. As time goes by, though, newly formed ions become less likely to form new particles from scratch and more likely to enlarge the sulphate particles that are already there – so the average size of the particles increases. Particles also start bumping into each other and sticking together. Since the biggest cooling bang for the buck comes from very small aerosol particles – about two ten-thousandths of a millimetre across seems to work well – these processes which encourage bigger particles mean less cooling per tonne of sulphur.

This explains some interesting things about the links between volcanoes and climate. At the more dramatic end, it explains why larger eruptions do not produce proportionately larger coolings. The more sulphur they inject, the bigger the particles, and bigger particles mean that they are scattered less efficiently and fall out faster. Tambora cooled the world more than Krakatau – but not by as much as you would expect if you just compared the sulphur emissions. At the more subtle end, it means that the smallest eruptions capable of getting any sulphur into the stratosphere at all have the greatest cooling potential per tonne of sulphur. In the past few years it has been suggested that a string of these small eruptions – including that of Soufrière Hills, as witnessed by Matt Watson – might have produced a veil which, while very low indeed in mass, was still strong enough due to its small particles to offset some global warming, and this might explain in part the lack of increase in surface temperatures seen over the first decade of the twenty-first century.[*]

When people first thought of producing veils on purpose, they imagined doing it the volcanic way, using sulphur dioxide – but

---

[*] This is one of many effects called on to explain the 'pause', or hiatus; some of them are dealt with, at least in passing, elsewhere in this book. The explanations include: changes in the distribution of heat between the atmosphere and the oceans (see Chapter Two); increased aerosols in the troposphere as a result of China's economic boom (see Chapter Ten); lower overall climate sensitivity (see Introduction). Some of the lack of change is almost certainly down to the fact that temperature records in the Arctic are incomplete, leading to warming there being under-represented in the global averages. The explanations are by no means mutually exclusive.

instead of releasing the gas all at once, they imagined it sprayed out at a constant rate, year after year. It soon became apparent that the tendency of additional sulphur dioxide to make existing small particles bigger, rather than creating new small particles, would be a problem, as would the tendency of particles to stick together. Optimum particle size would be hard to maintain; strong, lasting coolings might be hard to achieve.

Some thought this rendered the idea impractical. It seems more likely, though, that it just requires veilmakers to do more than simply mimic the actions of volcanoes. Although it would require somewhat more ambitious hardware, it is possible to imagine aircraft that would spray droplets of sulphuric acid – the substance to which sulphur dioxide eventually transforms itself – rather than sulphur dioxide, thus cutting out the intermediate step. Properly designed spraying systems built on that basis would be able to produce particles in the ideal size range, and once made, they would tend neither to grow nor to shrink, because there would be no extra source of sulphate ions to glom on to them, as there would be if the stratosphere were suffused with slowly oxidizing sulphur dioxide.

Keeping the particles small helps to minimize one of the side effects of a stratospheric veil, as smaller particles do not absorb or emit infrared very well, and thus warm the stratosphere less. But it does not solve the most obvious of the drawbacks: it still adds surfaces on which chlorine from CFCs can destroy ozone – indeed, it adds more surface area for a given mass. Studies using computer models indicate that a climate geoengineering programme that employs sulphate aerosols would, as Pinatubo did, thin the ozone layer globally by this mechanism, with particularly notable effects at the poles.

The effects of such a veil on the ozone layer are the subject of significant research, and not just using computer models. Part of the programme Watson leads studies individual aerosol particles trapped in laser beams to learn what goes on on their surfaces.

Eventually, the subject will need to be studied *in situ*, in the stratosphere – research that I expect both to raise the profile of climate geoengineering and to mobilize opposition to it.

Because the Montreal protocol curbed the use of CFCs and other ozone-destroying gases the level of chlorine in the stratosphere is now declining (it peaked in the early 2000s). If there had been no Montreal protocol and CFC production had continued to climb unabated, the world would currently face an ozone problem that was nasty and getting worse. Instead, thanks to prompt, coordinated action, it is passing through an ozone problem that is hardly perceptible in most places and that is in the process of being eliminated. There is still a seasonal Antarctic ozone hole, and a thinned ozone layer over the Arctic in the spring, but the effects of this disruption seem to be local and contained; of the changes that have swept across the Arctic over the past decades, and that continue to sweep across it – reductions in sea ice, much more reliable transportation, new levels of political representation for indigenous people, new post–cold war international politics – a modest increase in ultraviolet radiation doesn't even make the first division. In the Antarctic and the Arctic there is a strong expectation that the layer will recover as chlorine levels drop. A stratospheric veil of volcano-like sulphates would slow that recovery by making what chlorine there is more destructive; it might push the date that the ozone layer returns to its pre-CFC condition from the second half of this century to the early decades of the twenty-second.

If that still sounds worrying – or if future research shows that the net effect of sulphate aerosols on ozone might be worse than it looks today – there would be further options to explore. Though volcanoes use sulphate to cool the world, other aerosols are available, and might be preferable. Among those that have been discussed are oxides of aluminium and titanium, or even, rather beguilingly, very small diamonds. The surfaces of some of these putative particles are much less conducive to

ozone-destroying reactions, and thus might make a veil more
ozone friendly. Indeed, it is possible, if it were thought desirable,
that the particles' properties could be tailored so as to encourage
ozone formation, rather than depletion. (Shortly after the
discovery of the Antarctic ozone hole there was some discussion
of spraying chemicals that might interfere with the ozone-
destroying reaction into the Antarctic stratosphere. The approach
that was imagined, though, looked on closer inspection likely to
do more harm than good.)

Another reason for veilmakers to look at particles other
than sulphates might be the ratio of forward scattering to back
scattering. Sulphate particles offer quite a lot of forward scattering
for the amount of back scattering, thus diffusing more sunlight
than they reflect back out into space. That may, as we have seen,
be good for some plants, but it is bad for some solar-power
systems, and in thick veils it would give the sky a somewhat
'washed-out' appearance that many might wish to avoid.

There is also the question of how much the aerosol is likely to
heat the stratosphere. A warmer lower stratosphere might make
the tropopause more permeable, allowing more water vapour in,
which would have its own effects on ozone and other chemistry
and even on the likelihood of stratospheric clouds. It might also
have an impact on the thin, icy cirrus clouds that form in the
higher parts of the troposphere. Since cirrus clouds warm the
atmosphere below them (the infrared that they emit is a stronger
forcing than the sunlight they reflect), that would be a problem.*
These tropopause permeability issues could yet prove to be the
biggest drawback to such schemes.

Pinatubo showed that stratospheric aerosols could cool the
world – it did not show how to have them do so in a way that

---

* It has been suggested that because of this warming effect, deliberately disrupting cirrus clouds might be a way to cool the planet, but this form of
climate geoengineering has yet to be studied in any very great depth. Other
forms of geoengineering using clouds are dealt with in Chapter Ten.

best meets human needs. Looking for better ways requires the study of a whole range of aerosols and their thermal, optical and chemical effects, both in laboratories and, on a very limited basis, in the air. It also means studying their effects on human health – what goes up does, eventually, come down, and though the amounts of aerosol involved would in all likelihood be very small compared with the amounts of dust, pollen, pollutants and other things already blowing around the troposphere, knowing their health effects would be crucial.

It might be that sulphates would end up as the aerosol of choice; among other things, there might be an attraction to doing such things in a way that mimics natural processes quite closely. But if so, it should indeed be a choice, and a well-informed one, rather than just a pursuit of the first thing people thought of. To think that geoengineers would or should simply seek to replicate a volcanic veil in every detail is to mistake an inspiration for a design.

Promethean Science

In an essay about an earlier technology designed to lower the risks of bad weather – the introduction of lightning conductors in the eighteenth century – the historian Simon Schaffer develops the notion of 'Promethean science': 'an experimental enterprise that mixes a vaulting ambition to safeguard humanity against a major threat with the troubling hazards of following this science's recipes'. That veilmaking represents a vaulting ambition is not open to doubt. But what safeguards could it offer, and what troubling hazards?

This is the sort of problem where computer models, imperfect as they are, prove themselves useful. The lion's share of recent research into geoengineering by natural scientists has exploited this usefulness. The typical approach is to run the model with two, or sometimes four, times the preindustrial level of carbon dioxide and then run it again with, in addition to that carbon dioxide, a level of climate geoengineering that cuts down the sunshine coming in by just enough to offset the warming that the doubled or quadrupled carbon dioxide would be responsible for. Thus the models produce accounts of what happens on what we will call a Greenhouse Planet and an Engineered Planet; the researchers can compare these planets with each other and with the world as it is today – a Baseline Planet.

In 2011 Alan Robock, a professor at Rutgers University who studies the climatic effects of both volcanoes and nuclear wars, and Ben Kravitz, a student of his, led the creation of a project called GeoMIP. It offered a standardized way to compare Greenhouse Planets and Engineered Planets produced by various climate models from around the world. The results of the effort fed into the IPCC's Fifth Assessment Report, the relevant part of which was published in 2013. Below are the project's main findings to date, insights on which most of the models agree, pretty much, as do similar non-GeoMIP studies.

These insights are useful – but they can be misinterpreted and overinterpreted. In particular, people have a tendency to see what the models say about specific scenarios as being true of all approaches to veilmaking. But this is not the case. The scenarios chosen for GeoMIP are designed to elicit interesting results about the processes on an Engineered Planet. They are not realistic representations of approaches one would actually take if attempting geoengineering. Just as a real geoengineering project would not seek to be just like what is seen after a volcanic eruption, neither would it be just like what is seen in these models.

Take one of the most widely discussed conclusions from such studies: that an Engineered Planet that has the same average temperature as the Baseline Planet will not necessarily represent a world where the temperature is the same as the Baseline Planet's in every region. The Engineered Planet's poles will be a little warmer, its tropics a few tenths of a degree cooler. This is not a great surprise. Greenhouse warming at the poles is mostly concentrated in winter when the sun does not shine and stratospheric aerosols can provide no cooling. So to re-establish a pre-warming average temperature at the poles means cooling them a lot in summer, and if a cooling veil strong enough to do that is spread evenly over the planet it will cool the equator more than the greenhouse effect warms it. Thus, keeping the polar temperature stable 'overcools' the tropics and re-establishing the baseline temperature at the equator 'undercools' the poles.

This is a helpful reminder of a fundamental fact about geoengineering with veils. It is not an antidote to climate change. It is an additional form of climate change, one that has some effects that oppose those of the climate change brought on by greenhouse warming. But it is also open to overinterpretation. The overcooling and undercooling come about because the models of Engineered Planets in the GeoMIP project represent veilmaking in a deliberately uniform way, with the veil effect as marked at the equator as at the poles, in winter as in summer. A real geoengineering programme would not need to be constrained in that way. It is quite plausible to imagine a veil with some differences in strength according to latitude: inject fewer particles at the equator and more in higher latitudes and you could create a veil stronger at the poles than at the equator, and thus remove some or all of the undercooling/overcooling.

The mismatch between the climate change imposed by veilmaking and that imposed by a thickened greenhouse is more striking and worrying when it comes to the hydrological cycle. The evaporation and condensation of water all takes place in the troposphere. Because a stratospheric veil reduces the amount of energy coming into the troposphere, it reduces the energy available to drive that cycle. And the higher level of greenhouse gases on the Engineered Planet exacerbates the problem. By making the atmosphere more able to absorb infrared, carbon dioxide reduces the rate at which the air gets cooler with altitude – the lapse rate. That means that air near the surface does not rise with such force and the atmosphere sees less vigorous convection. Less water vapour is lifted into the sky.*

The Engineered Planet is thus drier than the Baseline Planet. In this, the models agree with the experience of 1992, when in the wake of Pinatubo there was low rainfall and low river flow. That said, the effect seen after Pinatubo was more extreme than you would expect as a result of an engineered programme running over a long period of time. This is because the continents cool (and warm) much quicker than the oceans, which have to absorb a lot of heat to change temperature even a little. So when something cuts down incoming sunshine suddenly, as Pinatubo did, it will cool the continents more than the oceans. That will suppress monsoons, driven as they are by the temperature difference between hot land and cooler water.

A slowly built-up geoengineered cooling would apply to land and water more equally, and that generalized effect on monsoons would be lessened (which is not to say that there would not be other effects of various sorts on some or all of the planet's monsoons). But even if Pinatubo provided rather more drying than an engineered veil might, veil-driven drying is still a cause for concern. The most serious of the risks posed by global warming in the coming century, to my mind, are those imposed on the agriculture of developing countries. A response to global warming that increases droughts seems a categorically bad idea.

---

* The same lapse-rate effect is seen on the Greenhouse Planet. But there the increase in evaporation due to higher surface temperatures more than counterbalances it; there is more evaporation and precipitation than there is on the Baseline Planet. If you were to increase the surface temperature without increasing greenhouse gases – if you had a climate-engineering technique that added to the sunlight reaching the surface, rather than reducing it – you would see a lot more increase in precipitation for a given temperature rise than you get if you raised the temperature through the addition of greenhouse gases.

When you compare the Engineered Planet with the Greenhouse Planet, though, rather than with the Baseline Planet, the situation becomes more complicated. To begin with, although the precipitation on the Greenhouse Planet is a good bit higher than it is on the Baseline Planet, thanks to the higher temperature, it does not go up everywhere. Most of the increased rainfall in a Greenhouse Planet falls on the oceans. This is not entirely inconsequential – it changes patterns of surface salinity a bit – but neither does it matter much to farmers. On the continents the effects of greenhouse warming are mixed, but broadly speaking models show the increased precipitation falling disproportionately on places that are already wet. Currently dry places often end up drier still, due to more intense evaporation. So Greenhouse Planets typically have strips of increased aridity across southern Africa, the Mediterranean basin, central America and the southwest states of the USA. In some places – notably parts of south and southeast Asia – rainfall increases, which is expected to be a problem. Torrential rains and floods are a significant source of climate risk.

In GeoMIP's Engineered Planets much, sometimes nearly all, of the increase in precipitation over the oceans is replaced by a decrease. Precipitation also falls in quite large parts of the temperate northern hemisphere. Typically, though, many places that are currently hot and dry see less of a worsening than they do on the corresponding Greenhouse Planet. The details differ a lot from model to model – but then models disagree quite a lot on precipitation anyway. As David Titley, who used to run the American navy's weather forecasting command, puts it, 'All models are wrong; but precipitation models are particularly wrong.'

Kravitz, now at the Pacific Northwest Laboratory, says that one of GeoMIP's more interesting results was the finding that, in general, Engineered Planets show less of the more-rain-for-the-already-wet, less-rain-for-the-already-dry pattern than is seen in

Greenhouse Planets. They also seem to show fewer very intense precipitation events, which are a major worry in the Greenhouse Planet projections. That said, there can still be problems in terms of less-rain-for-the-now-wet; there is, for example, a worrying tendency for Engineered Planets to provide less precipitation in bits of the Amazon basin than Baseline Planets do.

Further complications follow from the fact that the models are incomplete. Plants care about the moisture of soil, not precipitation per se, and soil moisture − which depends on both precipitation and evaporation − is not well captured in most of these models. And plants also care about the carbon-dioxide level. More carbon dioxide makes it easier for them to photosynthesize, which in turns makes it possible for them to get by with less water. This 'carbon-dioxide fertilization' is widely used in horticulture; commercial greenhouses often have extra carbon dioxide pumped into them. How much it adds to growth under real-world conditions is a matter of some debate.

So while plants in parts of an Engineered Planet will get less rain, the higher carbon-dioxide level in the atmosphere should allow at least some of them to make better use of what water they do get. The models of Engineered Planets in GeoMIP which were equipped to look at such responses all found that, despite some regional drying, the overall productivity of the plants on Engineered Planets was higher than on either Greenhouse Planets or Baseline Planets. This is a subject I will return to in a later chapter.

The various different factors at play − where the rains come, what water the plants need, how important it is to avoid very heavy precipitation − mean that it is hardly possible to say anything definitive about the impacts of the changed hydrological cycle that would be seen under a veil. The idea that the effects of lower average global precipitation would necessarily and uniformly be bad, though − which is the impression some people took away from early studies – is certainly simplistic and quite possibly flat wrong.

By way of contrast to the complexities of precipitation, one thing all models agree on is that once you have started building a stratospheric veil, you need to be careful about stopping. If the aircraft stop taking aerosols up to the stratosphere those that are there will soon fall out and the cooling effect will quickly fade away. In just a few years the planet will undergo most of the warming that the veil had been sparing it (though because the oceans are slow to warm, some residual cooling will last for longer). In general, faster warming makes adaptation harder, and very fast warming would be peculiarly damaging. The chance of such a 'termination shock' is often cited as a reason that geoengineering-by-veil would be a technology too dangerous to deploy. The possibility is indeed a worry – but not, I think, in itself a prohibitive one.

To begin with, the way the modelling is done tends to make termination shocks look particularly disturbing. This is because, in order to see what is going on, the models use scenarios in which the geoengineering is very strong – enough to counterbalance the warming caused by a doubling or quadrupling of carbon dioxide in the atmosphere. If you take away a very strong veil like that in a year or so you do indeed get a number of degrees of warming in a decade or less, which is very bad. But though GeoMIP models study such very thick veils, real-life geoengineering programmes would not necessarily produce them. If the veilmaking were a relatively modest affair, the termination shock would be more a termination shudder.

Then there is the question of whether, if you do have a thick veil, termination is actually likely. If it would indeed have disastrous results, why would anyone order it? If a particular country that had been doing all the veilmaking stopped doing so because of, say, a revolutionary change in its politics, would not another step up? There is an analogy here with GPS, the satellite-navigation infrastructure provided by America's armed forces. At a technical level the world only needs one such system, but the European Union is developing another, on the basis that something so vital cannot have a sole source. Though there is reason to suspect that there may be other motives at play too, that basic argument is a good one. To rely on just one country for something that the whole world depends on is foolish. In a world where one country was veilmaking, it would be prudent for others to match its capacity.

Redundancy along these lines would make a sudden termination through political caprice or institutional failure unlikely. That leaves a situation in which the benefits the world thinks it is getting from an ongoing veilmaking programme are suddenly swamped by a newly perceived problem – perhaps for everyone, perhaps for a specific region. It is hard to imagine such a development only making itself known after decades of geoengineering had led to a thick veil being put in place – bad side effects would seem much more likely to present themselves fairly early on. And even if there were a sudden and overwhelming change in the calculus or distribution of costs and benefits long after the programme had started, it would have to be of a pretty specific sort to require that all veilmaking cease forthwith, rather than that it be phased out gradually or scaled back to a less intense level.

Despite this, there are people who worry that the termination-shock problem means that if veilmaking is ever begun it will have to continue more or less indefinitely. Ray Pierrehumbert, a climate scientist at the University of Chicago who was one of the authors of the reports on geoengineering produced by America's National Research Council in 2015, is one such: 'When has humanity ever managed to sustain a concerted complex technological enterprise for centuries, let alone millennia?' he asked in an article about the report he wrote for *Slate*, a website. But the fact that sustained efforts have been rare in the past does not make them inconceivable. Humankind already has various continuing technological commitments. As we will see in a later chapter, without the technology that makes nitrogen-based fertilizers, the world would be very hard put to feed its current population, let alone its predicted one. But although people worry, rightly, about the world using too much fertilizer, no one really worries about the notional possibility that the chemical industry might stop making the stuff. The parallel is not exact: veilmaking is a bit more difficult than nitrogen fixation, and making fertilizer provides a prompt benefit to the maker – but it is still hard to see veilmaking becoming a lost art, particularly if many stand to suffer from the losing. A world that continued to find it necessary would in all likelihood continue to find it possible.

Another tricky problem that comparisons of Greenhouse Planets and Engineered Planets can raise is that of winners and losers. A key piece of research here is a comparison by Katharine Ricke and colleagues published in 2008. It employed the formidable resources of a system called climateprediction.net, which takes advantage of spare computing time on the machines of thousands of volunteers to run climate modelling experiments over and over again with very slightly different assumptions each time. In this experiment all the planets started off in conditions like those of 1990. Its Greenhouse Planets saw carbon-dioxide levels rise in accordance with a business-as-usual projection produced by the IPCC. Its Engineered Planets underwent the same increase in greenhouse gases, but with the addition of sulphates to the stratosphere aimed at keeping the global average temperature constant. The capabilities of climateprediction.net meant that this produced a whole gamut of Engineered Planets, some of which had a bit more of a veil, some of which had a bit less.

Ricke and her colleagues measured 'damage' by comparing the conditions in each of the model's 27 regions with the year-to-year variations in average temperature and precipitation in the 1990 Baseline version of the model. The year-to-year variations defined an envelope that could be seen as 'normal': when the average temperature, or precipitation, in a region on an Engineered Planet was within this envelope, there was no damage; when it moved beyond the envelope, damage was done to a degree proportional to the distance the temperature or precipitation had strayed. On the Greenhouse Planet temperature increases started to do damage in almost every region by the 2020s. On the Engineered Planets most regions stayed within their bounds a lot longer.

As the century of simulation wore on, though, what different regions experienced began to diverge. There were many scenarios in which one region would face no damage while a neighbour started to suffer. Take Asia in the 2070s. Most of the sulphate veils strong enough to keep China within the bounds of its Baseline 1990 climate were so strong that they chilled India significantly below its 1990s bounds; veils that left India within bounds saw China overheat. In neither case did Ricke and her colleagues calculate the damage to either nation to be remotely as great as it would have been on the Greenhouse Planet. But the divergent fortunes of the planet's two most populous nations obviously raised concerns.

The study also had global findings. One was that a sulphate veil could be thick enough to offset two thirds of the global average warming over the period while still being thin enough not to push precipitation beyond its accustomed bounds in more than a handful of regions. Another was that it identified an optimum level of geoengineering – one which, taking into account damage done by changes in both precipitation and temperature, minimized the harm done in comparison with that expected on the Greenhouse Planet. That optimum level would not have suited everyone; the damage done in West Africa was greater under that scenario than it would have been on the Greenhouse

Planet. But that led to perhaps the most surprising, and perhaps the most hopeful, of the study's results. A scenario was found in which the damage was decreased for many regions while being increased for none at all. The net benefit in such a scenario was less than it was in the global optimum. But no one was hurt at all, and the damage reduction compared with the situation on the Greenhouse Planet, though not felt by all, was widespread and significant.

Subsequent work, including some within GeoMIP, has mostly found similar results; relatively small amounts of veilmaking can produce planets on which it seems that everyone is better off than they would be on a planet with a similar greenhouse-gas level and no geoengineering. These are crude models, and quite crude analyses, and there are some studies which do not agree. The realities of stratospheric veils are doubtless far more complicated, and their effects may be surprisingly variable in both time and space. In 1816, after Tambora, there were summer days in New England where washing froze on the line – not something you would expect from the few degrees of change in average temperature that even a thick veil produces. But taken together, the research discussed in this chapter seems to offer four important conclusions.

It would be relatively easy to create a stratospheric veil strong enough to counterbalance part of the greenhouse-gas warming expected in this century. The degree to which that might slow or reverse the restoration of the ozone layer depends on the type of aerosol and the thickness of the veil; it seems likely that there are schemes in which such effects would not be prohibitive. A veil would not preserve the climate of a world in which greenhouse warming was not taking place: it would add a countervailing factor to the climate change, one that would change patterns of precipitation. Some of this composite climate change would still need adapting to, and some might do damage in some places. In many scenarios, though, especially those involving relatively small interventions, the benefits of a well-designed veilmaking programme could amply outweigh any damage.

Whether people can find such designs will depend, among other things, on how hard they look for them. Whether even good designs are implemented will depend on politics and public opinion. The evidence which suggests that this form of climate geoengineering could be a helpful component in strategies to reduce overall climate risk does not in itself mean it is the best available option for the world. Its hopeful message certainly does not ensure that such a geoengineering programme would be implemented in one of the benign ways that appears to be possible, even if it started off that way. A fundamental problem with intentional change is that the intentions themselves can change; ambitions can grow, beneficence can fade. It is not unreasonable that, faced with a Promethean science of vaulting ambition that seems, as David Nye puts it in his evocation of the technological sublime, to undermine 'all notions of limitation', people should worry about troubling hazards unrepresented in the world of models, experiments and notional optimizations.

But nor is it reasonable simply to ignore the findings from that world because of such worries. Overall, what those findings say about the possibility of reducing the risk of damage due to climate change means that geoengineering should be taken considerably more seriously than it has been over the past few decades.

5

Coming to Think This Way

Man is able to modify the influence of the climate that
he inhabits — to fix, you might say, the temperature
most convenient to him.
            Le Comte de Buffon, *Histoire Naturelle* (1785)

Even if one were to know for sure that intervening in the
climate would reduce the risks the world faces, it would not
follow that such intervention was necessarily a good idea. There
might be opportunity costs: climate geoengineering could make
other things that would bring greater benefits less likely to a
degree that outweighs any good it does of itself. There might
be slippery slopes: the ability to intervene for good might be
inextricable from an ability to intervene in ways that do harm,
and efforts to do good might make those harmful interventions
much more likely, even inevitable.

And there are deeper reservations. There are things that seem
likely to do good that people refuse to do because they see them
as at odds with their core beliefs. One of the greatest gags in
Blackadder to the contrary, Sir Thomas More did not simply fail
to realize that he could recant his Catholicism, kicking himself
for the oversight as the flames licked higher. He preferred to die
as himself than give up on something central to his life and be
forced to live as someone else.

I think there are people who, though they would not

necessarily put it this way, would rather see greater suffering due to climate change than have human empire dominate the previously natural world to the extent geoengineering would seem to entail. There is a fear that such a transfer of power would mark a fundamental loss of both what is natural and what is human; it brings to mind Christ's question to his disciples: 'For what is a man advantaged, if he gain the whole world, and lose himself, or be cast away?' The author and activist Bill McKibben has talked of climate change as the 'end of nature'. The deliberate climate change that geoengineering entails might seem to fit that phrase even better, seeming to mark, as it does, the final, irreversible replacement of the authentic world with a fake one. If that is how it seems to you I think I can understand your rejecting the idea, even if the rejection comes with great costs. I think, though, that we might both see a problem in your stance if the costs were to fall not just on you, but also on others who would choose differently.

While I think I appreciate the power of this deep objection I do not think it should rule out all consideration of climate geoengineering. One reason is that I don't think nature can be ended so simply or so finally. Another is that while today's geophysical predicament is unprecedented, people have grappled with the idea of how best humans can relate to their climate for centuries, and they have often imagined changing it. Although it is common to hear climate geoengineering portrayed as a new idea, it is the idea that humans should not seek to change the climate that is the novelty, at least in the western intellectual tradition. I do not take this to mean that the way people used to think on these matters was right and the way people think today is wrong. But I do take it to mean that it is worth appreciating something of the range of views that has been taken in the past to help inform those that could be taken in the future. Having looked at the contemporary science of climate geoengineering, let us turn to its prehistory.

In the Enlightenment it was the peak of reason to believe that climate – which was, in effect, nature itself, but in localized and sensible form – both shaped humans and could be shaped by humans. The eighteenth century's explosion of exploration, speculation and tabulation used climate to explain the differences between people found in different places: the cold of the North encouraged languages with many consonants, since people did not want to chill their teeth; the oppressive heat of the Orient encouraged despotism; the changeable weather of littoral Europe led to fickleness and its cousin, innovation. It also used the actions of these people to explain their climates, at least to some extent. It was commonly held that the climate of Europe – taken to be, by and large, a good climate – had been much improved by the clearance of savage forest.

Before the Enlightenment, in this view, human changes to the climate had been largely incidental, if providential. After the Enlightenment it was possible to imagine them being brought about intentionally through ingenuity and the increasing power of the state. Draining swamps, managing forests (that provided crucial supplies of wood both for burning and for warship-making) and regulating rivers were all seen as making the climate subject to, and the business of, government. When the upper Rhine was disciplined from its meandering, anasto-mosing natural state in the early nineteenth century it was an act of state power designed to serve political and environmental ends in a way that hardly acknowledged a distinction between them. The rectification of the Rhine dried the land and thus improved the climate; it also inscribed in the landscape itself a purportedly immutable border between France and its German neighbour.

Where European government was imposed beyond Europe, climate was frequently used as an explanation for why the natives were the way they were, and the natives' perceived short-comings as users of the land were often pressed into service as an explanation for the climate's inhospitality. A better, more orderly, more European society would bring with it a better nature – a better climate.

Take America. In the seventeenth and eighteenth centuries Europeans saw America's climate as uncivilized, its winters too cold and its summers too hot. But that did not mean it was immutable. Thomas Jefferson, along with others less notable, believed that the spread of the right type of European civilization across America could in and of itself improve the climate. It was a stance which sat comfortably alongside his comrade Tom Paine's battle cry that 'we have it in our power to change the world' – and alongside the discovery by Benjamin Franklin, with his kites and keys, that Promethean science could call down and tame lightning. A century later, the land speculator Charles Dana Wilber encouraged farmers to head west on the basis of a very Jeffersonian belief: that 'rain follows the plow'. Such ideas were not held universally; there were contemporaries who dismissed both Jefferson and Wilber. But a contested idea can still be highly influential.

Island colonies were places that prior thought, from Plato's *Republic* to More's *Utopia*, had deemed particularly suitable locations for idealized forms of government, and for some that came to include the government of the climate. On Mauritius, St Helena and elsewhere, achieving a balance of forest, farmland and plantations that ensured ample rainfall came to be seen as an integral part of the management of the people and of the place.

In a world that believed in managing climate it was natural that, during the climate emergency that followed the eruption of Tambora in 1815, many called on government to intervene and bring things back to the way they had been. The famine that hit Ireland in Tambora's chilly aftermath explains the urgency with which an essayist in the *Dublin Chronicle* called for the Royal Navy to be put to work moving icebergs south in order to alleviate the situation. In France, the Minister of the Interior sent a bulletin to the prefects of the country's *départements* ordering them to survey all the actions that might be taken to ameliorate the 'marked cooling of the atmosphere, abrupt changes in the seasons and hurricanes' that were plaguing the country.

As the nineteenth century drew on, evidence for other, greater climate changes began to appear, and again administrators were called on to respond. By the middle of the century European explorers had discovered cave paintings, dried lakebeds and other signs indicating that the Sahara had once been wetter and greener. Joined with a self-serving, and misleading, reading of classical texts that portrayed the region as the 'breadbasket of the Roman Empire' was used to justify French colonial rule; the land's fallen semi-arid state was taken to be the result of bad landscape management by its post-Roman inhabitants. Left in Arab hands, things would get even worse. 'The Sahara, this hearth of evil, stretches every day its arms towards us,' warned a panicked pamphlet in 1883; 'it will soon enclose us, suffocate us, annihilate us.' Ambitious landscape interventions were undertaken to avert this; colonial records report the planting of millions of trees. Even more dramatic moves were suggested. Ferdinand de Lesseps, the architect of the Suez Canal, imagined further canals with which to turn the heart of the Sahara into an inland sea, tempering the desert climate; Jules Verne, who had previously speculated about changing the tilt of the Earth's axis to melt the ice of the Arctic, made conflict around the creation of a Saharan Sea the central idea of his last novel.

By the time Verne's book, *The Invasion of the Sea*, was published in 1905 notions of man's relation to his climate were far from what they had been a century earlier. The growth of industrial cities, whose climate was clearly human and vile – fetid water, soot-dark skies, air that raked at the lungs – made the idea that progress in and of itself brought better climates harder to believe. And changes in the way people sought to explain the world and the people it contained left less room for climate.

The historians Fabien Locher and Jean-Baptiste Fressoz point out that in the nineteenth century sociology, which focused on the internal dynamics of societies rather than on their external environments, and anthropology, which began seeking out biological differences, provided new accounts of why people in different places lived in different ways. And the previously unimagined germs that Louis Pasteur and Robert Koch brought to humanity's notice in the 1870s and 1880s led to accounts of disease that greatly reduced the role previously ascribed to the malign influences of bad climates.

New explanations focusing on races and germs came to be built into the way societies were shaped, as well as the way they were talked about. They brought with them new measures of hygiene, both personal and racial, and new anxieties about policing how people washed and with whom they bred. This added to the explanations' importance; knowledge that becomes tied up with the exercise of power becomes powerful in and of itself.

Shorn of its social role as a shaper and product of human affairs, a co-production of humanity and nature, the climate became just another part of the increasingly mechanized picture the sciences had of the natural world, and at the same time a matter of private rather than public concern. If you want a date for the last phase of this transition, try 1883. It was the year that the term 'eugenics' was invented, foreshadowing the most atrocious governmental forays into the control of genes; it was also the year the modern domestic thermostat was patented, as a purely technical mechanism for controlling climate, albeit on the smallest of scales, that carried no social or governmental implications whatsoever – just numbers on a dial.

Martians and Moral Equivalents

To the modern mind, it seems obvious that there was another reason Enlightenment ideas about the climate being subject to human control fell from favour in the latter part of the nineteenth century: they weren't true. Yes, draining swamps can improve public health, and the presence or absence of forests on tropical islands matters when it comes to rainfall and erosion. The idea that a good civilization brings about a good climate, though, is hard to sustain. Human influence on the European climate over the thousand years prior to the Enlightenment, models suggest, probably led to slightly colder winters; low-albedo forests absorb more winter sunshine than snow-covered pastures, heaths and fields, and Europeans replaced the former with the latter. But that change was neither recognized nor beneficial. There was nothing the prefects of France could do to stop the storms that Tambora set in motion; no amount of iceberg wrangling by the floating forests of the Royal Navy would have changed the fate of the starving Irish one whit.

All true, but beside the point. History is replete with examples of things people believed that were not true and which, to those with today's understanding, must surely have seemed untrue at the time. People have often, and persistently, acted in ways they think will have a given effect despite the fact that, as science now understands the world, such actions must routinely have failed to achieve anything like their stated goals. So persistent are such beliefs that 'because it was wrong' is for the most part a woefully inadequate explanation when faced with the question of why a specific group of people gave up a particular belief or set of beliefs.

That said, by the late nineteenth century the strange and re-markable grandeur of the ice ages was making it harder to believe that humans were much of a factor. Having covered Europe and America alike, they were evidence that the Earth had a global climate, not just a set of local climates. Changes in the climate of the Earth as a whole needed to be understood at the scale of the Earth as a whole. As we have seen, it was in exploring possible answers at that geophysical, whole-planet scale that Arrhenius made his calculations of the greenhouse effect, and that the link between the global climate and volcanic eruptions was first explored. And for people working on the scale of a whole planet, a scale that included the rest of the solar system was not too much of a stretch (Arrhenius, for one, always saw himself working at what he called a cosmic scale). Thus the rhythmic oscillations in the Earth's path around the sun were investigated as possible influences on the Earth's climate, and the conditions of other planets were studied as possible analogues from which to learn.

One outcome of this was a particularly influential flight of geoengineering fancy: the story of the canals of Mars. Observations of Mars with the improved telescopes of the late nineteenth century revealed what appeared to be a complex system of linear markings. Today these are accepted as having been, in effect, optical illusions. Then they were a cause célèbre; Percival Lowell, an American astronomer, took them to be canals made by a planet-spanning civilization, setting off a fascination with the idea of alien intelligence.

Why should there be canals on Mars? To Lowell it was obvious. He believed Mars to be an older planet than the Earth, and thus both a drier one – he subscribed to a theory that all planets tended to dry out as they aged – and one with inhabitants further along the evolutionary road of progress than humans had yet travelled. Humans were already using canals to join oceans and sunder continents; Martians were advanced enough to build them on a planetary scale in order to irrigate their increasingly arid planet. The motives of the Martian canal builders, Lowell thought, were as clearly seen through the telescope as the surface of the planet itself. In his refusal to distinguish astronomical observation from sociological explanation, if nothing else, he prefigured the hybridization of the human world and the earthsystem that characterizes the Anthropocene.

Lowell's canal dreams inspired many further speculations. H. G. Wells suggested that if the Martians could remake one planet in their quest for a better environment, they might also turn their technology to conquering another. The Martians in his *War of the Worlds* were more evolved than the men of the nineteenth century in their intellect and their technology, but not in their moral sympathies. They thought no more of pushing aside civilized Europeans than those Europeans thought of dispossessing the aboriginal peoples of the antipodes. Others thought that the Martians' technical prowess must be accompanied by some moral progress. A globe-spanning irrigation network, they reasoned, must mean a globe-spanning government, and indeed it might have helped to bring it into being. In 1906 William James, the psychologist and pragmatist philosopher, gave a speech at Stanford entitled 'The Moral Equivalent of War', a powerful phrase which was brought back to prominence by Jimmy Carter when trying to confront the energy crisis of the 1970s. James wanted to abolish war but to keep many of the virtues that war brought with it; he put particular value on the shared devotion to a higher cause it required and inculcated. He thus argued that wars between men should be replaced by a 'war on nature' aimed at improving the planet for the good of all. The Martian canals could be seen as an example of just this sort of thing; their global nature showed that the Martians had given up war against each other in favour of war against their deserts.*

No one, as far as I know, thought that the Martians were wrong to be waging this war – that they should simply accept the fate their worsening environment held for them. What more noble employment could there be than grand ceaseless work for the common good? But the noble struggle had an

---

* David Scott Dolan, an Australian scholar, notes that James and Lowell enjoyed each other's company when they met while both seeking cures for nervous exhaustion on the Riviera in 1900. Given that it would be quite unlikely for Lowell to go any length of time without speaking of Martians, James could easily have been thinking of their war on their planet's desert when he called for an end to human warfare.

air of tragedy, too. Robert Sherlock's *Man as a Geological Agent*, published in 1922, prefigures more recent Anthropocene discussions with its impressive account of the sheer scale of the Earth's mines, cities and canals. At its conclusion, the canal builders of Mars are invoked to put those human rearrangements into context – and to be put into cosmic context themselves. 'The canals of Mars, if they really exist . . . are far greater than any our engineers have imagined. The battle of the Martians with Nature has been on a much more gigantic scale than Man's conflict, and yet we hear that the Martian is on the point of extinction, and Mars of becoming totally lifeless. Even on Mars the mighty engineering works seem merely to scratch the skin of the planet, and the final result of Martian activity on the solar system seems likely to be infinitesimal.'

Sherlock shows that at least some of those thinking about engineering as a planetary phenomenon used Lowell's ideas as a benchmark, and for much of the twentieth century speculations about human redesign of the Earth are much more similar in spirit to Lowell's megaprojects than to the Enlightenment belief that civilization and climate were providential co-productions. Early- and mid-twentieth-century ideas about humans changing the climate tend to focus on specific projects aimed at specific ends: diverting ocean currents with canals and dams; rearranging river flow; creating inland seas, as in the Sahara; even putting mirrors into space to give the Earth rings like Saturn's, rings that would shield some places from the sun and reflect more of it to others. In the Soviet Union, where James's notion of 'war on nature' had an established place in state ideology, projects to re-route whole river networks added to large-scale destruction of the environment by other means, though few of them were carried out in full. In the West, though, most geoengineering ideas were hardly even plans; they were speculations explored in the same spirit of planetary playfulness as that in which people had speculated about the motives of Martians, the sort of fanciful idea that some geologists and geophysicists referred to as 'geopoetry'. Just such disinterested playfulness is to be found in *The Foreseeable Future*, a 1953 compendium of technological speculations by the great British physicist George Thompson. In imagining the future of the climate, Thompson suggests, I think for the first time, that the injection of a few million tonnes of some sort of absorbing or reflecting particles into the stratosphere could be used to cool the Earth's surface, should people wish to do so. He offers no reason why they should – modern concern about global warming had, at the time, yet to get under way. He is just playing with the idea.

In 1842, poised between the worlds of Jefferson and Lowell, Henry David Thoreau reviewed *The Paradise within the Reach of all Men, without Labor, by Powers of Nature and Machinery*, a utopian work by John Adolphus Etzler. Etzler, a German immigrant influenced by the French anarchist Charles Fourier (and a friend of John Roebling, architect of one of the earliest masterpieces of America's technological sublime, the Brooklyn Bridge) was seized by a belief in the power of wind, water and sun to revolutionize transport and industry and 'free [men] from almost all the evils that afflict mankind . . . except death', allowing all humanity to 'enjoy a new world, far superior to the present, and raise themselves far higher in the scale of being'. In his review, Thoreau distinguishes two types of reformer: 'One says he will transform himself and then nature and circumstances will be right . . . The other will reform nature and circumstances, and then man will be right. Talk no more vaguely, says he, of reforming the world – I will reform the globe itself.' In a broad-brush way, the distinction can be used to separate the views of human control of the climate at the end of the eighteenth century from those at the beginning of the twentieth – and to distinguish two different approaches today.

In the Enlightenment model the influence on climate

flows from the way humans organize their lives. It requires no technological expression in and of itself, merely the appropriate disposition of the other factors of the economy. This is, in essence, the moderate green view of today: build the right energy infrastructure, embody the right virtues of conservation and efficiency, and all will be well. In the later model, climate has become an objective part of geophysics, and its modification is to be understood in that light, more quantitative than moral, a matter of instrumental procedures applied directly to the world, a reform not of society but of the globe itself. This is how people now see geoengineering.

What I want you to consider is that there could be a third option – one in which the question of how to 'reform the globe itself' is asked in a way that requires the respondent to think about the sort of world in which such a question could be justly answered. It is an option which rejects the idea that if only humans could be made right then the 'nature and circumstances' of the planet would come right in response, and which also rejects the idea that the planet should be blithely reworked to suit the current wishes and interests of those able to do so. It is one that asks instead how the challenge of remaking the planet can be used to bring about a world that could be trusted with the power to meet that challenge; how society can create a fulcrum that allows planet-changing leverage to be applied with justice and prudence. I have no programme to bring this about. I realize that it is, perhaps, as utopian a dream as Etzler's. But I still think it deserves to be considered, if not as a possibility, at least as an inspiration – my inspiration, over the course of writing this book; yours, perhaps, after reading it.

The Day Before Yesterday

In the years after World War II, two new styles of thinking about climate intervention were fostered. In one of them human action on the environment is not the moral equivalent of war, but simply its continuation. By the 1950s the relationship between military might and physics that created the atomic bomb had expanded to include much of geophysics, and the possibility of reshaping many aspects of the earthsystem to military ends cropped up repeatedly.

In his book *Arming Mother Nature: The Birth of Catastrophic Environmentalism*, Jacob Darwin Hamblin looks at the ideas for 'environmental warfare' examined by a NATO committee under the chairmanship of Theodore von Kármán, one of the key figures in twentieth-century aeronautical engineering, and led after his death by Edward Teller, one of the creators of the hydrogen bomb. The committee's brief was to find geophysical ways to wage war that would supplement the atomic, chemical and biological modes already provided by other scientists – to find ways to weaponize everything from earthquakes to the weather to the Van Allen radiation belts discovered by the first satellites. Similar considerations were entertained by the JASON group, comprised mostly of academic physicists who provided the Pentagon with big ideas and audited the big ideas coming from elsewhere. If the planet came to feel fragile in the 1960s and afterwards, Hamblin argues, it was in part because there were people actively working on techniques for breaking it.

There were ways in which this military speculation, too, was playful. You cannot read 'How to Wreck the Environment', a 1968 essay by Gordon MacDonald, a UCLA physicist and a member of the JASON group, without feeling the delight he takes in playing with ideas for geophysical warfare as wild as the deliberate triggering of solar flares. But MacDonald was well aware that not all games are nice games. His essay, written as a warning, talks about the possibility of a hole in the ozone layer. It was not an entirely new idea: as we have seen, Sydney Chapman, who first explained the chemistry of the ozone layer in the 1930s, considered the possibility of making temporary holes in it for the benefit of astronomers. But MacDonald was talking about the possibilities of killing people with such holes. And not all the speculation remained speculation; ideas about environmental warfare from the 1950s and 1960s shaped the defoliation campaigns of the Vietnam War that sought to deny the enemy forest cover, and the cloud-seeding operations that were designed to turn the Ho Chi Minh Trail to mud. The horrors that chemtrailers imagine governments perpetrating in the sky are fantasies; but they are not entirely without historical precedent.

At much the same time, the prospect of climate change brought with it, in its early days, a new geopoetry of remediation. Roger Revelle, the man who had first drawn attention to the 'geophysical experiment' of increasing carbon-dioxide levels, included a mention of the effects that experiment was expected to have on the climate in a 1965 report on environmental pollution drafted for President Johnson – the first greenhouse-warming alert ever presented to a head of state. But Revelle made no suggestion that carbon-dioxide emissions should be reduced as a way of dealing with the problem. Instead he described a technical fix – floating ping-pong balls on the surface of the ocean to make it more reflective.

A decade later an Italian physicist, Cesare Marchetti, outlined a scheme in which much of the world's industrial carbon dioxide would be piped directly from power plants to the ocean depths, rather than released into the atmosphere. Under the pressures of the abyss the carbon dioxide would become a liquid, denser than water, which would pool harmlessly in depressions in the sea floor. Marchetti's 1977 paper on this scheme was the first to use the term 'geoengineering'. It was published in *Climatic Change*, a then-new scientific journal edited by an ambitious, broad-ranging and politically engaged climate modeller called Stephen Schneider. Schneider had a taste for things that would provoke debate; he engaged with ideas about geoengineering more than any other leading climate scientist of his generation.

While Marchetti imagined ocean-floor lakes of carbon dioxide,

Freeman Dyson, a mathematical physicist with a rich imagination and a pronounced speculative bent (and a pre-eminent JASON) wrote of planting vast forests of sycamore to suck the stuff directly from the atmosphere. A little earlier, Mikhail Budyko, a Russian climate scientist whose work was driven, in part, by the perceived need to understand how rerouting rivers in the Arctic would change the climate more generally, suggested that, if the world needed cooling, aircraft burning sulphur-enriched fuel might be sufficient to the need. By the 1980s Marchetti's term 'geoengineering' had come to encompass all such technological schemes to counteract human intensification of the greenhouse effect, whatever specific technology they might employ.

When, in 1988, worries about greenhouse warming broke out of academia and into the world of public affairs, you might have expected such speculation to grow in volume and seriousness. And there were some signs that it might. Just days before the eruption of Mount Pinatubo, a conference in Palm Coast, Florida was discussing ways in which grand engineering projects might be used to slow, halt or even reverse global warming. At the same time a report by America's National Academies of Science called 'Policy Implications of Climate Change' devoted an entire chapter, drafted in large part by Schneider, to geoengineering. It discussed, among other things, orbiting mirrors in space, the possible use of artillery pieces to launch sulphates into the stratosphere and the use of iron to fertilize the southern oceans, increasing the growth of carbon dioxide–hungry photosynthesizers. At the 1992 American Association for the Advancement of Science meeting in San Francisco, scientists gave the issue a public airing, as did magazines and newspapers reporting on the meeting[*]. With the future of the climate now a hot topic and Pinatubo's cooling effects in evidence, you might have expected the trickle of interest to grow into a flood.

---

[*] The first interview I ever did on the subject, I believe, was during that meeting, in a McDonald's on Powell Street, with cable cars click-clacketing past.

Instead it dried up altogether. In 1965 it seemed reasonable that the only remedy Revelle would include in a report to the president mentioning the greenhouse effect was an ocean covered with ping-pong balls; who in his right mind would have thought of suggesting to Lyndon Johnson that the basic course of industrial capitalism needed to be altered? But thirty years later, rearranging the economy and wider society so as to eliminate emissions had become the whole agenda.

The Rise of Carbon Dioxide Politics

The worldwide neglect might have been down to the idea simply seeming too far out, and that surely played a role. But as someone writing about science at the time, and with a pretty high tolerance – to be honest, appetite – for the far out, I found that throughout the 1990s I was able to abuse the generosity of various editors to get to scientific meetings on subjects quite as far out as geoengineering: the threat which asteroids posed to the Earth, and techniques for reducing it; the challenge of interstellar, as opposed to merely interplanetary, space missions; the possibility of colonizing Mars. Indeed, there were even meetings on ways in which, subsequent to such colonization, the Martian climate might be engineered to human advantage, using tailored aerosols and greenhouse gases to warm and thicken the atmosphere to the point where water might flow. Such 'terraforming' owes an obvious debt to Lowell – except that where Lowell saw Martians solemnly trying to postpone their planet's inevitable demise, the terraformers were imagining ways to bring a dead-already world to life. It was in these debates that I was first exposed to some of the questions that animate this book, questions not just about how to reshape a planet, but about what right one might have to do so.

But for more or less a decade after that pre-Pinatubo get-together in Florida I came across no meetings dedicated to engineering the Earth's climate, as opposed to that of some other planet, and I don't think I missed any. Ideas about the terraforming of Mars made the cover of the prestigious journal *Nature* in 1992 – but from 1991 to 2001 it published not a single scientific article mentioning climate geoengineering.

What distinguished geoengineering from those other ideas was that it had to do with something that actually mattered. There were no asteroids on impact courses that the Earth needed to worry about, and no planets around alien stars with a pressing need for probing. The colonization and terraforming of Mars could wait for another day, or decade, or century. But from the late 1980s on there was widespread agreement that the climate had already changed, and that the risks posed by further change needed addressing forthwith.

This sense of urgency, rather than encouraging a debate on geoengineering, suppressed it. The world turned its back on playful speculations about the future to focus on the situation at hand, and climate moved with precipitate speed from being not seen as an issue at all to being a problem for which there was a recognized response laid down in international law.

Though climate scientists had been talking of greenhouse warming for decades, and had done so with increasing alarm since the early 1980s, the issue only really emerged on the political stage in 1988. That year's El Niño saw temperature records broken all over North America; Jim Hansen testified to the reality of climate change caused by industrial emissions in front of the US Senate; a large meeting of scientists and environmental groups in Montreal addressed the question head on for the first time. The next year the IPCC was set up to assess the problem and to draw the developing world into the discussion. In 1992 almost all the world's governments signed the UNFCCC at the 'Earth Summit' in Rio de Janeiro, taking on themselves the duty of 'avoiding dangerous climate change'; by 1997, less than a decade after the scorching summer during which Hansen gave his testimony, the Kyoto protocol to the UNFCCC had committed the world's developed countries to specific carbon-dioxide reductions. Although some nations, notably the USA, never ratified the protocol, it went into effect in 2003.

One reason the process was so quick was that its focus was narrow: carbon dioxide. This made obvious sense to scientists, as carbon-dioxide emissions represent both the largest current human interference in the climate system and the one that is growing fastest. It also made sense as practical politics. Climate is huge, amorphous, irreducibly complex. Carbon dioxide is a molecule. It is pretty obvious which of the two is going to be an easier subject for agreement. Groups that saw the politics, morality and metaphysics of the climate problem very differently from each other could nevertheless reach agreements about carbon dioxide; it could bring together natural scientists, green activists and non-governmental organizations, developing nations, international diplomats and even economists.

Climate scientists knew that carbon dioxide was only part of the climate-change story. But they knew it was the biggest part, in the watts-per-square-metre language of radiative forcing. And they believed that if the world was going to act on climate, which they thought it should, it would need fairly simple goals to subscribe to. They also had a guiding principle that by and large the best thing for the natural world, and the human world which is built atop it, was to minimize the change it was subjected to. In discussions between scientists in the late 1980s the idea of minimizing change led to a proposal that the target for action should be the rate at which things were changing, rather than the total amount of change foreseen – adaptation, of people and of ecosystems, being in principle easier when change is slower. But that goal was rejected in favour of one that would attempt to restrict the total amount of change. It seemed clearer and sharper; it was also a goal that computer models could more easily translate into the language of carbon-dioxide emissions.

Some of the scientists probably also appreciated that a

definition in terms of carbon dioxide gave them a particular power. Carbon dioxide is a technical matter, the sort of thing that fits well inside the realm of science, the sort of thing that scientists have authority to talk about. If climate change is seen primarily as a matter of carbon dioxide, then scientists will have a particular standing in the debate; knowledge that becomes tied up in the exercise of power becomes powerful in and of itself. In the environmental movement climate scientists became treated as the voice of the planet, with a unique ability to say what the natural world needed – to set the 'planetary boundaries' beyond which human intervention in the earthsystem should never stray. With the debate framed this way the scientists' belief that the planet should stay as close as possible to its pre-industrial carbon-dioxide levels took precedence over other concerns; the massive changes that meeting those needs would force on to the social and economic parts of the earthsystem were secondary. The world should change so that the planet should not.

Political environmentalists also tend to like the natural world as it is. But while natural scientists feel free, at least in their professional lives, to more or less ignore the social and economic world, political environmentalists are committed to changing it into something less obsessed with growth, less exploitative, more just, more sustainable. They often believe that the link between economic growth and the potential for happiness, a link that is held to underpin the capitalist system, is largely imaginary, and always believe that there are better ways of living than those the current system tends to endorse. They see contemporary life as alienated from a nature, and from aspects of a human nature, that it should embrace. The act of conforming to the confines of that embrace and accepting limits which they see nature as setting on human life is held to be responsible, morally upright and, in all likelihood, the basis of lasting happiness. They are very much reformers of Thoreau's first type: reform of the self and the community is at the heart of the project.

Many social and natural scientists working on the environment have sympathy with at least some of this agenda, not least because people often become environmental scientists in their twenties because they have become environmentalists in their teens. This does not mean their work is necessarily biased; one can be objective without being neutral. It can, though, complicate things.

Carbon dioxide suited scientists because it seemed like a straightforward measure of the problem. It suited greens because it was a pretty good proxy for the industrial society against which their movement was a reaction. The international negotiations that set up the UNFCCC showed that it suited developing countries because it was primarily a developed-country issue; at the time of Rio, the vast majority of all the industrial emissions since the eighteenth century had come from Europe and America. The UNFCCC embedded that idea in a key clause referring to the signatories' 'common but differentiated responsibility' for climate change. The differentiation can be seen in the Kyoto protocol's architecture: developed countries accepted restrictions on their emissions but developing countries did not have to. This makes ever less sense – at some point in the 2020s, on current trends, the total amount of carbon dioxide emitted by China over the past fifty years will surpass the total for America over its entire history. But it is the way the agreement came to be made, and an agreement that required things of the developed world but not the developing world was a novelty that developing world politicians warmed to.

The diplomats entrusted with negotiations, too, found focusing on carbon dioxide congenial. It made climate change a thing-there-ought-to-be-less-of problem, and by the early 1990s people negotiating treaties had started to feel at home with things-there-ought-to-be-less-of; they knew how agreements about things-there-ought-to-be-less-of worked, and were pretty good at reaching agreement on them. For the 1987 Montreal protocol the thing-there-ought-to-be-less-of was

CFC production; diplomats had come up with a clever way of delivering the less that was needed, one that differentiated the obligations of developed and developing countries, and offered assistance from the rich to the poor, both things which would be themes of future climate negotiations. In arms-control accords like the Strategic Arms Reduction Treaty of 1991 and the Intermediate-range Nuclear Forces Treaty of 1987 the things-there-ought-to-be-less-of were particular types of weapon, and those deals worked too. The language of UNFCCC documents is strikingly reminiscent of such agreements, with their focus on calibrated levels of cuts and on monitoring and verification mechanisms designed to show that the cuts have actually been made.

While they hadn't applied it to nuclear weapons, economists, too, thought they had a good approach to things-there-ought-to-be-less-of: create a market in them. At the time of Rio, America had started using this approach to cut sulphur emissions from coal-fired power plants. Operators got a limited supply of permits to pollute. Those who could cut their pollution cheaply could make money selling the permits they didn't need to companies who found the process more difficult. Economists liked this sort of scheme because it encouraged the most cost-efficient reductions; people who find it easy to reduce did so the most. Carbon markets had a political appeal, too, in the context of the 'third way' favoured by Bill Clinton and Tony Blair; they used a 'right-wing' mechanism, the market, to deliver a 'left-wing' goal, environmental improvement. The market approach was implicit in the UNFCCC, and explicit in the Kyoto protocol five years later. It has, so far, been an abject failure. Europe's Emissions Trading System, the world's largest carbon market, has crashed twice. Attempts to introduce a national market in the USA were one of the notable disappointments supporters of Barack Obama suffered after his election in 2008.

This illustrates a general problem with focusing so much of the world's climate action on carbon dioxide: it has proved a much easier substance to agree on than to control. The mechanism set up under the UNFCCC to help developing countries emit less carbon dioxide has had very little effect, as have all the other provisions of the Kyoto protocol. Global emissions have increased at as high a rate as anyone imagined. Though carbon-dioxide emissions in some developed countries have declined a little, this has been more to do with the fact that energy-intensive industries have moved to the developing world, taking their emissions with them. While Britain emitted 15 per cent less carbon dioxide in 2005 than it did in 1990, the carbon dioxide emitted in producing all the products Britain consumed was 19 per cent higher; it was just that more of the production, and thus more of the carbon-dioxide emission, was going on elsewhere.*

There are many reasons why deep global cuts in carbon dioxide are difficult to achieve, some of which were dealt with in the introduction. Two, though, stand out. One is that fossil fuels are built into the foundations of the industrial and economic system, which means cutting emissions is hard – especially so since the costs of the cuts are concentrated on a powerful sector of the economy. The other is that cutting carbon dioxide provides no short-term benefits. Because what matters to the climate is the total amount of carbon dioxide in the atmosphere, not the rate at which it is increased, carbon dioxide cuts made today will have more or less no effect on climate for thirty years. History hangs over everything. This, perhaps more than anything else, is what makes climate negotiations difficult; costs people bear now lead to benefits that other people will see in the second half of the century.

These are fundamental problems, and they would not go away if people focused less exclusively on carbon dioxide; but an approach that did not make it the be all and end all could

---

* The carbon dioxide emitted by making products in Britain that will be consumed elsewhere is pretty trivial in comparison.

see progress on other fronts that the primacy of carbon-dioxide politics has sidelined. For example, there has been relatively little action on other human interventions in the climate. Cumulative human carbon-dioxide emissions are currently adding about 1.8 watts to the Earth's radiative forcing. But humans add half as much again by other means. They emit HCFCs, the ozone friendly replacements for CFCs that are themselves strong greenhouse gases. They emit methane, both industrially and from farmlands, and nitrous oxide, both of which are greenhouse gases. These gases are covered by the Kyoto protocol, but they are a long way from being its focus. Humans also emit soot, which absorbs sunshine and emits infrared, warming the atmosphere and the surface — with a particularly dramatic effect when it settles on snow. Action on these other greenhouse gases and 'short-lived climate forcers' has started to become an international focus since the Copenhagen climate summit's breakdown in 2009. It could have started to do so well before then.

A more significant victim of the politics which equated climate action with the reduction of carbon-dioxide emissions was the effort to find ways for people to adapt to changed climates. Adaptation has some great advantages over emissions reduction (known in policy circles as mitigation): because many societies are not particularly well adapted to their present climates, helping them to prepare for worse future climates can do good right here and now. And because there are lots of heres, and because now starts now, adaptation can deliver benefits even if enacted piecemeal, without a global agreement.

Despite these strong arguments, adaptation was sorely neglected in academic studies of climate change, in climate politics, in climate activism and, as a result, by many countries for much of the 1990s and 2000s. It was seen not as a vital counterpart to mitigation but as a counsel of despair, as throwing in the towel, as a poor second choice to be considered only if the main aim — control of the atmosphere's carbon-dioxide level

– proved unattainable. Almost none of the people who wanted to talk about adaptation ever suggested that it did away with the need for mitigation. Still, they found themselves, as one put it to me, as popular as people who keep farting at the dinner table. This persistent unwillingness to think about adaptation – and to address the question of how to pay for the large-scale adaptation that will be needed in some developing nations – badly undercut the world's response to climate change.

The last victim was the sort of climate geoengineering I have described in the past few chapters. The only type of climate geoengineering that made sense in a world where climate was defined as a carbon dioxide issue was some sort of technology for capturing carbon from the air, as discussed in the introduction and more fully in Chapter Eight. If carbon markets worked, such techniques would be moneymakers.

But a world of carbon-dioxide politics had no place at all for geoengineering technologies that sought to limit the amount of sunshine coming into the earthsystem. They were like adaptation – indeed, in many ways that is a good way of thinking about them, as a world-spanning, insufficient and partial adaptation to the problem of too much heat. And so the worry that applied to adaptation – that it took people's attention away from the real problem of carbon dioxide – applied to such ideas with peculiar force. With sunshades in the sky, the carbon-dioxide levels that were the one thing everyone could agree on would lose salience. What sort of monster would want to undercut the very basis of climate politics in such a way?

6

Moving the Goalposts

It's coming through a hole in the air,
From those nights in Tiananem Square
It's coming from the feel
That this ain't exactly real,
Or it's real, but it ain't exactly there.
Leonard Cohen, 'Democracy' (1992)

Lowell Wood is an intriguing figure, a slouching bear-like man happy to deploy his intellect with a force proportionate to his bulk. He spent most of his career as a designer of nuclear weapons at Lawrence Livermore National Laboratory, east of San Francisco Bay, where he was a protégé of Edward Teller. Teller, who had an insider's clout in the political arena, was something of an outcast in the world of physics. In 1954 he had told an inquiry that he thought Robert Oppenheimer, the scientific leader of the Manhattan Project which had developed the Hiroshima and Nagasaki bombs, should no longer have a security clearance; many of those who, like Teller, had worked under Oppenheimer considered this a betrayal and shunned the betrayer. But Teller inspired a quasi-filial devotion in his hand-picked young colleagues at Livermore, and encouraged them to think widely about ways that physics (including, quite often, the physics of nuclear explosions) could be applied to problems other than the making of weapons. He and they were indulged in all this by a laboratory happy to spend money on whatever distractions might refresh its key workers' interest in life when they were taking a break from the business of death.

Insulated from the politics of mainstream science and from the need to publish papers that would enhance their careers, and certain in their belief that science could give humans near limitless power – did the weapons they designed not recreate the fires at the heart of the sun? – in the mid-1990s Teller and Wood started to make some calculations about how to change the climate. In 1997, with a colleague called Roderick Hyde, they produced the only major paper on stratospheric veilmaking published in the 1990s, though they brought it out as a working paper inside the lab rather than sending it to an academic journal.

The paper was full of physical insight into how various sorts of aerosol could be used; as well as looking at sulphates, it looked at scattering particles made of metal, at ions trapped in tiny lattices of carbon that might fine-tune the wavelengths they had their effect on, at systems where the scattering was done by mirrors in space rather than by aerosols in the stratosphere. The paper also looked at similar systems which might be used to warm the planet should there be an ice age on the horizon – the wizards of Livermore had a grand enough conception of humanity's place in the universe that they were able to worry about geological time scales as well as current policy predicaments. The ice-age spin also sent the message that the Livermore work was important beyond the context of man-made climate change – on the importance of which some on America's political right, with whom Teller was closely aligned, were eager to cast doubt. With an eye to the same constituency, the paper also insisted that sunlight-scattering schemes needed to be studied because, compared with the reduction of carbon-dioxide emissions, they were very, very cheap.

Wood presented the paper's conclusions at a meeting on climate issues held in Aspen, Colorado in 1998. David Keith and Ken Caldeira were sitting at the back of the hall, and as they now tell the story, they scoffed at the presentation like smart-arse schoolboys. Keith's background was in physics; when setting out as a researcher in the early 1990s, he did highly regarded research on new ways of trapping individual atoms. But, always keen on environmental issues, he went on to apply himself to a peculiarly wide range of climate-related research. He did engineering-based policy analysis, he worked on calculations of the rate at which heat flows from the equator to the poles, he designed a satellite instrument that could nail down with unprecedented precision the rate at which radiation entered and left the earthsystem. (Thanks largely to NASA internal politics, fifteen years on that instrument has yet to fly.)

Keith considered himself far better versed in climate science than Wood. The same applied to Caldeira, whose work centred on earthsystem modelling. He had come to it by a curious route that included stints on Wall Street and a fair amount of political activism against nuclear weapons. On returning to higher education in the late 1980s, he had found an intellectual home in the small, eclectic and rather fabulous Department of Earth System Science at New York University. (How eclectic? Well, you remember that conference on interstellar space missions I mentioned? That was them.)

Keith and Caldeira told each other that if Wood were a proper climate scientist, he would appreciate the degree to which analysing just the energy going into a system and the energy coming out, while always necessary, was never enough. When Wood talked about a stratospheric veil or a space-based mirror reflecting away enough energy, in watts per square metre, to balance the warming due to a doubling of carbon dioxide, he was ignoring the fact that the two forcings, while quantitatively the same, were qualitatively quite different. Greenhouse warming worked 24 hours a day; mirrors worked only in daylight. Greenhouse warming was strongest in winter and at the poles – the time and place that mirrors had least effect. Those mismatches felt like a big deal. Not appreciating them marked Wood down as a cocksure interloper, an impression reinforced by his overreaching conclusion – that Promethean physicists could deliver a climate optimized for human beings with scant regard to what the sun and the earthsystem might have to say on the matter. That they could fit the planet with a thermostat.

Caldeira, as it happened, also worked at Livermore, in a climate modelling division that could be traced back to Teller's post-war interest in predicting, and possibly controlling, the weather. But he had, and wanted, no contact with the weapons-designing part of the laboratory; Teller, Wood and Hyde had not consulted him when working on their ideas.* To investigate the flaws he felt sure he would find in Wood's analysis, he and a colleague, Govindasamy Bala, used an up-to-date climate model to compare a Greenhouse Planet, an Engineered Planet and a Baseline Planet. So completely was climate geoengineering off the agenda that no one had actually done this before.

Caldeira and Bala found what subsequent models such as those of GeoMIP found. The match between the cooling provided by less sunlight getting through the stratosphere and the warming furnished by greenhouse gases, while not exact, was actually pretty close. The differences that Keith and Caldeira had prided themselves on appreciating back in Aspen turned out to matter very little. The ocean, which is slow to warm, cares not at all whether energy moves in or out by day or night, and sees little difference between winter and summer. And a great deal of climate change is moderated by the ocean.

Caldeira was not exactly converted to climate geoengineering.

---

* This may seem odd, but Livermore is a somewhat odd institution. Different security clearances and cultures set up distinctions between the nuclear-weapons designers 'inside the fence' and the other researchers there. Also, consulting others who have thought about the issues is not always something physicists in the mould of Teller or Wood are keen to do. They would often rather reach their own conclusions from first principles; this can bring fresh insights, and it can also bring avoidable error.

But he realized, slightly shamefacedly, that he had been too ready to dismiss its potential, and began to look into it more systematically. Over the next few years he ran further model experiments, looking at the effects on the hydrological cycle and on ecosystems. It was the sort of science that appealed to him – big-picture stuff a little way from the mainstream, where asking the right questions and getting first-order answers to them would let you survey a whole new field of enquiry. The same taste later led him to look at what dissolved carbon dioxide might do to ocean chemistry – that is, at the problem, only then coming to the fore, of 'ocean acidification'.* That dissolving carbon dioxide in water produces a weak acid is basic chemistry, and that doing so on a global scale might change the pH – a measure of acidity – of the oceans had been obvious since Revelle's days, but such pH changes were generally seen as a minor matter. As a doubling of the carbon-dioxide level went from being, as Caldeira put it, 'a nightmare to avoid' to 'the least we can get away with', ocean acidification took on a new importance, and Caldeira was one of the first to recognize it. It has since become the topic of a great deal of research. The extent of the ecological effects it will bring is still hard to gauge – the pH of seawater is already variable in many places, and both organisms and ecosystems have some capacity to adapt to changes in the level. Nevertheless, Caldeira and others have come to believe that the process imperils many of the world's coral reefs and a number of other ecosystems too.

Caldeira's geoengineering work was not picked up by other climate-modelling groups, but it was noticed by a man of great influence: Paul Crutzen. Crutzen's work on the chemistry of the ozone layer had won him a share in a Nobel Prize – only

---

* To avoid confusion, it is worth pointing out that the oceans will not in fact become acidic: acidification is a relative term. The earth's oceans are slightly alkaline – the opposite of acid. Adding carbon dioxide makes them a bit more acid; that makes it an 'acidification'. But it does not mean that the seas could end up being acidic; they will stay alkaline.

the second ever awarded for work on the atmosphere – and underpinned the Montreal protocol. He would have been respected in atmospheric science for that alone, but he had an understated, rather Dutch, slightly Yoda-ish charisma that added yet more to his influence, both in Europe and in the United States. He had political instincts that served him beyond academia, a feeling for the stuff that mattered and for how to get people to see that it mattered – hence his successful introduction of the term 'Anthropocene' into the scientific lexicon in the early 2000s. In 2006 he announced, in an essay in Stephen Schneider's journal *Climatic Change*, that climate geoengineering had become something that mattered, and that scientists had to start doing more research into it.

The essay was simple, thoughtful and convincing. It started off lamenting the fact that carbon-dioxide emissions were not being reduced. Taking up Caldeira's work, it then noted that some sort of stratospheric veil might provide an alternative response – adding that in Crutzen's opinion, such action would not necessarily put the ozone layer in severe danger. Such a scheme could never replace emissions reduction, Crutzen went on; the reduction of emissions offered the surest way to a stable climate, and it was the only way to curtail ocean acidification.

This stress on acidification was one of the clever things about the way Crutzen framed his argument. The fact that sunshine geoengineering can do nothing about this non-climatic effect of increased carbon dioxide is often treated by critics of the idea as a fatal shortcoming. Crutzen saw that it could be a political asset. Because sunshine geoengineering efforts would do nothing about ocean acidification, they did much less to undermine the politics surrounding the reduction of carbon-dioxide emissions. Scientists working on stratospheric veils to stabilize temperatures could reassure people that even if their work bore fruit, the separate problem of acidification meant that there would still be an overwhelming need to control carbon-dioxide emissions.

The most important thing about the essay, though, was not what it said about ocean acidification or anything else. It was that everything it said was said by Paul Crutzen, saviour of the ozone layer. 'The messenger was the message,' Schneider told me later. Specifically, Crutzen's cachet in the realm of stratospheric chemistry protected anyone wishing to follow up his call for further research from being seen as an ozone vandal. More generally, Crutzen's imprimatur provided some protection against the suspicion that any approach to climate change other than one focusing on the reduction of carbon-dioxide emissions might play into the hands of Teller's friends on the right, committed to avoiding the costs of such reductions.*

An interest in the possibilities of sulphate aerosols as cooling agents is obviously not the same as denying the role of carbon dioxide in the climate – indeed, such denial should leave one uninterested in taking any action on the climate at all. But because both involve, in different ways, the decoupling of carbon-dioxide emissions from the climate, both can lend strength to arguments against reducing carbon-dioxide emissions. The possibility that people could flip from a position of saying that there was no problem at all to saying that there is a problem but the answer is climate geoengineering, not emissions reduction, worries many of the academics who work on geoengineering – most of whom, I think it is safe to say, share the broadly left-wing sympathies common among those who work on climate issues. That such a pivot is oddly easy, given its apparent incoherence, can be seen in *Superfreakonomics*, a book by Chicago economist Steven Levitt and journalist Stephen Dubner. They begin their chapter on climate change with arguments purporting to show

---

* Teller was closely associated with the George C. Marshall Institute, a think tank responsible for some of the more disgraceful *ad hominem* attacks on climate scientists who were demonstrating the importance of carbon-dioxide emissions in the 1990s – including Ben Santer, a colleague of Teller, Wood, Caldeira and Bala at Livermore.

that carbon dioxide is not that much of a problem; they then move seamlessly on to the idea that stratospheric aerosols are a far cheaper solution to the problem than emissions reduction. The main source for their discussion of an aerosol scheme was Nathan Myhrvold, a friend of Bill Gates who was once Chief Technology Officer at Microsoft. Myhrvold has looked at the possibility of a scheme for delivering sulphur to the stratosphere through balloon-supported pipes; Lowell Wood was one of those who worked on the idea with him.

Such arguments and connections make geoengineering very politically suspicious in some eyes. If it had not been for Crutzen's blessing, I suspect the suspicion would be even more ingrained, and fewer scientists would be active in the area; that is why a number of Crutzen's colleagues tried quite strongly to dissuade him from publishing the essay. But Crutzen, a cunning man, was not only, or even primarily, interested in providing support for those researching the topic. His essay could be read as a straightforward suggestion that climate engineering was something that had to be considered and looked at, even if never put it into action. It could also be read as a warning that unless action on emissions reduction got serious, such extreme action might become either a vital necessity or very politically expedient. Crutzen knew that, whatever he wrote, most scientists working on the environment would look at that possibility with fear and distaste. He hoped that by making the prospect of sunshine geoengineering seem more real, his essay might spur the action that would make it less likely. Today, Crutzen says that the second interpretation was the one he intended; he wanted to scare people. But both readings were possible, and both messages were taken away.

One other striking thing about that essay: if emissions reduction looked weak and feeble to Crutzen in 2006, by 2014 it looked almost 300 billion tonnes of carbon dioxide weaker and feebler still.

From Plan B to Breathing Space

After Crutzen's essay, publications on climate geoengineering shot up in number. Members of the world's various climate-modelling groups started to look at Engineered Planets with their computers, an effort that eventually led to the structured comparisons of the GeoMIP programme. Some atmospheric chemists, alarmed by the idea of messing with the stratosphere but intrigued by Crutzen's partial blessing, began to look at the implications for ozone. Volcanologists brought to bear their expertise on the aftermaths of eruptions. Cautiously, organizations began to host meetings on the topic.*

Around the same time, Bill Gates took on his role as the field's sugar daddy. When he started to move into a phase of his life where Microsoft was not the be-all and end-all, Gates took to informing himself about subjects that interested him by getting smart, wide-ranging experts to come and talk to him. This brought both Keith and Caldeira into his circle as informants on climate and energy issues. Since both were interested in geoengineering, the topic came up, and Gates found it interesting, too. Learning that there were no government funds committed

---

* One of the first, at NASA's Ames Research Center, which is across the bay from Livermore and not far from Stanford, where Caldeira now works, added a barbarous new term to the subject's lexicon. The meeting was sanctioned by the then director of Ames, Pete Worden, a former Air Force general who had worked on the Star Wars programme with Teller and Wood, and combined a similar penchant for provocative technological boosterism with a rather better sense of what you can get away with, if only just, in a big bureaucracy. The bureaucrats below him, though, were keen to avoid the words 'geoengineering' and 'NASA' being linked, especially in the title of a conference. When Caldeira, one of the organizers, heard of their concerns, he responded in partial jest that they should just replace 'geoengineering' with the dreariest management-speak circumlocution they could come up with – 'something like "solar-radiation management"'. The term stuck and, abbreviated by the initials SRM, it has become ubiquitous in the literature. I hate it. It sounds like something you'd find stencilled under a stopcock sticking out of a bulkhead on Spaceship Earth. It's no way to talk about the world.

to such research, Gates and some former Microsoft colleagues set up the fund that Keith and Caldeira make grants from (it is not used for investments). Keith and Caldeira would have been leaders in the field simply on the basis of their work, but having this fund at their disposal has undoubtedly given them extra heft. It has allowed them to support work that would otherwise not be supported, and create space for discussions that might otherwise not have taken place.

By 2008, what the journalist Eli Kintisch has called a 'geoclique' had begun to develop around Caldeira and Keith. Subsets of the clique would turn up at the increasing number of conferences and conference sessions devoted to the topic. Reports on the subject – including the most influential one, the Royal Society's 2009 'Geoengineering the Climate: Science, Governance and Uncertainty' – would typically have Keith, Caldeira and some other members of the clique among their authors. The clique was not limited to climate scientists; there were some engineers, social scientists, policy people and journalists, too. As well as Kintisch, a reporter for the journal *Science*, there was also Jeff Goodell, a reporter for *Rolling Stone*, and me.*

I tend to think of the clique's members as anyone I saw at three or more geoengineering conferences around that time. Defined that way it was far from unanimous in its views. Alan Robock, one of the leaders of GeoMIP, was and is resolutely sceptical about the idea that stratospheric veils are a promising intervention. His most widely read article on climate geoengineering is 'Twenty Reasons Why Geoengineering May Be a Bad Idea'. Clive Hamilton, a philosopher and public intellectual from Australia whom I first met at the Copenhagen conference, deeply distrusts the hubris, reactionary politics and anthropocentrism he sees as endemic in discussions of geoengineering. James Fleming, America's leading historian of meteorology, can be relied on to

---

* Kintisch's book *Hack the Planet* and Goodell's *How to Cool the Planet* came out in 2011.

point out the degree to which discussions ignore the history of previous work on modifying the weather and climate by the military and others – attempts which he sees as ever germane in a those-who-forget-history-are-doomed-to-repeat-it way. The debate within the clique has mostly been cordial, but it is not without sometimes-tiresome trottings-out of long-established disagreement and occasional flashes of genuine rancour.

The constituency for debate has expanded steadily since the early days of the geoclique; the first international academic conference on geoengineering research, held in Berlin in the summer of 2014, attracted about four hundred.* GeoMIP has brought in a number of early-career natural scientists. The bulk of the growth in interest, though, has been in the social sciences and humanities; the Berlin conference saw considerably more papers delivered on law than on hardware.

The key ideas in the debate have changed little during its (still moderate) expansion. One spur of particularly well-rehearsed argument is the risk that simply talking about climate geoengineering will lead to less climate mitigation – the same concern that contributed to the marginalization of adaptation efforts in the past. This is often called the 'moral hazard' problem. In economics, the moral hazard of an insurance policy is the extent to which it encourages the insured to take more risks. Thus, subsidized flood insurance encourages people heedlessly to build on flood plains; likewise, the knowledge that banks have a very good chance of being bailed out if things go wrong encourages bankers to take self-serving risks.

In geoengineering, 'moral hazard' has been used to describe the expectation that if cooling technologies seem a real possibility, people will put less effort into reducing carbon-dioxide emissions. This fear is clearly related to the persistent nightmare of a 'superfreak pivot' in which the American right moves overnight from a position of not wanting to act on climate at all

---

* Disclosure: I was on its advisory board and chaired parts of the proceedings.

to wanting to act only through some sort of crash geoengineering programme. But moral hazard is a broader issue than that. The superfreak pivot imagines a new enthusiasm for geoengineering; the moral-hazard worry doesn't require enthusiasm for geoengineering, merely a slacking off on emissions reduction because geoengineering is thought to be an option. If people believe that there's a plan B that could work, they will pursue plan A with less vigour.

This does not seem much of a practical worry in the near term. The political incentives, as opposed to the policy aims, that drive mitigation's Plan A are to please the voters who identify as green and the lobbies of companies that make renewable energy hardware. Climate geoengineering does neither of those things, a reality which for the moment renders moot its 'incredible economics'; if it is not the answer the constituents want, it doesn't matter if it's cheap. It is worth noting that around the time George W. Bush's administration was abjuring the Kyoto protocol, it quietly quashed a report on geoengineering as a possible cheap alternative that had been produced by its own Department of Energy; to produce such a report at that time would have looked more superfreaky than the government wished – why inflame green opinion yet further? If a government that is not pursuing mitigation dares not discuss geoengineering in public, the obverse is also true. The country that has been spending the most money on geoengineering research while I have been writing this book is Germany, which has a number of well-funded interdisciplinary projects looking at the natural science, social science and policy implications of various approaches to climate geoengineering. Germany is also home to the world's most ambitious strategy for moving away from fossil fuels towards renewable energy, the *Energiewende*.

Even if it does not currently look politically germane, though, the moral-hazard point has to be taken seriously. It does seem likely that, other things being equal, an increased role for climate geoengineering in climate policy would reduce the pressure for emissions reduction. This is not necessarily a negative. For moral hazard to be a problem, you need to show that the behaviour it is encouraging is actually a bad thing. Climate geoengineering research thus raises a real moral hazard only if it reduces action on mitigation without making progress of its own in reducing the risks of climate change. If research on geoengineering pays off in providing a way of slowing the warming that can be deployed in a satisfactorily safe and just manner, then reduced interest in mitigation could be rational and morally justified.

Unfortunately, many people thinking about geoengineering talk about it in a way that tends instead to increase the moral hazard: use-only-in-case-of-emergency. A 2008 report by Novim, an American think tank, framed climate geoengineering as a possible response to 'climate emergencies'. The idea was that the right response to 'normal' climate change is the reduction of carbon-dioxide emissions, but that research into sunshine geoengineering techniques was needed because of the risk of a climate emergency requiring a prompt response above and beyond those reductions – a runaway emission of methane from polar permafrost, perhaps, or a reordering of the ocean currents, or a collapsing ice sheet. A similar way of thinking could also be seen in the influential Royal Society report a year later, and in a report for the US Bipartisan Policy Center two years after that.*

The use-only-in-case-of-emergency idea has various attractions, especially if you are a researcher who wants to study geoengineering without being seen as an obstacle to climate action or an enabler of the superfreak pivot. It highlights the fact that solar geoengineering, like a volcanic eruption, could start to cool the climate in months, whereas emissions reduction takes a

---

* By my attendance-at-three-meetings definition, at least half the authors on each of these three reports were members of the geoclique; Keith and Caldeira contributed to them all.

number of decades to do anything of note. But it acknowledges the primacy of such reductions in the absence of an emergency.

Treating climate geoengineering as a last resort not currently required thus allows researchers to put on record their misgivings about making any practical use of their studies while continuing their work on the subject; it absolves them from having to get into discussions of how it might be justifiably and appropriately deployed. It is a particularly strong version of the Plan A/Plan B approach, one in which Plan B is not just defined from the outset as less attractive but actually sealed away under glass.

The reason I think this approach exacerbates the moral-hazard problem is that it gets rid of prohibitions against talking about geoengineering without providing an incentive to work hard on finding ways in which geoengineering could actually be made to work – by which I mean not just how it could deliver climate benefits, but how it could do so in a just and responsible manner. My feeling is that most of the moral hazard posed by geoengineering is experienced up front; the very fact that there is geoengineering research is what creates the hazard. But not all geoengineering research is equally well suited to reducing further risks. Risk reduction depends on finding – indeed, designing – approaches that might work and approaches that definitely don't work. But a lot of geoengineering research simply looks at some simplified scenarios more or less as if they were natural phenomena – something a bit like an El Niño, perhaps. It is designed to let people say more about geoengineering as it plays out in a few scenarios rather than to help them think of ways to do it better, or to show which ways of doing it definitely shouldn't be tried at all. And it is that sort of how-to-do-it-or-not knowledge that the future might actually welcome. As the philosopher Stephen Gardiner has pointed out, doing things the way they are currently done increases the risk of bad outcomes being visited on future generations – because doing the research reduces the urgency of mitigation – while also leaving them facing all the risks that would come from actually deploying the mooted but untested geoengineering technologies – because the research is not aimed at understanding and reducing those risks. The future ends up not better protected, but doubly screwed.

There is an alternative to the Plan-B frame for looking at climate geoengineering – one first raised in the scientific literature by Tom Wigley, an Australian climate scientist who has been at the heart of the field since the 1970s. In a paper that came out just after Paul Crutzen's, Wigley outlined what could be called the 'breathing-space' approach – that of using climate geoengineering to slow the rate of warming for five decades or so, thus allowing more time for the development and deployment of fossil-fuel-free energy technologies more advanced than today's.

The breathing-space approach, which would mean trying to design systems that might be used in the fairly near term, seems much more radical than the in-case-of-emergency approach, which means just keeping an eye on the issue. But it has numerous advantages. In-case-of-emergency more or less requires that climate engineering be introduced quickly and at full force if it is introduced at all – it's an emergency, dammit! It also requires that it be introduced when the climate system is showing some particularly upsetting degree of instability – which means that the political system will be doing so, too. Panicky upset hardly seems the best setting for rushing to use a radical world-reshaping technology. The breathing-space approach, on the other hand, allows an incremental introduction of the technology, with a much lower initial forcing and a slow build-up.

It also allows people to plan an exit strategy from the start. The obvious one would be, at some point in the future, once emissions were more or less zero, to start withdrawing carbon dioxide from the atmosphere, thinning the veil – slowly and incrementally, just as you built it – as you do so. This synchronization would tie together the two ideas that have been lumped together as 'geoengineering' since the early work of Budyko, Marchetti and Dyson but that often seem to have little in common: the sunshine approaches and the carbon approaches.

The relationship between the two approaches is one of the bones of contention that the geoclique gnaws on repeatedly. Some people working on carbon-dioxide approaches resent the fact that most discussions of geoengineering deal with veilmaking and similar sunshine approaches first or even exclusively (the fact that their ideas are dealt with only in a later part of this book might strike them as making the point). Others object to carbon-dioxide reduction being treated as geoengineering in the first place, as the term can alienate greens who would normally be enthusiasts for, say, forms of soil management that increase carbon stocks.

The breathing-space scenario is one in which the two approaches are shown to be complementary, not opposed.* It also smoothes over the Plan A/Plan B dichotomy. It does not undermine it – emissions reduction is still seen as the primary process for which the breathing space is needed. But breathing space makes sunshine geoengineering an adjunct to mitigation, rather than an alternative to be avoided, or else to be rushed out in case of failure.

But although breathing space seems to me a more sensible approach to geoengineering than waiting for an emergency, it does make some problems more acute. An emergency might never come; but breathing space suggests at least some geoengineering come what may. And so it requires you to come up with answers to how that process is governed.

As we have seen, it is plausible that there may be levels of

---

* As Tim Kruger, who works on geoengineering at Oxford University, has pointed out, this can be taken as another blow to the 'incredible economics' idea; if you think about building a shield and later removing the greenhouse gases that made it necessary as a single process, climate geoengineering doesn't look anything like as cheap.

climate geoengineering in which every region is better off than it would be with no such engineering. But the same studies also show that almost every level of geoengineering would be suboptimal for most regions on the planet. All would be winners – but all would be in a position where, if the system were deployed differently, they might expect to win more. If they were to win more, though, others would at best win less – which can feel like losing. And some might actually lose in absolute terms. Who decides? And how are they held accountable?

This is commonly referred to as the 'whose-hand-on-the-thermostat' problem, and you can see why, but it's not an image I am very happy with. For one thing, the idea of a thermostat gives a wildly overstated sense of how precise and uniform control over the climate might be, and how domesticated the world would get. More fundamentally, the image of a hand on a thermostat divides the development of the technology – which is seen purely as a mechanism of control – from the agency that uses it, which it implicitly encourages people to see as any one of a number of pre-existing institutions with leave to act in the world, such as nation states.

It is possible that that might be the way that geoengineering comes about; that it could be a technology developed in purely instrumental ways, to be used by whatever agency managed, through some political procedure or use of power, to get its hands on it – perhaps, even, to be fought over. Much better, though, for the decision-making process to grow up alongside the technology for delivering on those decisions, with the possibilities opened up by the two processes influencing each other. Much better, rather than treating geoengineering as a technocratic way of avoiding politics, to use it as a way of reinventing politics. Exploring the potential of geoengineering could spur and shape the development of a new way of making planetary decisions. The aim should not be the development of a thermostat alone; it should be the development of a new hand to use it. It should be to fashion from the institutions and hopes of the human world a fulcrum well suited to the pressures and demands of levers that can move the earthsystem.

Expanding the Boundaries

At the 2010 UNFCCC negotiations in Cancún, Mexico, I watched the nations of the world endorse the Copenhagen accord, a sketchy and unsatisfactory document produced by a handful of great powers at the Copenhagen conference of the previous year. In doing so, they formally committed themselves to limiting climate change to two degrees Celsius above the pre-industrial average. The two-degree limit has been much discussed since the 1990s, when various policy shapers adopted it as an alternative to the earlier idea of trying simply to limit the rate of climate change. It is not a boundary firmly rooted in science. For some places and some species, change of less than two degrees could be calamitous. At the same time, a change of two degrees need not be necessarily or universally catastrophic; it might prove to be within the scope of adaptation without leading to undue suffering. As ever, uncertainties abound. But the two-degree boundary is the one on which the process settled.

If the earthsystem turns out to be less sensitive to greenhouse gases than climate scientists have come to believe, this boundary may not end up broken. If so it will be a matter of good luck, not good judgement. The emissions-reduction targets the various countries have lodged with the UNFCCC in the run-up to its next big meeting, to be held in Paris at the end of 2015 – targets to which they are not bound and which they are far from certain to meet – are too loose to make such an outcome likely under most assumptions about climate sensitivity. As a result, it is increasingly common to hear eminent people in the field of climate change say that the two-degree target simply cannot be met. Some see this as realism; some see it as a counsel of despair.

I see it as untrue. In a breathing-space scenario a relatively modest amount of climate geoengineering – a much thinner veil than that of Pinatubo, though much longer lived – would stand a good chance of keeping the global-temperature increase well below two degrees Celsius for most of this century, and perhaps, depending on the eventual rate of emissions reduction and the sensitivity of the climate, for much longer.

If you have no truck with solar geoengineering and still take the two-degree limit seriously you have to be willing to contemplate geoengineering with carbon instead. Half the computer models looked at for the most recent IPCC report said that if the climate were to be kept below the two-degree limit, emissions would have to be negative by 2100. Humans would have to be actively taking carbon dioxide out of the air, rather than just refraining from putting any more in.

So if people are serious about the two-degree limit, one or another form of geoengineering needs to be treated as a real possibility. The risks, costs, politics and practicalities need to be debated in a process which admits that there is more to climate change than emissions reduction and adaptation. This would be a big shift. The challenge is made even bigger by the fact that current climate-change discussions are small – in scope if not in attendance. The UNFCCC process started in the 1990s staggers from meeting to meeting, from Copenhagen to Cancún to Paris, wrapped up in its own procedures and minutiae, largely ignored by those with real power in the governments involved. Hobbled in various ways from the outset, it is more worthy of respect than many who have not observed it would believe. But it is not up to the task.

It would be amazing if it were. What humans are doing to the climate is unprecedented. The effects will last for centuries, even millennia. Why would one expect the problem to be adequately handled by a set of policies cobbled together over a few years and enacted by a world political system to which climate concern is more or less alien? To appreciate how differently people have thought about climate change in the past – as part of the evolving moral order of the world, as something that could be dealt with by an ocean of ping-pong balls, as a mandate for the wholesale re-design of industrial civilization – is not to say that any one of those views was right. It is to say that, over a couple of centuries, all those views and more were expressed and felt. To look at that diversity of views and still think that the way people decided to deal with climate change when scrambling to build a basis for political action in the early 1990s can be treated as the one true path for the centuries to come – that it should be followed more or less as is in times that lie as far in the future as the eras of Roger Revelle, Percival Lowell, Ferdinand de Lesseps and Thomas Jefferson lie in the past – feels like folly.

It is on this basis that I think a more ambitious and wide-ranging discussion of geoengineering as part of the human response to climate change is vital. It is obviously feasible to design systems for deliberately altering the climate, and international discussions should recognize that. Exploring how best such techniques might be used would be a useful, even transformative, expansion of their remit even if, in the end, no engineering scheme was ever put into practice. Dwight Eisenhower said that 'whenever I run into a problem that I can't solve, I always make it bigger. I can never solve it by trying to make it smaller, but if I make it big enough I can begin to see the outlines of a solution.' Even if climate geoengineering is never used to provide some of Wigley's breathing space, discussing it might help clarify the inadequacy of the current debate, and provoke an appetite for new ways forward. If it is used, such discussions are the place where the building of the hand that uses it might begin.

Increasing the range of ideas available and deepening the pool of thinkers involved would not only improve climate politics; it could also improve discussions of geoengineering. At the moment, climate geoengineering is a very small field. There are very few funded programmes; hardly any senior researchers anywhere in the world devote most of their time to the subject; a number of leading lights are retirees. It is conceptually small, too. New ideas are quite rare, and a lot of published research is at best incremental. For a field devoted to something that still strikes many as truly wild, it can feel oddly tame, as though the very fact they are talking about such an outré topic makes the speakers do so in a conservative way.

People at meetings such as the one in Berlin in 2014 worry a lot about whether or not to deploy geoengineering, which is right. But they tend to do so in a constrained form, one which assumes that people already know what it might be to do geoengineering (for the most part this is taken as being the bringing about of something that looks like one of the GeoMIP scenarios) and why one might do it (to take control of the climate and stop it from changing). Everywhere you see problems that crop up in particular geoengineering scenarios – such as 'undercooling', or termination shocks – being taken as intrinsic to geoengineering per se.

Both natural and social scientists tend to make this sort of mistake – to talk as though what geoengineering is has already been decided, rather than treating it as something still up for grabs. Broadening the debate on geoengineering – a goal which more or less everyone already in the debate says they share – is treated as a matter of bringing more and different expertise to bear on geoengineering as it is currently imagined, rather than reimagining it. It feels disturbingly like the building of an inverted pyramid – an ever-wider superstructure of policy speculation and social considerations balanced on a barely expanding natural science base and a fixed set of views as to the possible modes of its application.

One effective way to broaden the debate would be to think up more and different forms of it to discuss, and to discuss them in the real context of possible action – the context of hands as well as thermostats. Introducing geoengineering options into wider debates on climate policy could help in that. So could data derived from experiments designed to gauge the effects of proposed engineering techniques in the real world.

This is something David Keith is working on. Since 2013 he and his friend and mentor James Anderson, who oversaw the U-2 flights into the Antarctic ozone hole in the 1980s, have been developing plans to explore the effects of possible veilmaking aerosols on stratospheric ozone by squirting very small amounts of various compounds into the stratosphere and measuring their chemical effects over subsequent days. The idea is not just to quantify what the effects of a given type of veilmaking aerosol might be, but to try and work out what processes are going on, and thus how the damage done by a veilmaking programme could be lessened, or, indeed, done away with all together, by tailoring the size distribution, or altitude of injection, or composition of the aerosols used. This is not an attempt to 'find out what geoengineering does'; it is an attempt to produce some of the knowledge that will be needed if people are to decide what sort of geoengineering there could be.

The two broadenings – new settings for debate and new data to debate in them – are linked. To do geoengineering-oriented fieldwork of the sort Keith is proposing, or the cloud-whitening experiments other people are looking at – which we shall come to in Chapter Ten – requires the setting up of procedures to discuss and approve such work. Such field trials would thus provide a spur and a setting for a wider institutional discussion of what the ultimate purpose of such knowledge might be, and how it could best be used.

What it might be and what it might mean to mount a real geoengineering programme are not set in stone – there are all sorts of ways in which it could be done, and it could be aimed towards any number of goals. It could be used to respond to an emergency, sure, but it might also be a source of breathing space, or a counterweight to some other action, or a limit to the rate of change. Expanding the range of possible scenarios could and should go hand in hand with research aimed at finding out what effects can be achieved, what side effects can be averted, what supplementary benefits might be designed in, what legal constraints already apply or should apply, what political contexts might confer legitimacy, what obstacles may prove insurmountable, what risks unacceptable, and how research can be governed responsibly.

Such expansion would be a fine consequence of bringing discussion of geoengineering into the mainstream of climate politics. It is not the only outcome imaginable, though. There is, as always, the fear of the superfreak pivot – that, once unleashed, the idea of geoengineering will be so appealing that it will be pursued uncritically as the only game in town. Closely related (and to my mind much more likely) is the concern that it could simply prove polarizing, one of many issues within climate politics over which there is no agreement, another buttress to the gridlock – an outcome that would be unlikely to engender fresh thinking about the field or new ways forward in climate politics.

One step that might avoid that impasse would be to use the process of widening the discussion of geoengineering to move away from talk of using it as a system of control. For one thing, such talk is unrealistic; control in any normal sense of the term is, at the moment at least, not on the table. Whether through sunshine or through carbon, geoengineers can hope to do little more than design ways to influence average climates in moderately predictable ways – to influence the boundaries within which the system tends to stay. That would be a remarkable achievement, but it would not be control.

The fact that geoengineering could not in fact control the climate is sometimes used as an argument against it. But its inability counts as a failing only if you accept the premise that control is the aim. I don't think you should. The technocratic aspiration towards control has run through science and its applications for centuries; it informed significant strands in the development of sociology, psychology and biology. In many of those applications it has proved, as a goal, to be flawed, unattainable or both. In some it has been very damaging. Leaving aside attempts to control the genetic endowment of individuals or the racial purity of nations, talk of 'human engineering' of the sort that was heard among scientific managers and biologists alike in the early twentieth century is hard to square with respect for human autonomy.

There are better things to take than control. Foremost among those, to me, is care. In an essay on Mary Shelley's *Frankenstein* – a novel written in the darkling shadow of Tambora and sub-titled, we should remember, '*The Modern Prometheus*' – the French thinker Bruno Latour argues that Victor Frankenstein's crime is not that he brought forth a strange ungodly creature, but that he abandoned it when he should have cared for it; that he should have loved his monster. The same could be said of the technologically transformed earthsystem that the human world has created. Do not reject or recoil from it, tempting as that might be; it is here, and needs to be dealt with. So care for it instead: for the people who live in it; for the creatures that partake in it; for the great wounded wonderful thing itself. Geoengineering could be the means of that care.

An approach to geoengineering that centred on the possibilities it offered for taking care of people, rather than taking control of the climate, would be a better, saner and more fruitful thing. A utopian impulse attractive enough to be welcomed by many, it could still be robust and provisional enough that it can get along without universal assent. People who wish to control the earth-system are set for conflict both with others who want to do the controlling differently and with those who want no one to have control over them at all. People who don't subscribe to the need to care will not necessarily wish to stop others from doing so.

★

Making geoengineering a bigger part of climate discussions, and taking the prospect of its use seriously, sounds difficult, dangerous and, to many, unpalatable. It normalizes what some would wish always to remain an outrage. A world in which such things are discussed is a world less natural, a world that can no longer simply be restored to the way it was. Such a world would sit in an earthsystem not merely distorted by human action, but intentionally reshaped to human design – and subject to further reshaping, should human intentions shift. That is a big change.

But such big changes are not unprecedented. Humans may not yet have started to intentionally change the climate, but they have put their minds to changing other parts of the earthsystem. They have re-routed the great cycles of the planet's metabolism – the systems that govern life's supplies of carbon, nitrogen, sulphur and other vital elements – to a degree that was unimaginable a century ago, and which many people still do not appreciate. By extending their martial abilities they have introduced to the world they live in the possibility of an apocalyptic end. Some of these changes, I think, are in themselves forms of geoengineering, though few understand them as such. Others seem to me to be a crucial context for attempts to imagine what geoengineering might be. They allow us to imagine who its engineers might be, and what care they might take.

We will come back to the possibilities of climate geoengineering, and its integration into politics, at the end of this book. Before that, though, to have a richer, rounder view of such a possibility, it is time to look at some of the interventions humankind has already made in the earthsystem.

PART TWO
Substances

7

Nitrogen

There can be little doubt that from now on synthetic
nitrogen compounds will continue to be drawn in
increasing amounts from the atmosphere into the life
cycle . . . This extraordinary development is something
much more than a fundamental new departure in
industry. It represents nothing less than the ushering in
of a new ethnological era in the history of the human
race, a new cosmic epoch. In the short span of a dozen
years – geologically speaking in an instant – man
has initiated transformations literally comparable in
magnitude with cosmic processes.

Alfred Lotka,
Elements of Physical Biology (1925)

The most striking deliberate human intervention in the earthsystem – and the one that I think has the best claim to be seen as
a non-climate form of geoengineering – lacks almost any hint of
the sublime. No awe-inspiring volcanoes; no nifty stratospheric
jets; no angsty symbolist sunsets: just bags of fertilizer. But
the human capacity to make that fertilizer, a breakthrough
technological capacity carefully nurtured by various national and
global interests, has changed the size and shape of the human
world as much as any development other than agriculture itself
and the harnessing of fossil fuels. It is having a deep impact

on the earthsystem as a result – and also changing how some researchers think about such impacts.

Nitrogen makes up most of the Earth's atmosphere. And all life on Earth needs nitrogen. But almost none of the planet's life forms can use the copious nitrogen in the air, because it exists in the peculiarly inert form of two identical atoms firmly fastened one to the other. To be incorporated into DNA and proteins – and without such incorporation there can be neither DNA nor proteins – those atoms must be prised apart and bound to some other element: 'fixed', as scientists say.

On the early Earth, some four billion years ago, nitrogen was fixed by bolts of lightning. Then as now they tore apart the molecules in their path, liberating nitrogen atoms to join with oxygen atoms torn from carbon dioxide molecules. The nitrogen oxides thus created were almost certainly the main supply of fixed nitrogen for the planet's earliest life.

Before very long, though, life found a way to fix its own nitrogen. Evolution has made bacteria brilliant biochemists, and some of them found a way to twist nitrogen molecules and insert hydrogen ions into the openings thus produced, turning the nitrogen into ammonia. In its chemical properties, ammonia is quite different from the oxides made by lightning. But nitrogen in ammonia is still considered fixed, because if you have ammonia to hand it is comparatively easy to make all the other nitrogenous molecules life needs, just as it is if you start off with nitrogen oxides.

It is also possible to turn fixed nitrogen back into the inert form – a process called denitrification – and various microbes mastered that trick, too. And so a biogeochemical cycle began to turn, one of the suite of nested, interlocking cycles by which the substances from which life is made flow through the earthsystem. The nitrogen-fixers pulled nitrogen out of the atmosphere and incorporated it into the biosphere; they and other organisms made use of it; denitrifiers returned it to the inert atmospheric stock.

And, for the next 3.5 billion years or so, that was more or less the whole story. The earthsystem's life forms became more complex and interesting; the uses to which fixed nitrogen was put became more various, and the food chains it passed along became complex webs. But however complex the web, the nitrogen-fixing microbes provided the way in and denitrifying ones provided the exits. When plants evolved about 400 million years ago, some of them came up with ways to encourage nitrogen-fixing bacteria to live in and around their roots, improving their supply. That allowed the amount of organic matter on the planet to increase. But the basic dynamics of the nitrogen cycle were more or less as they had been before.

Then the twentieth century happened, and through a mixture of deliberate planning, unthinking enthusiasm and force of circumstance, the human race took over.

The nineteenth century saw Europe's imperial powers extend their reach all around the world. European maps, trade and armed force penetrated beyond the other continents' coasts, where they had been established for some time, into their deepest interiors. Europe's cities grew to a size unprecedented in human history and the railways that sliced between them moved people faster than humans had ever moved before. Rivers were tamed; farms brought forth new plenitude. Nights once lit only by candle, lightning and moonlight were made bright by gas and electricity; words that could once only be carried by breath or paper crossed oceans as quickly as they used to cross streets. Canals cut through continents; quarries ate through hillsides; mines hollowed out mountains.

And at the century's close there was a sense of weary, uncomfortable completion. The unknown was exhausted, the known triumphant, and the world felt jaded as much as gilded; *fin de siècle* had arrived. The frontier that defined America's challenge and potential, the demographer and historian Frederick Turner told his compatriots, had closed. The scramble for Africa, Europe's last great theft, was almost over. The British Empire on which the sun never set lacked for that very reason a sunset beyond which to sail. Riches continued to accumulate, quarries to be quarried, mines to be mined. But the age was now one of intensification not extension, one constrained by unaccustomed limits.

It was thus in keeping with the times for Sir William Crookes, perhaps Britain's leading chemist, to draw attention to a limit of planetwide importance. In the 1860s Crookes had discovered a new element, thallium. In the 1890s he had become one of the first to identify helium in samples of Earthly material – the element had only previously been known to be present in the sun and stars. But the admonitory and inspiring presidential address he gave to the British Association for the Advancement of Science in 1898 began with a subject much more mundane: the British Empire's insufficient supply of manure.

Crookes was worried about the world's supply of wheat. The number of people wanting to eat it was increasing, as was the amount of it many of them wanted to eat. The amount of land on which it grew was not increasing, at least not by much; the world had been mapped and its fertile lands more or less all expropriated by the mappers. And there were no acceptable alternatives, at least not for white people: 'Other races,' Crookes allowed, 'vastly superior to us in numbers, but differing widely in material and intellectual progress, are eaters of Indian corn, rice, millet and other grains. But none of these grains have the food value, the concentrated health sustaining power of wheat, and it is on this account that the accumulated experience of civilised mankind has set wheat apart as the fit and proper food for the development of muscle and brains.' So the amount of wheat that each parcel of land provided had to increase: it was 'vital to the progress of civilised humanity'.

Fifty years earlier, the great German chemist Justus von Liebig had convinced the world that for every crop, under every set of circumstances, there is a 'limiting' nutrient – one which, if supplied in greater quantities, will on its own spur growth. For wheat, under most of the circumstances farmers dealt with, the limiting nutrient was fixed nitrogen, which was also a vital constituent of explosives, crucial both to mining and to the manufacture of munitions. If more nitrogen could be ploughed into the world's wheat fields there would be bread for all who deserved it. If not, Crookes warned, the world faced a 'catastrophe little short of starvation . . . and even the extinction of gunpowder!'

The ability of bacteria to fix nitrogen had been discovered a decade before Crookes's speech, and that advance might have seemed a hopeful sign. But people had been using bacterial nitrogen fixation long before they knew that they were doing so. Clover, legumes and other plants that have particularly good relationships with the nitrogen–fixing bacteria in their roots were already used to reinvigorate depleted soils. Such husbandry had allowed an intensification of farming in the nineteenth century, but the process could not be carried on indefinitely. The more of your land you cover with crops to replenish its nitrogen, the less you have on which to grow more wheat. And there was no obvious way to get the bacteria to hurry up. As Crookes noted with regret, 'the nitrogen which with a light heart we liberate in a battleship broadside has taken millions of minute organisms patiently working for centuries to win from the atmosphere'.

Another option was to return more nitrogen to the soil in the form of food waste and animal and human manure. Liebig had noted with frustration the way in which the 'manners and customs peculiar to the English people' made them too fastidious to gather up their city shit and return it to the land, preferring to flush the riches robbed from the soil of England, its empire and its trading partners down the Thames, the Mersey and the Tyne. Karl Marx saw the 'metabolic rift' between the country where food was grown and the towns where it was consumed, its nutrients never to return to the soil, as one of the ways in which industrial society had been alienated from its productive roots (quite literally, in this case). Crookes acknowledged Liebig's point, but felt that manure, like bacteria, would prove insufficient to the task he foresaw.

A third option was to mine the nitrates. A quirk of the earthsystem's workings had made islands off the coast of Peru a phenomenal source of guano. The Humboldt current brought cold, nutrient-rich waters from the deep ocean to the surface there, supporting lots of fish and thus lots of birds; very low rainfall meant the birds' nitrogen- and phosphorus-rich droppings were not washed back into the sea, but instead built up in deposits tens of metres in depth. By the middle of the nineteenth century hundreds of thousands of tonnes of guano were being exported every year, at the miserable expense of tens of thousands of Chinese workers treated as slaves.

As the best guano islands had started to get played out, a partial replacement had been found in mineral nitrates from the Atacama Desert, the richest source of such deposits on the planet. The export of these riches dominated the regional economy and triggered wars. Spain tried to take Peru's guano islands in the 1860s. In the 1880s, supported by Britain, Chile seized the Atacama nitrate deposits from Peru and Bolivia. But even though Britain had thus assured itself preferential treatment, by the 1890s the best guano was largely gone and the nitrates were going. Crookes warned that ever-increasing demand would use up the desert's best beds in just a few decades.

The worry that the supply of food could not keep up with the growth of the population was not new. Crookes was speaking exactly a century after the parson Thomas Malthus had first voiced concerns on the matter in his *Essay on the Principle of Population*. Nor was it new to worry that a mineral resource might run out. As Crookes noted, Stanley Jevons had warned decades before that Britain could run out of coal, arguing that the more efficiently the stuff was used, the greater the demand for it would be. Crookes just combined the gloomy ideas of Malthus and Jevons: the food supply might fail to keep up with population growth because a vital natural resource was being depleted. It was a then-new insight that seems terribly familiar today, most famously as the 1970's argument of *The Limits to Growth*.

But though the origin of his idea was gloomy, Crookes himself was anything but. He was saying only that the world could run out of wheat – not that it must run out, and not that it would run out. The advancement of science to which the British Association was dedicated, Crookes told his audience, could solve the problem. Turning inert atmospheric nitrogen into fixed oxides or ammonia is, after all, just a chemical reaction, and the mastering of ever more difficult and more profitable chemical reactions had made chemistry the great industrial science of the closing century. If science could fix nitrogen industrially, the problem would be solved.

For the previous century the nitrogen-fixing reactions required had 'eluded the strenuous attempts of those who have tried to wrest the secret from nature', Crookes admitted. This did nothing to weaken his belief that the secret would be wrested away from nature at some point. But it did suggest that just waiting for ingenuity to take its course would not be enough. The risk of wheat shortages meant that the nitrogen problem differed 'materially from other chemical discoveries which are in the air, so to speak'. Men of science must come together to make it a priority.

Today most scientists making similar claims about an urgently needed technological breakthrough would turn to governments for support. But before the twentieth century governments had very little role in science. Attempts to direct scientific investigations to specific ends were actively resisted: a friend of Crookes, William Noble, ran a 'Society for Opposing the Endowment of Research'. There were no great foundations like that of Rockefeller or Bill and Melinda Gates to turn to when you had an idea about how the common good could be advanced. The chemical industry did not do large-scale research and development in the modern way (though in Germany it was beginning to). It was basically down to scientists such as those in Crookes's audience: 'It is the chemist who must come to the rescue.'

And lots of chemists tried. Some of the world's most celebrated – Wilhelm Ostwald in Germany, Henri Louis Le Châtelier in France – worked on the problem in the subsequent decade. Crookes himself was among those who suspected that copious new supplies of hydroelectricity might be brought to bear on the problem. The resources of the new plant at Niagara Falls were used to power thousands of electric arcs that produced the sharp geranium smell of ozone – another molecule produced through the torture of air – as they tried to fix nitrogen the way that lightning does. Norwegian dams were used to the same end, and rather more successfully. But even with cheap electricity, the low yield of fixed nitrogen made the capital costs prohibitive.

The breakthrough was made by Fritz Haber, a professor at the University of Freiburg, working with the German chemical company BASF. Eschewing the electrical route, he and his excellent laboratory technicians created a system that produced ammonia by passing a continuous stream of hydrogen and nitrogen over a hot catalyst at very high pressure. Carl Bosch of BASF scaled the process up into something with industrial possibilities. In 1919 Haber received the Nobel Prize in Chemistry for his efforts; Bosch was similarly recognized in 1931.[*]

Their technology changed the world far more deeply than Crookes, or anyone else, would have imagined. Crookes was

---

[*] Haber was not himself seeking to produce fertilizer, or gunpowder; he wanted to win an academic argument with a rival, Walther Nernst, about thermodynamics. But his desire was fulfilled in a context where academic ambitions could be coupled to industrial interests.

talking about fixing a couple of million tonnes of nitrogen a year to assure Europe's wheat supplies for the foreseeable future. Today industry fixes over a hundred million tonnes a year, comfortably more than all the Earth's nitrogen-fixing soil bacteria put together. This extraordinary industrialization of the planet's metabolism changed the conditions under which the human race lives. It adapted the earthsystem to a quadrupling of the human population and gave humans a dominant role in one of the earthsystem's basic biogeochemical cycles.

To call this geoengineering is to use language that Crookes, Ostwald and Haber would not have recognized, and that many people might be moved to reject. But I think it fits. Consider the similarities to the debates about geoengineering the climate. What Haber and Bosch did, and what the others wanted to do, was in part a response to a specific global threat – the inability to feed the growing population. The nature of this threat was not obvious in everyday life, any more than climate change is, but the scientific elite had identified it.

Unlike most of today's climate scientists (but like those who see a need to take climate geoengineering seriously), Crookes and his contemporaries saw no political solution to the problem; Crookes was not about to suggest that everybody agree to eat less, or that white people give up wheat in favour of rice, or that the Empire renounce cordite. Instead, he and men like him looked to newly powerful science. If geoengineering means a large-scale, purposeful technological manipulation of the earthsystem towards a given end, then what happened to the nitrogen cycle in the century following Haber's breakthrough is the best and most dramatic example that history provides. The lessons to be learned from a century in which nitrogen fixation transformed agriculture and human history, recasting environmental thinking in the process, deserve a chapter of our attention.

The Making of the Population Bomb

This transformation of the nitrogen cycle took some time to get into its swing. While artificial fertilizers were taken up keenly in some rich countries, such as the Netherlands, it was a different industrial breakthrough that came to the immediate aid of the world's wheat crop: the internal combustion engine. In so doing, it created the context – agricultural, economic and intellectual – in which fertilizers would come into their own.

When, in 1931, an economist from Britain's Ministry of Agriculture revisited Crookes's speech at another meeting of the British Association, he noted that in the intervening decades, global yields of wheat had risen by only 11 per cent – a far smaller improvement than Crookes had deemed necessary. The most important factor in the wheat market was not better yields, but new land: land which had previously been too marginal to farm had been opened up by new types of farm machinery.

In 1924 the western prairies of Canada had boasted five combine harvesters. In 1929 they had 7,250. By then America was producing over 200,000 tractors a year. Similar change was afoot in Argentina and Australia. The amount of labour needed for farming was falling, and that had allowed the area under cultivation to increase by more than 20 per cent in less than 20 years. The labour done without was not only human. When a tractor was bought, land which had previously been needed as pasture for the horses that pulled ploughs, carts and cultivators could be turned over to crops. That could give a farmer a quarter more land. In the generation after the First World War America's population of draft animals fell by 15 million.

By the end of the 1920s, the problem in many places was one not of scarcity, but of oversupply. Grain prices were plummeting and farmers were in debt. And so the tractors were worked yet harder. Twenty thousand hectares of previously virgin land in the plains of the United States were being ploughed under every day. Little was done to enrich, irrigate or protect the new fields.

In the early 1930s drought set in. The wind took the ploughed and weakened topsoil; the rain, when it returned in sudden spates, cut through what was left behind, leaving it gullied beyond any further use. Hundreds of thousands of American women, men and children left the parched lands; clouds of soil rose three or four kilometres into the sky.

No environmental issue had ever seemed as pressing as the Dust Bowl. As Hugh Bennett, a pioneer of soil conservation, put it, 'when people along the seaboard of the eastern United States began to taste fresh soil from the plains 2,000 miles away, many of them realized for the first time that something somewhere had gone wrong with the land'. Books appeared with titles like *Deserts on the March* and *The Rape of the Earth*. Estimates suggested that in many American states more than half the soil that had been there when settlers and slaves first arrived had since been stripped away – sometimes a lot more than half.

Understandably downplayed during the Second World War, these worries returned in amplified form in two Malthusian bestsellers published in 1948, *Our Plundered Planet* – written by Fairfield Osborn and endorsed on the cover by Albert Einstein and Eleanor Roosevelt – and *Road to Survival* – written by William Vogt and introduced by Bernard Baruch, financier and confidant to presidents. Both books decried the fact that erosion was eating away at the Earth's ability to support its people. Agricultural yields were falling and natural resources becoming exhausted even as the number of people who needed food grew ever faster.

The second bit was true. There were more people on the planet than ever before, and the rate at which the population was growing was higher than ever before. When Malthus wrote his essay on population in 1798, the world is estimated to have held just under a billion people. When Crookes made his speech in 1898, it held about 1.6 billion. Fifty years later, when Vogt and Osborn were published, the population was more or less that of 1798 and 1898 added together: 2.5 billion. This gave new potency to Malthus's eighteenth-century concern that the growth of the population, which feeds upon itself and thus always tends to accelerate, would outstrip the growth in the food supply, which Malthus could not imagine growing in a similarly self-fuelling way. Vogt and Osborn, though, went beyond Malthus. They argued not just that food supply could not keep up. They argued that it could not increase at all – indeed that, thanks to erosion and other forms of soil degradation, it was decreasing and would continue to do so.

The idea that America was using up the endowments that had made it great had gained currency in the late nineteenth century; it can be seen as another expression of the *fin-de-siècle* feeling of limit and loss. Osborn's uncle, who, like his nephew after him, was the president of the New York Zoological Society, had been one of those who gave it voice. (He was also a leading eugenicist, and one can detect a similar sensibility behind both concerns – a sense of the noble and singular at risk from growth beyond control.) Osborn the younger and Vogt took that idea of the best being behind them and globalized it, tying together fears of environmental degradation around the world and stories of desertification and decline throughout history, stories to which the experience of the Dust Bowl added new weight.

Both books are informed by a scientifically inflected sense of how the interconnectedness of the world is at once spiritual, mechanical and energetic. All forms of energy, from that of sunlight to that of spinach to that which lifts aeroplanes into the sky, could be quantified in terms of calories. Man, writes Vogt, 'quite as much as any Ford or Packard, requires fuelling'. Poor farming put the very body of the nation at risk. 'Not only is soil washing into the sea,' Vogt wrote, 'but . . . bread and pork chops and potatoes. The Gulf of Mexico . . . is stained with the substance from which our children build bone and muscle and blood.'

As well as expressing the metabolic links between humans, crops and the soil, the books also stress the interconnectedness of human life through politics, trade and war. They boast a sincere internationalism, a belief in solutions at a global scale. But the global solutions are hard to stomach. Whether or not erosion and depletion could be held in check, many poor countries simply had too many people in them. Something had to be done. 'The pressure of population,' Osborn coyly admitted, 'cannot be solved in a way that is consistent with the ideals of humanity.' Vogt explained the implications of that inconsistency with spectacular bluntness. British withdrawal from India (where the natives, he noted, bred 'with the irresponsibility of codfish') 'may well result in the reversal of the population trend that this country so badly needs'. The 'horror of extensive famines in China within the next few years' might be 'not only desirable but indispensable'.

How does one write such things? Well, there have always been men who think it better to let others starve than to change the system that starves them. The numbing effect of the un-precedented slaughter of the Second World War might perhaps make such a laissez-faire attitude to mass mortality easier to embrace – even if, for Vogt, that slaughter had been insufficient. ('Unfortunately,' he wrote, 'in spite of the war, the German massacres, and localised malnutrition, the population of Europe, excluding Russia, increased by 11,000,000 between 1936 and 1946.') The world beyond America looked both broken and out of control. The British Empire was ready to collapse. China was in turmoil. And the war-ending atom bombs seemed to have burst the very bounds of nature. Whole cities – perhaps whole countries – were now susceptible to annihilation. In late 1945 opinion polls found that a quarter of Americans said they expected nuclear science to destroy the world. Humankind seemed at the same time more powerful than ever and more terribly beset by threats and limits – it had come to the coffin corner within which it still lives.

By showing that the world really could end – though the early atom bombs were not up to the task, the hydrogen bombs which followed in their thousands were easily capable of finishing off civilization – nuclear weapons not only brought in their own new existential anxiety, they also upped the ante for all other fears. If there was one way in which humans could literally end the world, then surely there were others.

It can have been no coincidence that Vogt chose Baruch, a key figure in the nascent world of nuclear diplomacy, to write his introduction: Baruch's call for international efforts to achieve a 'favourable biophysical relationship with the Earth' would have been read in the context of his attempts to reach international agreement on the fruits of that other branch of physics. Vogt alludes to a similar eschatological equivalence when he imagines the quandary facing a scientist who creates a cure for malaria but is unsure whether he should tell anyone: why save millions from disease only to see them and their children inevitably starve? The cure, Vogt's imagined scientist realizes, had 'a power that was perhaps as dangerous as the atomic bomb'. In the early 1950s the two imagined dooms were linked explicitly in the title of a pamphlet by General William Draper which drew heavily on Vogt: it was called 'The Population Bomb'. (The title was later recycled by Paul and Anne Ehrlich, ecologists at Stanford, for a 1968 book.)

Osborn became a leader in the post-war movement to control population, a movement which was, like his uncle's conservationism, tainted by an ineradicable eugenic heritage. So did Vogt. That movement, well funded by charitable donors and widely endorsed by scientists, many of whom maintained some level of eugenic belief, led to terrible coercions in the form of compulsory abortions and forced sterilization both in America (specifically, among Native Americans) and around the world. Its message that the earthsystem could simply not carry the number of people it would soon be faced with – 'It is obvious,' wrote

Vogt, 'that fifty years hence the world cannot support three billion people at any but coolie standards' – was a deep influence on people like the Ehrlichs in the next generation of American environmentalists. It was also wrong. The Earth, it turned out, could hold many more people than they imagined, and the means by which it would do so were being developed as Vogt and Osborn wrote.

In 1943 the Rockefeller Foundation set up a 'Mexican Agricultural Program' to investigate how the country could feed itself efficiently and avoid stripping itself of resources. Vogt visited it soon after it was opened, emphasizing the importance of soil conservation and organic farming which eschewed industrial inputs. The second part of his advice was ignored. One of the Rockefeller workers, Norman Borlaug, a gifted breeder of plants, was struck by the poverty of Mexico's soils, especially their lack of nitrates – a poverty he went on to see as basic to soils in most of the countries of the tropics and subtropics. The local crop varieties, he knew, were adapted to this poverty; they yielded little and grew sparsely because they had little to work with. But he thought he might be able to breed versions of those crops which would be able to stomach a richer diet while at the same time resisting the diseases to which full, crowded fields fall prey. And if he could succeed, he knew that all the nitrogen those greedier plants needed could be easily supplied – the world was awash with industrially produced nitrogen. It was deciding the fate of nations as he crossed his first wheat strains.

Defusing the Population Bomb

Crookes, Haber and the other chemists of the twentieth century all talked approvingly of nitrogen fixation as a way of supplying food. But they did not hide the fact that it had another purpose that might prove more pressing. In 1910, 13 per cent of the Chilean nitrates imported into the United States were used as fertilizer; almost twice that amount was used in the chemical industry, three times as much or more in the manufacture of explosives. When the Royal Navy cut off Germany's supplies of Chilean nitrate in the First World War the Haber–Bosch process allowed Germany to continue furnishing itself with gunpowder and other ammunition, very likely prolonging hostilities. After Germany lost, the victors gathered in Versailles insisted on compulsory licences to the Haber–Bosch process as part of their spoils.

Some of the nitrogen was indeed used for agriculture. By the time Carl Bosch received his Nobel Prize* in 1931 he was the head of IG Farben, the vast company into which BASF and Germany's other big chemical companies had been rolled up. The company was producing explosives, but it was also producing hundreds of thousands of tonnes of nitrogen fertilizer a year at its four-and-a-bit-square-kilometre Leuna plant in Saxony-Anhalt. And not all of nitrogen's explosive uses were hostile; the new abundance of dynamite helped miners dig deeper. But mostly it was a means of killing.

According to one estimate, armaments made possible by the Haber–Bosch process have accounted for 150 million deaths over the course of the twentieth century. The unprecedented slaughter of the Second World War would have been unthinkable without it. According to a later chemist and Nobel laureate, Linus Pauling, over the course of that war a quantity of explosives equivalent to six million tonnes of TNT was used. The Hiroshima and Nagasaki bombs represented less than half of one per cent of that total. Everything else came almost entirely from the Haber–Bosch process. The allies used thousands of tonnes of high explosive in their raids on the Leuna plant alone.

Artificial fertilizer use was taking off in America at the time

---

* There is an obvious irony in the award of Nobel Prizes – created from the money Alfred Nobel made as a developer of nitrogen-based explosives and armaments and intended, at least in part, as expiation for the damage those explosives did as weapons of war – to two men who made the scope for killing with such explosives far greater than it had been in Nobel's day.

that Borlaug was starting his work on Mexican crops in part because waging war demanded more food from fewer farmers and in part because so much nitrogen was being fixed for ammunition that sparing some for those farmers was easy. By the time Borlaug came up with his first high-yielding rust-resistant nitrogen-hungry superwheat in 1949 there was no trouble fertilizing it. The dwarf varieties he bred later, which get their reduced stature because they put less energy into growing stalks and more into growing seeds – which contain protein, and thus nitrogen – used fertilizer even better.

Continued support from the Rockefeller Foundation was harder to guarantee than a ready supply of fertilizer. As Nick Cullather documents in *The Hungry World*, his magnificent history of food and development, some of the foundation's leaders, notably John D. Rockefeller III himself, were much taken with Vogt and Osborn's new Malthusianism, and felt that any attempt to improve things in the short term – for example, by scientifically improving crop yields – would lead to yet greater starvation further down the line. To this, E. C. Stakman, Borlaug's boss in Mexico, and Warren Weaver, the head of the programme in New York (and one of the most fascinatingly influential people in the history of twentieth-century science) argued that science could stretch the Earth's capacity. Weaver pointed to the vast mismatch between the Worldfalls-worth of sunlight hitting the planet and the number of calories needed to feed the world. Yes, the planet was finite in its endowment of material riches. But within that constraint, its capacity to support life could be enhanced. The belief common to Vogt and Osborn and many of their colleagues and followers that there was a definitive, fixed 'carrying capacity' for the environment was wrong. Circumstances could alter cases.

Weaver's view prevailed, up to a point; the Rockefeller Foundation decided to play both sides of the Malthusian street. It would fund population-control programmes, but it would also continue to fund work on agriculture; to get the desired result in the race between the population and the farmer, it would try to both restrain the first and speed on the second.

By the 1950s, farmers in much of the industrialized world were using artificial nitrogen, and Mexico was providing a model for how it could be used in developing countries, too, if the right crop strains were provided. The model had its drawbacks. Though the Rockefeller project's original focus has been on village agriculture, Borlaug's work concentrated on wheat, not corn, the local staple. This was one of a number of ways in which it benefited large farming concerns and thus American business interests. Another was that the close-packed, high-yielding fields that the new crops made possible required more pesticides, adding a new ecological burden and a new cost. Small farmers, whose livelihoods had been one of the original focal points for the Rockefeller programme, could often not afford to reap its benefits. But yields increased a lot. That was important not just for humanitarian reasons, but also for political ones.

In the 1950s and 1960s the Asian peasant, endlessly threatening to become a communist, was the focus of a great deal of Washington's geopolitical concern. The introduction of high-yielding crops and the nitrogen fertilizers they required into the Indian subcontinent in the 1960s was designed as a way of stopping the spread of Russian and Chinese influence there. When William Gaud, administrator of the US Agency for International Development, hailed India's famine-busting 1968 wheat crop as a 'Green Revolution', it was in explicit contrast to the red revolutions America was fighting against elsewhere on the continent. This is one of the reasons I feel it is fair to treat the takeover of the nitrogen cycle as a form of purposeful geoengineering. It was not simply a matter of individual farmers all over the planet doing something which improved their livelihoods. It was to a significant degree a deliberate act of policy, based on the belief that a more fertile planet would be a better one in a wide range of humanitarian, geopolitical and economic ways.

At roughly the time the Green Revolution was being given its name, the amount of nitrogen fixed by the Haber–Bosch process surpassed that fixed by all the microbes in the world's soil. American fertilizer use had grown from its 1940s level of some 500,000 tonnes of nitrogen a year to 8 million tonnes or so. In Europe the level was about 10 million tonnes. And though American and European use never got all that much higher – both use about 11 or 12 million tonnes a year now – elsewhere things were just getting started. Between 1980 and 2000 the world's fertilizer companies fixed as much nitrogen as had been fixed in all the rest of the twentieth century. The Chinese nitrogen fertilizer industry, set up in the 1970s, now fixes almost 50 million tonnes of nitrogen annually. Overall, the Haber–Bosch process now fixes over 140 million tonnes a year. About 20 per cent of it is used by the chemicals industry and to make explosives; the rest is used to feed the world, most of it by farmers in developing countries.

The spectacular increase in nitrogen use has been accompanied by spectacular increases in yield. British winter wheat provides an example, with applications of nitrate fertilizer going up almost tenfold in the fifty years after the end of the Second World War, and yields going up by a factor of four, to over 8 tonnes per hectare. Elsewhere the effects have been even more impressive. While most of the nitrogen is used to grow cereals, roughly a tenth is used for growing oil crops, a tenth for fruits and vegetables, and a tenth to enrich pastures and grow fodder.

There are costs to this, the first of which is measured in energy. The earthsystem's biogeochemical cycles do not freewheel; it takes energy to run them, just as it takes energy to drive ocean currents and winds. Re-engineering the nitrogen cycle on a planetary scale has required a redistribution of energy which, while small by the earthsystem's standards, is appreciable by humankind's: about 1.5 per cent of the world's industrial energy use is given over to nitrogen fixation. That would have been equivalent to almost a fifth of the global energy used in Crookes's day, which is one reason why Crookes would never have foreseen nitrogen fixation going as far as it has gone. Nitrogen fixation, mechanization and the use of pesticides have made agriculture an extremely energy-intensive operation; indeed, it is now the most energy-intensive sector of the economy in developed countries. Producing a dollar's worth of food typically takes more than five times as much energy as producing the same value through manufacturing.

This energy intensification of agriculture has fundamentally changed human relations with the natural world. On pre-industrial farms the energy produced, in the form of food, was typically ten times the farm's energy input in terms of the labour of humans and animals. Today, the food energy produced on farms is typically less than the energy needed to power the machines, near and far, used in growing and harvesting that food, from tractors to nitrogen-fixation plants. The energy used to create the context in which sunlight is turned into calories of carbohydrate, fat and protein is greater than the amount of sunlight thus transformed that actually ends up eaten.

This sounds almost absurd: why use more energy to make less energy? The answer lies in what the farmers use less of as a result: labour and land. Making more food used to require more of one or other of those things, and that limited how much more food you could hope to make. As it became possible to replace labour and land with energy, the ability to expand the absolute amount of food that the world could grow increased. In the second half of the twentieth century it became possible to make a lot more food by making farming more energy intensive. And as energy was becoming cheaper and more abundant than ever before, that made a lot of sense.

This energy transformation is not the whole story of twentieth-century agriculture, and the mere provision of calories does not in itself abolish famine or hunger – people need the political and economic ability to get the food they need. But vastly increased productivity proved Vogt and Osborn wrong about Malthusian famines. Farmed in the way they were farmed at the beginning of the twentieth century, today's 1.5 billion hectares of cropland would feed about three billion people eating a diet typical of 1900 (which is to say, an insufficient one) – close enough to Vogt's vision of a world of 3 billion eating the most meagre 'coolie' meals. Instead, however, that land feeds seven billion. The riches are not evenly spread, and there are many who are hungry – 840 million people, not far off the entire population of the world in Malthus's day. But there is enough food overall that a more equitable distribution would banish such hunger. And even in the absence of equity, the chronically underfed are both an ever-smaller proportion of the population and ever fewer in number.

And at the heart of this transformation is the re-engineered nitrogen cycle. Vaclav Smil, who has done more than anyone to measure and communicate the scale of the nitrogen revolution, calculates that 50 per cent of the nitrogen atoms in the world's food can in principle be traced back to a Haber–Bosch converter. And you are what you eat. The fibre of your muscle fibres; the synapses of your brain; the chromosomes on which your genome is stored: all are built with atoms pulled from the atmosphere in a factory. Vogt was being somewhat fanciful (though in a way I appreciate) when he suggested that the flesh and bones of America's children were to be found staining the waters of the Gulf of Mexico. I am not being remotely fanciful when I say that the world's industrial-nitrogen metabolism passes through your every cell and underlies your every thought. It is a physical truth.

Far from Fixed

That said, the Gulf of Mexico is still stained, and that is still a worry, and that worry is still linked to soil fertility and nitrogen use. Since the 1960s, every summer brings with it a bloom of photosynthetic algae, mostly to the west of the mouth of the Mississippi. The decomposition of this rich broth of algae uses up all the oxygen in the water, leaving no scope for animals to eke out a living. The area of the bloom becomes a dead zone, free of fish, its floor a desert. In the past decade the area of the dead zone has averaged about 15,000 square kilometres.

The algae bloom because of excess fertilizer from the heartlands of American agriculture. From the 1960s to the 1990s, the flow of nitrogen out of the Mississippi and the Atchafalaya, which drain into the Gulf, doubled. The dead zone, at first an occasional thing, became an almost annual event – only in drought years is it now not seen – and doubled in its average size between the 1980s and the 2000s. It is perhaps the most visible part of the high price the planet has paid for the twentieth century's re-engineering of the nitrogen cycle, but it is only a small part of the billable total. In almost every area of environmental concern, from air pollution to biodiversity, nitrogen pollution is now a major issue.

The problems are not all due to the Haber–Bosch process. Other changes in agronomy have boosted the amount of nitrogen fixed by some of the crops themselves. And when you burn fossil fuels, especially at high temperature, some of the nitrogen in the air burns, too, thus creating the oxides of nitrogen collectively known as NOx. As fossil-fuel burning rose over the twentieth century so did the amount of nitrogen fixed this way. Today the amount of nitrogen fixed by combustion in this way is equivalent to about a quarter of the amount fixed by the Haber–Bosch process.

What the interventions all have in common is that they strengthen the fixing side of the cycle. Meanwhile, little has been done to encourage the other side – the various forms of microbial denitrification that return nitrogen to its inert form. Even without any conscious effort on the part of humankind, the denitrifying bacteria responsible for that side of things have upped their game, since they have more fixed nitrogen to work with. But they cannot make good all of humankind's work. And so, with more coming in and less going out, the world's stock of fixed nitrogen has been growing.

At this point, the word 'fixed' – which dates back to the worldview of the alchemists – becomes misleading. For biology and chemistry, 'fixed' is fair enough; it nicely captures the idea that nitrogen which was just floating around has been knitted into organic matter. But from the environmental point of view, it is what happens after that initial fixing that matters. Fixed nitrogen does not stay put; it moves around physically, from soil to water to air and back; and it changes its chemical proclivities too.

Because it is this mobility and activity that concerns them, environmental scientists don't talk about fixed nitrogen; instead, they talk about reactive nitrogen – nitrogen that flows in a cascade from one form to another. And they talk about it a great deal. If it were not for climate change, the build-up of various different forms of reactive nitrogen in the world's soils, water and air would probably be the single biggest topic of environmental conversation and concern. In 2011 the *European Nitrogen Assessment*, a 200-author, 600-page report not unlike those of the IPCC, estimated the damage done by reactive nitrogen in Europe at somewhere between 70 billion and 320 billion euros a year. You can't scale that up to the whole world directly, because Europe suffers from some nitrogen-related problems more acutely than other places, and gets off more lightly with others. But a comparable worldwide figure would definitely be in the hundreds of billions of dollars, and quite probably over a trillion.

Some of this damage is done by reactive nitrogen acting as the fertilizer it was intended to be. Fertilizer helps some organisms more than others; one of the things Borlaug set out to breed into his Green Revolution crops was the ability to be the sort of plants that gobbled up nitrogen with gusto. This uneven response means that nitrogen tends to reduce the biodiversity of the soils to which it is applied; it boosts populations of the relatively few plants and microorganisms particularly good at making use of it at the expense of the less apt, more diverse majority. To the extent that biodiversity provides resilience – a wider range of creatures in the soil should give it more ways to respond to change – this can be a problem.

As the Gulf of Mexico's dead zone shows, the problems of oversupplied nutrients go beyond the place where they're applied. There have now been more than 540 such dead zones identified, from the Sea of Cortez to the Sea of Japan, from Shetland to Tasmania. The nutrients do not have to build up to dead-zone levels to perturb the environment – they can, for example, damage reefs by encouraging the growth of algae rather than coral larvae. On the other hand, they can sometimes be of benefit to people. When the Aswan Dam was built in the 1960s it reduced the flow of sediments down the Nile into the Mediterranean enough that the catches of fishermen in the delta began to collapse. Run-off from the fertilizer now used copiously on Egyptian farms downstream of the dam, though, has gone some way to restoring the sea's productivity. It would be Pollyanna-ish to see this unintended amelioration of unintended damage as an unproblematically happy outcome. But it may still be welcome to the fishing communities which have kept their livelihood as a result.

Reactive nitrogen in the water does not need to get to the sea to have an effect. Many rivers and lakes near farmland are choked with weed and algae as a result of overfertilization. Nitrates in drinking water can also pose a health hazard. In developed countries, water-treatment systems minimize the risk this poses, but people in developing countries drinking untreated water from land where a lot of fertilizer is used are at increased risk of various ailments, including colon cancer.

Not all the reactive nitrogen leaves the soil by way of water.

Some sticks around in mineralized forms for ages. And some turns into gas – mostly ammonia, with some NOx and some nitrous oxide (N_2O, which, despite the O, isn't classed as a NOx, and behaves rather differently*). And once reactive nitrogen gases get into the atmosphere – whether from the soil of a farm, from the chimney of a power station or from the exhaust pipe of a car – a great deal more chemistry ensues, leading to a lot more harm.

A suite of reactions between the various forms and combinations of nitrogen and oxygen produces ozone. Ozone up in the stratosphere, screening out ultraviolet light, is a boon; ozone in the troposphere is a bane, doing damage to people's lungs and plants' leaves. Other reactions produce organic compounds and various aerosol particles, which are also bad for human health. The authors of the European Nitrogen Assessment estimated that reactive nitrogen in the air is cutting six months to a year off the life expectancy of the whole population of central Europe (those to the west, who get fresh air from the Atlantic, do better). The health costs they attribute to diseases brought on by poor air quality dominate their estimate of the damage done by nitrogen. Worldwide, outdoor air pollution is thought to cause perhaps four to seven million deaths a year; reactive nitrogen and the things it produces account for a significant fraction of them.

When the nitrogen in the atmosphere comes back to the surface, either dissolved in rain (which it makes more acid than it would otherwise be) or in aerosols or gases, it starts fertilizing things again. More or less all terrestrial ecosystems – heathlands, wetlands, managed and unmanaged forests, tundra, coastal dunes, the lot – are now fertilized in this way, though not all are

---

* NOx molecules are single nitrogen atoms with oxygen atoms stuck to them; they are reactive and short lived. In N_2O one nitrogen atom is bound firmly to another nitrogen atom on one side and an oxygen on the other side, which makes for a much more stable arrangement.

affected equally.* Models strongly suggest that this is increasing the productivity of many of these ecosystems; better-fed plants photosynthesize more. It is one of the reasons why satellites have seen the average colour of the planet get a bit closer to the colour of a leaf over the past decades. As on farms, the increase is selective; ecologists expect biodiversity to drop even as productivity rises, with some species marginalized or pushed out of their habitats altogether. How big an issue this may turn out to be, it is hard to say.

Reactive nitrogen which has returned to the surface, though, it is no more likely to stay there than it was in the first place. Just like fertilizer applied on purpose, reactive nitrogen deposited from the air can get back out of the soil and into the water or air again. And again. Until stored away in some ocean sediment beyond the reach of any recycling mechanism save that of plate tectonics, or returned inert to the atmosphere whence it came, reactive nitrogen can go on causing mischief in all sorts of guises.**

Some reactive nitrogen, though, manages to stay above the fray, at least for a century or so. This is the nitrogen released in the form of nitrous oxide. It is too stable to take part in the reactions that use up the NOx in the troposphere, and because it isn't soluble, it doesn't fall back to the ground as rain. But this doesn't mean that it has no environmental effect. Nitrous oxide is a powerful

---

* This unintended fertilization story applies to farms, too, including those which make a point of not using industrial fertilizers. Organic farms throughout the world are fertilized against their will by nitrogen that can be traced back to power stations, cars, and industrially fertilized land upwind, and though the amount of nitrogen is small compared with that applied by non-organic farms, it is probably large enough to have some effect on the yield. People following strictly organic diets still have once-industrial nitrogen knitted into their cells, though to a much lesser extent than the rest of the population.

** Once back in the atmosphere, a given nitrogen molecule could in principle come straight back into play by being re-fixed, but the odds are against it. Even at the accelerated rates of the twenty-first century, only a ten millionth of the atmosphere's stock of nitrogen gets fixed per year.

greenhouse gas; molecule for molecule, its effect is calculated to be about 300 times that of carbon dioxide over a century. The IPCC estimates that human activity – mostly farming – has raised the atmospheric level of the gas by about a fifth. At the moment, the amount of heat trapped in the atmosphere by humankind's cumulative emissions of nitrous oxide is about a tenth that trapped by carbon dioxide. Like most of the nitrogen production, most of the added nitrous oxide has come about in just the past few decades.

Because nitrous oxide is stable enough to stay around for decades, it has time to drift up to the stratosphere – where, in the presence of ultraviolet light, it finally breaks down into NOx compounds that attack the ozone. The first studies of chemical harm to the ozone layer by Paul Crutzen and Harold Johnston looked at just this problem, though they imagined the NOx coming from fleets of supersonic airliners. The airliners never materialized; the world's concern shifted, rightly, to CFCs. But now CFC emissions have been slashed and their presence in the stratosphere has plateaued; emissions of nitrous oxide, on the other hand, are continuing to grow. As a result, nitrous oxide emissions, mostly from farmlands, are now doing more to slow the ozone layer's recovery than emissions of any other anthropogenic chemical.

How to Spot a Geoengineer

It is the surest of bets that Sir William Crookes gave no thought to the state of the stratosphere when he called on his colleagues to ensure that white men could continue to enjoy the advantages of wheat and cordite. He was speaking before the stratosphere had been discovered. And it is a pretty safe bet that he gave little more thought to most of the other side effects of increasing the planet's stock of reactive nitrogen; as with the stratosphere, such worries depended on ideas that the scientific world had not yet had.

While there was a sense, at the beginning of the twentieth century, that the world was becoming circumscribed, the environmental ramifications of that closing down had hardly begun to dawn. Barry Commoner's insight that there is no 'away' into which things can be thrown on a finite planet was still a lifetime off. The idea that bits of the earthsystem might be flooded with reactive nitrogen because a low rate of denitrification was causing the whole nitrogen cycle to back up would have been alien to Crookes and his contemporaries, who had only a vague notion of biogeochemical cycles.

And had Crookes known – had he the gift been given to see the world as science now sees it? Surely he would have urged care of various sorts. Perhaps he would have sought to reserve the technology's use to a single country. But would he have said that his vision of pulling down fertilizer from the sky – of making bread from air, as a eulogist of Fritz Haber put it – was not worth the damage it would do? Would he have been like that malaria researcher Vogt imagined in *Road to Survival*, gripped by the fear that his cure for the disease would, by allowing more life, lead to death and misery on the scale of atomic warfare?

I doubt it. That was not the sort of attitude that a scientist at the heart of empire would have found it easy to come by. He seems much more likely to have said that if there were problems, they should be foreseen – just as he had foreseen a world wheat shortage – and the appropriate solutions sought. It would be to men of science, again, that he would think the world should turn.

There are those who would wish he had thought differently. Any organic farmer who thinks all other farmers should follow her lead implicitly wants a world without industrial nitrogen fixation. What might such an imagined world be like? Given that, today, half the population is effectively fed by industrial nitrogen, you might think that such a world would have a considerably lower population, something many would consider a plus; I have had it put to me by environmentalists that it was in allowing another doubling of the population that the nitrogen revolution did its greatest harm. Few would advocate measures designed quickly to reduce today's population by half; but the idea of never having added the second half apparently sounds attractive to many. The pressure that humanity puts on its environment, they think, would surely be less in such a world.

Not necessarily. The demographic transition which begins when declining death rates allow long-stable populations to grow quickly and ends when a subsequent decline in birth rates evens things out was well under way in much of the world well before the nitrogen revolution came into its own. When Vogt and Osborn were sounding the alarm about population in the 1940s there were already two and half billion people on the planet. The number of children those people had drove population dynamics for the rest of the century. And their fertility did not depend on their beliefs about future crop yields; it is a fair bet that if there had been no nitrogen geoengineering going on, they would have had more or less exactly the same number of children as they did. Which is to say that if the population were to have stayed close to its mid-century level, it would not have been because of fewer births. It would have been because of more deaths.

Unwilling as they would have been to die, or to see their children die, without synthetic fertilizers those people would have sought other ways to increase the available food. The Danish economist Ester Boserup argued in the middle of the twentieth century that the world Malthus described, in which agricultural production was perpetually as high as it could be – and perpetually at risk of being outpaced by the population – did not look very much like the world she actually observed. It seemed to her that, instead, people varied the amount of food they produced to match their populations. If the people of the mid-twentieth century had not been able to increase production through additional nitrogen, they would have sought other means.

The first half of the twentieth century illustrates one of the outcomes that would have been likely on a planet of enforced organic farming: a lot more land would have been put to the plough. Calculations of what it would take to feed today's population with agricultural yields per hectare no better than those of the pre-nitrogen 1940s show that the amount of extra land needed would have been huge – 17 million square kilometres, an expanse roughly the size of Russia.

Perhaps such clearances would have been possible. Before the scale of the nitrogen revolution had been fully appreciated, people arguing against Malthusianism were happy to suggest huge changes in land use that sound remarkable fifty years on. 'Vast areas . . . are given over to unproductive tropical forests,' wrote John Maddox, the editor of *Nature*, in his 1972 book *The Doomsday Syndrome*. 'The Amazon and Congo [river basins], for example, include enough land at present not used for agriculture to provide 1,000 million acres [four million square kilometres] of cultivable land, enough to feed 1,000 million people or more.' Four million square kilometres is half the total combined area of the Amazon and Congo rainforests; the space for wilderness in this world of organic farms would have been pinched.

The fact that, if nitrogen had not been available, farms would have spread to all sorts of places in no way implies that those places would have been good farmland. As in the 1920s and 1930s, when debt and the internal combustion engine tore up America's plains, the attempt to open up new farmland would have failed in some places, maybe many – including, in all likelihood, Maddox's favoured Amazon and Congo basins. It would have created further dust bowls, denying local farmers their crops and families their food, creating landscapes unfit for man or beast.

A doubling of the farmed area is probably unrealistic; although synthetic fertilizers are responsible for a lot of the gains in yield over the past sixty years, they are not the only factor, and if they had not been available other factors might have played more of a role. But even if the added land required were only half the area of Russia, it would still be a lot. And it would require more farmers, too. The yield gains nitrogen fertilizers provided did not depend on field labour, and sometimes actively decreased the need for it; for example, they have encouraged the separation of livestock and arable farming, because manure from beasts is of little value if something similar and ready for spreading is easily bought. This 'metabolic rift' – to borrow from Marx – between the two arms of farming has, along with mechanization, allowed both to be practised on much larger scales than hitherto, with far fewer workers needed for the amount of food produced.

High-intensity agriculture without industrial nitrogen would have to rely more on mixed farms. The working practices of the best such farms are more complex than big simple monoculture and so there would almost certainly have to be more workers on them. Employing more of the workforce in agriculture would slow growth in more economically vibrant areas, and thus slow economic growth overall. While farm labour can undoubtedly be satisfying, the alacrity with which a lot of people give it up when they can see an alternative suggests that a need for much more such employment would not necessarily be a net source of human happiness.

And my portrayal of this world – a world with far less wilderness and wildlife, with some severe regional environmental crises, with more people poor and leading constrained lives which experience suggests they might rather not have led, with less of the urbanization that serves to fuel both economic growth and, often, personal emancipation – represents an attempt on my part to lean quite heavily towards generosity in taking a Boserup-inspired approach to a world without industrial nitrogen fixation. It may be that the Malthusians would have been proved right, and mass mortality would have held the population to two thirds or three quarters of what it is today.

In avoiding such an outcome, it seems to me that the nitrogen revolution did something very like what Tom Wigley has imagined climate geoengineering doing: it provided breathing space. It used fossil-fuel energy to create room in the earthsystem for the global demographic transition to play itself out without disrupting huge amounts of land and incurring all sorts of other of damage. In so doing, it caused a lot of disruption. But it reduced the risks of much worse outcomes.

It could surely have been done better; in speeding up and re-routing the nitrogen cycle, humans have undoubtedly made a mess. But the intervention and the mess are not indistinguishable, and this seems to me a last way in which human intervention in the nitrogen cycle deserves to be seen as geoengineering. In debates on geoengineering the climate, one often hears the idea that geoengineering is already under way, in that humans are already doing huge amounts to the Earth's surface, atmosphere and biogeochemical cycles. This is sometimes used as an argument for doing purposeful geoengineering, on the basis that it would just be a continuation of what is going on already, but more thoughtfully and to a better end. It is sometimes used as an argument against geoengineering, on the basis that the mess shows that humans can't handle such planetary processes, and that trying to fix the problems of this 'geoengineering' with deliberate geoengineering is likely to be both fruitless and wrongheaded.

But, as David Keith has remarked, equating making a mess with doing engineering is an affront to engineers. Engineers, like all sorts of other people, often make a mess; but it is not all that they do, and not what they set out to do. And it is not necessarily an unavoidable side effect of what they are doing. Once the problem is pointed out, it can often be put right.

There are a lot of people working on putting right the nitro-gen mess right now. Their objective is not to do away with industrial nitrogen fixation, but to limit its scope and side effects and render the benefits it has supplied sustainable. There are technical means of doing this to some extent – for example, by setting up denitrification systems at animal feedlots which produce vast amounts of manure that has no market, and thus giving the back-to-the-atmosphere leg of the cycle a boost. Such boosts, though, cannot be the whole story; it is not possible simply to speed up global denitrification through a technological innovation akin to Haber–Bosch.

This reflects the basic principle that to exert its influence on the workings of the 120,000-terawatt earthsystem, the 15-terawatt human world needs leverage. In the nitrogen cycle – and in the carbon cycle, which we will look at in the next two chapters – that leverage comes from concentration. Humans get their impact by intervening at the point in the cycle where the element is most concentrated. The atmosphere is the most concentrated reservoir of nitrogen. Stages in the cycle where things are diffuse and dilute are far harder to get to grips with – and whereas nitrogen makes up 80 per cent of the air, reactive nitrogen in the environment is measured in parts per billion.

So while scientists working to reduce the adverse impact of reactive nitrogen on the environment need various technologies, what they need more is the attention of their peers and the public, and a role in policy.

It was from one of these scientists, Mark Sutton of the Centre for Terrestrial Ecology in Edinburgh, that I first came across the idea that the nitrogen cycle was a worked example of geo-engineering. Learning about the history of the nitrogen cycle, as laid out in this chapter, convinced me that he has a point. But so did watching him in action, coordinating scientists to produce that European Nitrogen Assessment, convening meetings to look at specific aspects of nitrogen policy as it applies to farming and nutrition, contributing to subcommittee after subcommittee of an alphabet soup of international agreements and protocols, cobbling together alliances through which the problems could be addressed.

In doing all this, it seems to me that Sutton and a network of colleagues around the world are continuing the work that William Crookes, the Rockefeller Foundation, Lyndon Johnson and the central committee of the Chinese Communist Party began. The articulated intent to change the world at the scale of countries, continents and humankind as a whole that raised intervention in the nitrogen cycle to its current scale can still be seen when Sutton and his colleagues look at ways of strengthening the Convention on Transboundary Air Pollution, or argue for change to the way that farms are subsidized in the European Union, or urge the affluent to reduce their meat consumption, or discuss bringing nitrous oxide under the Vienna Convention – which already protects the stratosphere through the Montreal protocol – or look at practical reformulations for fertilizers to be sold in Malawi.

These people do not seek to stop human interference in the nitrogen cycle – none of them would dream that the modern world could function without Haber–Bosch. They mean to reduce the overall scale of the human intervention by reducing the amount of nitrogen wasted, and thus the amount that has to be produced. But this is not a simple pulling back; in seeking to enhance the care and thoughtfulness with which humans manipulate the nitrogen cycle, the incentives they have and the safeguards they observe, they are increasing the depth and intelligence of the intervention – moving it into tax systems and treaty obligations, but also into health advice and social norms – even as they seek to reduce its overall impact. Their attempts to distinguish engineering from making a mess, to clean up the system their forebears put in place while not sacrificing the progress it made possible, are deliberate and global in intent. And they are inspiring.

One way to recognize geoengineering, it seems to me, is to see dedicated geoengineers at work.

8

Carbon Past, Carbon Present

*As an eight-year-old boy, I walked through the dunes
with my father. 'Look around you,' he said. 'All you
can see is really there. Yet, behind each thing lies a
mystery, hidden from view. If you try hard you may
get a glimpse of that world, but never will you be able
to see it all.'*

Peter Westbroek,
Life as a Geological Force (1991)

Every story has another story flowing past it, heading in the
opposite direction. The detective's story of whodunit, which
starts with a murder, is the criminal's story of being undone,
like as not to end in another death. An escape to the wilderness
holds alongside it the story of a wilderness no longer wild, seen
through eyes that can but be civilized; a voyage to the moon is
a vision of the Earth. The nineteenth-century *fin-de-siècle* story
of human empire coming to its limits was countered, even at
the time of its telling, by stories in which those limits were
themselves becoming a thing of the past.

Radioactivity, in which individual atoms announce themselves
to the world with palpable bursts of energy, was discovered in
1897. In 1903 Ernest Rutherford and Frederick Soddy came
up with an explanation of the phenomenon: atoms of one
element were turning into atoms of another, liberating some of

their internal energy in the process. This 'transmutation' (Soddy, a chemist with a mystical streak, liked the word; Rutherford, a no-nonsense physicist, was wary of its alchemical overtones) showed that that internal energy must be prodigious. Sir William Crookes told British newspaper readers that the energy in a gram of radium would serve to lift the whole Royal Navy a kilometre or so into the air.

Rutherford tended to play down the prospect of ever liberating this energy from its atomic confines. Soddy was more of an enthusiast. In the remarkable last chapter of his 1909 book *The Interpretation of Radium* – a chapter which inspired H. G. Wells's 1914 novel *The World Set Free* and, directly or indirectly, many more of the twentieth century's nuclear dreams and nightmares – he laid out what it could mean to tap into the remarkable energy in the atoms of radium and, in principle, all other elements.

For all the world-spanning industrial pomp of human empire, Soddy told his readers, life in their Edwardian civilization was little better than that of an 'aboriginal savage, ignorant of agriculture and the means of kindling fire'. Research into radium, though, showed that 'the hard struggle for existence on the bare leavings of natural energy . . . is no longer the only possible or enduring lot of Man'. There was 'no limit to the amount of energy in the world available to support life, save only the limit imposed by the boundaries of knowledge . . . So far as the future is concerned, an entirely new prospect has been opened up . . . a more exalted material destiny than any which have been foretold.' The Promethean powers on offer might even be applied to the climate, making it possible to 'transform a desert continent, thaw the frozen poles, and make the whole world one smiling Garden of Eden'.

The twentieth century did indeed see a new age of unprecedented abundance when it came to energy. But contrary to what Soddy – and generations of later scientists similarly awed by the power of the atom – expected, the transforming energy was not that of the atom. Ernest Gellner, the great social theorist, used to mock Marxism's 'wrong address theory' of the rise of nationalism, in which 'the spirit of history made a terrible boob' in the nineteenth century and delivered the raised collective consciousness intended for the working class to nations instead, thus strengthening many of the shackles that socialism had been intended to break. There is something oddly similar about the way that the unprecedented amounts of energy that Soddy and Wells imagined flowing from the atom, and which they saw as leading to world government, the abolition of war and cornucopian plenty, were instead delivered to and through the industrial economy that already existed, powered from then until now by the carbon and hydrogen in fossil fuels, overseen from then until now by nation states.

If you think this means that the twentieth century was just business as usual on the energy front, rather than a radical trans-formation, consider the sheer scope of what fossil fuels went on to do. Excluding, for the moment, the burning of wood and other biomass, in 1900 the world's total primary energy supply was 33 exajoules, derived almost entirely from wood and fossil fuels.* In 2000 it was about 400 exajoules, and well over 85 per cent of it was still coming from fossil fuels. Thus, in a century which saw the population double twice, the amount of energy derived from fossil fuels doubled four times – as did, far from coincidentally, world gross domestic product (GDP). Coal, oil and gas got ever more abundant and ever cheaper; from 1920 to 1970, oil prices dropped by a factor of 3 as oil production rose tenfold. Natural gas, not a fuel at all to Soddy, grew to make up more than 20 per cent of the total energy mix by the end of the

---

* An exajoule, a billion billion joules, is not a unit for which one can easily develop a feel. But then, no units for energies this great are easy to relate to. If it helps, one exajoule is equivalent to the energy in 164 million barrels of oil, or 278 terawatt hours.

century – powering, among other things, the industrial fertilizer plants so vital to the second of the population's two doublings. Atomic power, on the other hand, barely made it to 5 per cent.

Those impressive figures understate the true scale of the century's energy transformation. Both the amount of useful work that could be done with a given amount of energy and the types of work on offer also expanded, multiplying the effects of the phenomenal growth in the total energy used. When Soddy and Rutherford were arguing about alchemy, most fossil fuels were still driving the pistons of steam engines; but the much more efficient steam turbine was beginning to make its presence felt, as was the much lighter internal combustion engine. Those technologies grew ever more capable and ever more affordable, allowing the work of machines to replace or enhance that of draft animals, women and men. Electrification allowed the energy derived from fossil fuels to be used more cleanly, more efficiently and for far more diverse industrial and domestic purposes. In 1900, according to Vaclav Smil, the world consumed about 0.2 exajoules of electricity (roughly what Massachusetts uses today). In 2000 the figure was 2,300 times higher.

More fuel and better technologies made possible what had been barely imaginable. The paramount economic role of coal in the first half of the century gave miners a power over the economy that no workers had ever held before, a power which changed the fortunes of organised labour and the basis of political life in industrialized countries. Relations between countries changed too. No one has used fossil fuels to lift whole navies, as Crookes imagined radium doing. But they have raised into the air armed forces far more fearsome than any seen on a nineteenth-century ocean, and fossil fuels, along with fossil-fuel-fixed nitrogen, made possible slaughter on a scale quite as large as that of the wars that Wells foretold. The most far-flung of Soddy's dreams was that atomic power would open the way to other planets. But when, 60 years after his book came out, engines with a power measured in gigawatts lifted *Apollo 8* off its launchpad, they did so by burning kerosene little different from that in the paraffin heaters of Soddy and Rutherford's lab. As for thawing the frozen pole, fossil fuels have taken the matter in hand without even trying.

Why then introduce Soddy's Promethean atomic speculations, if they were so misguided (or, at least, misdelivered)? One reason is to stress the point that, for all that the twentieth century felt in ways like a closing, it also presented itself as an opening – and nothing spoke of that novelty more than radiation and the atom. Acknowledging the degree to which those atomic dreams have been fulfilled by fossil fuels provides a way to appreciate again a central fact of that century – that fossil fuels have remade the world to a degree that has no historical parallel. Another reason is that I think it is important to appreciate what atomic dreams meant to the imagination. Just as the atomic bomb shaped the way the twentieth century thought about catastrophe, so the possibility of deploying the atom's potential for peace shaped the way it thought about plenty. Soddy was among the first to articulate those dreams.

And there is a third reason to start this chapter with Soddy. He did not just talk about what research into radium meant in terms of expanding access to energy. He also talked about it in terms of expanding the human experience of time – time past and time future. And it is into the realms of distant time that I want to take this chapter's discussion of the carbon cycle, a discussion that focuses not on carbon geoengineering per se, to which we will turn in the following chapter, but on the broader ways in which humans have been shaped by, and have started to reshape, the carbon cycle – the part of the earthsystem that mediates the deepest connections between life and climate.

The story of humankind and the nitrogen cycle fits pretty well into the confines of the twentieth century. At the beginning of the century humans played a minor role in the nitrogen cycle.

At the end of the century they had taken it over. But the story of humankind's relationship with the carbon cycle stretches hundreds of millions of years back in time and millennia into the future. It thus requires us to plumb depths of time for which the world had no measure before radioactivity.

Part of the sense of closing off felt at the end of the nineteenth century could be laid at the door of physics. From the eighteenth century on it had been widely accepted that the Earth had been born in heat – perhaps from the collision of a comet with the sun – and had been slowly cooling down ever since, in rather the same way as it was accepted that the atmosphere got cooler and thinner all the way up to space. In the nineteenth century the second law of thermodynamics, the one that says that temperature differences are necessary if useful work is to be done, gave a new long-drawn-out doominess to the idea. As hot things cool down, the useful temperature differences between them and the rest of the world are worn away – a law that binds solar systems as firmly as it does steam engines. And so a new end of the world was imagined: a 'heat-death' in which everything hot has cooled, everything cold has warmed and no meaningful activity of any sort is possible – an entropic nullity foreshadowed in the closing pages of Wells's *The Time Machine*, with its bloated, dying sun reddening the sky of a largely lifeless Earth.

In *The Interpretation of Radium* Soddy wrote that this thermo-dynamic pessimism had been a good reason for nineteenth-century physicists to steer clear of saying much about what their work meant for the human spirit: the conclusions were too depressing. 'To what purpose,' he asked, 'is the incessant upwards struggle of civilisation which history and the biological sciences has [*sic*] made us aware of, if its arena is a slowly dying world, destined to carry ultimately all it bears to one inevitable doom?'

Radioactivity changed that by revealing not just new sources of energy, but new expanses of time through which it could be employed and enjoyed. The reason that radioactivity had hitherto been invisible to science was that nature released the vast energies of the atom at an extraordinarily slow rate. 'Radioactivity,' Soddy wrote, 'has accustomed us in the laboratory to the matter-of-fact investigation of processes which require for their completion thousands of millions of years. In one sense the existence of such processes may be said largely to have annihilated time. That is to say, at one bound the limits of the possible extent of past and future time have been enormously extended. We are no longer merely the dying inhabitants of a world itself slowly dying.' The second law of thermodynamics was not exactly repealed; things were still going downhill. But the hill was vastly larger than had previously been thought.

Faced with these new expanses of time, Soddy became fascinatingly fanciful, imagining vast cosmic cycles of creation and destruction both in the time to come and in the time already passed, invoking racial memories of lost civilizations which had harnessed the power of the atom (his remarks on using atomic energy to melt ice caps and bring life to deserts were presented as speculations as to what might have been done in some earlier lost age, rather than as hopes or predictions for the future). Others, though, brought a more practical approach to the new possibilities.

Thermodynamics had not only specified a gloomy heat-death end-point for the Earth's cooling; it had also provided qualitative estimates of the cooling's pace. These suggested that the Earth was perhaps 100 million years old. Similar calculations suggested that the sun (also assumed to simply be a hot thing cooling down) was not much older. Thus, thermodynamics limited not just the future, but also the past – to the consternation of geologists and palaeontologists, who thought the evidence of the rocks they studied and the fossils within them pointed to a much older planet. In 1904 Rutherford suggested that radioactivity might be a way to settle the dispute. By 1911 Arthur Holmes, a young English geologist, had used a radiological dating method based on the ratio between uranium and lead (which some uranium decays into) to assign an age of 370 million years to a rock from the Devonian age – far from the oldest period in the record. The world was clearly a billion years old, maybe more.*

If the twentieth century seemed constrained and crowded in many ways, it was to be unparalleled in the expanse of time that it had to look back on. Dating methods based on ever more subtle and precise ways of monitoring the slow processes of radiation continued to push back the age of the Earth until, in the 1950s, it reached some 4.5 billion years, where it has stuck. The history of life is almost as long.

The carbon cycle began to turn in something like its modern way pretty early on, when bacteria learned to use the energy in sunlight to fix carbon. Fixing carbon, like fixing nitrogen, means taking an inert gas from the air and turning it into something biologically useful. The inert atmospheric carbon is carbon dioxide; fixing it means removing the oxygen, a process called reduction. Once reduced through photosynthesis, the carbon can be used to make the vast range of sugars, proteins, fats, nucleic acids and all the other 'organic' molecules of which living things are made. All this reduced, fixed carbon is then available for other things to eat.

To reduce any substance requires a source of electrons. Early photosynthesizers pulled the electrons they needed from various chemical compounds, but after a while one group learned the impressive trick of pulling them from water, a process that produces free oxygen as a by-product. This changed the way

---

* Radioactivity did not just provide the dates that proved the previous ideas wrong – it also explained, in large part, why they had been wrong. The Earth is constantly furnished with new heat to lose as radioactive elements in its mantle decay; the sun creates as much energy through radioactive transmutations as it loses through heat.

the earthsystem worked more than any other evolutionary innovation in the history of life. It opened up the metabolic possibilities that more or less all life visible to the naked eye now depends on, and it set up the carbon cycle as it is seen today. Photosynthesizers produce reduced organic carbon and oxygen from carbon dioxide and water; almost everything else uses that oxygen (or some derivative of it) to turn reduced carbon back into carbon dioxide and water, a process called respiration. Almost all the carbon fixed by the photosynthesizers is thus quickly re-released through the respiration of the things which eat the plants, or which eat the things that eat the plants. A small fraction of the fixed carbon, though, will escape being eaten, and get buried in sediments instead.

For most of the Earth's history, almost all of this went on at sea, though there were lichens on the continents, and algae in lakes and rivers. A bit less than 500 million years ago, though, plants took to the land for the first time. By the time of the Carboniferous, the continents – which were at the time assembling themselves into a single massive supercontinent, Pangaea – were rich with forests and swamps. And in those swamps, where oxygen was scarce, it was possible for reduced carbon to stay reduced and be turned to coal. Millions of years of sunlight were locked into trillions of tonnes of rock.

And there it stayed. As Pangaea broke up and the Atlantic Ocean separated the Americas from the rest of the word; as the dinosaurs roared their dominance and as plants first decked themselves with flowers; as the Himalayas rose from the floor of the ocean to scrape the tropopause: for all those hundreds of millions of years, trillions of tonnes of coal stayed stored up. And at the current rate of use it would take humans less than a millennium to burn the whole lot.

Alfred Lotka was one of the first to see the remarkable contrast between the length of time the Earth had spent making coal and oil and the length of time in which humans would be able to burn the whole reserve up. His *Elements of Physical Biology* (1925) contains a remarkably far-sighted appraisal of the situation:

> [The] process of fossilization is slow and would not, in itself, in any short period, materially affect the carbon cycle. It has, however, furnished the occasion for a phenomenon which, judged in a cosmic perspective, represents a purely ephemeral flare ... but which to us, the human race in the 20th century, is of altogether transcendent importance: The great industrial era is founded upon, and at the present day inexorably dependent upon, the exploitation of the fossil fuel accumulated in past geological ages.
>
> We have every reason to be optimistic; to believe that we shall be found, ultimately, to have taken at the flood this great tide in the affairs of men, and that we shall presently be carried on the crest of the wave into a safer harbor. There we shall view with an even mind the exhaustion of the fuel that took us into port, knowing that practically imperishable resources have in the meantime been unlocked, abundantly sufficient for all our journeys to the end of time. But whatever may be the ultimate course of events, the present is an eminently atypical epoch. Economically we are living on our capital, biologically we are changing radically the complexion of our share in the carbon cycle by throwing into the atmosphere, from coal fires and metallurgical furnaces, ten times as much carbon dioxide as in the natural biological processes of breathing. How large a single item this represents will be realised when attention is drawn to the fact that these human agencies alone would, in the course of about five hundred years, double the amount of carbon dioxide in the entire atmosphere, if no compensating influences entered into play.

I do not know what 'practically imperishable resources' Lotka expected to find unlocked in time to make up for the exhaustion of fossil-fuel reserves. It is possible he meant atomic energy, a subject he was definitely versed in; it is also possible he was thinking of some sort of renewable energy. Later in his book he talks of wind power representing 5,000 times the energy being derived from coal, and of ocean currents and hydropower being similarly immense in their potential, language that recalls the renewable utopian thought of John Adolphus Etzler. I do know that imagining that humans could double carbon-dioxide levels in just half a millennium was a dramatic claim at that time, even if today it is an underestimate. It could be done in a century.

The Anthropocene
That humans can, in lifetimes, make changes over which the Earth unaided would labour for millions of years is at the heart of the idea of the Anthropocene. Since Paul Crutzen brought the idea to prominence in 2000, the Anthropocene has been taken up as a framing device by a wide range of people interested in global change. Some natural scientists see it as a way of stressing the depth of human involvement in the earthsystem. Some social scientists and historians find it a useful way to break down barriers between the social and natural sciences they consider outmoded. Rather as conversation about Lowell's Mars had to combine ideas about the planet's physical condition and its inhabitants' intentions, so conversations about the earthsystem need to find a new way of talking about the intermingling of the human and the previously-seen-as-inhuman; the Anthropocene, the very name of which makes human action central to the sciences of the current planet, looks like an arena in which that great realignment might take place.

At the same time, many resist the idea. Some see it as agitprop, a dramatic gesture with which some scientists are seeking to exploit the power they enjoy as the namers of names to add weight to their views about the depth and danger of human involvement in the earthsystem. There is some truth to this. Crutzen is a canny man, and acts with an agenda; so do those who have followed him. That said, the fact that a way of looking at the world is expedient to its proponents does not in itself invalidate it.

Some object to the idea as legitimizing the level of human interference in the earthsystem when the aim must be to reverse it. If talking about the Anthropocene shifts environmental debates towards the best strategies for managing human dominance over the earthsystem and away from a steadfast refusal to countenance such an idea, a fundamental back-to-nature strand in environmental thought would seem to be sundered. To the extent that there is any point in talking about the Anthropocene, in this analysis it is in the context of retreating from it and back into some more natural epoch as quickly as possible.

And some see the Anthropocene as a category mistake; shoe-horning the enormity of what humans are doing to the planet into the strict and objective confines imposed by the rules of geological nomenclature is, they say, a disservice both to the disciplinary niceties of geology and to the discourse surrounding human influences on nature. The effect of humans on the earthsystem is something bigger, stranger and richer in meaning than the moving of a sea level, the raising of a mountain range or the impact of an asteroid. Again, it's not a bad point.

The fact that there is much to be said both for the adoption of the Anthropocene and for various arguments against it underlines the degree to which the idea is both important and paradoxical. The paradox, in a nutshell, is this: humans are grown so powerful that they have become a force of nature — and forces of nature are those things which, by definition, are beyond the power of humans to control.

Proponents of the Anthropocene can point to all sorts of ways in which humans now outstrip nature: their dominance of the nitrogen cycle; the amount of soil they move around with bulldozers and ploughs, greatly accelerating the transfer of sediment to the sea; their creation of completely new ecosystems which mix together species from multiple continents; their killing off of species at a greatly accelerated rate. They point to aspects of human activity that will be preserved over the Soddy-esque periods of time that geologists pay heed to – the strange interminglings of metal and stone that make up a city like Amsterdam, Alexandria or Adelaide will be recognizable in the fossil record for as far into the future as their lithified remains exist.

But if there is an Exhibit A in the case for the Anthropocene, it is the carbon-dioxide level. Before the Industrial Revolution came along, the carbon cycle was pretty much in balance. Terrestrial plants were taking up over 100 billion tonnes of carbon a year through photosynthesis, with another 50 billion tonnes fixed by the microorganisms of the ocean; respiration was returning carbon dioxide to the atmosphere at the same rate. The three easily accessed pools of carbon through which the cycle ran – the atmosphere, the organic material of the biosphere, and the dissolved carbon stocks of the oceans – were pretty much fixed in size. The biosphere contained between three and five times as much as the atmosphere, most of it in the soils rather than in the living plants. The oceans contained some 20 times as much again – 40 trillion tonnes of the stuff, mostly in the form of bicarbonate ions. Carbon moved between these pools all the time, but the flows were equal in each direction.

The human addition of 10 billion tonnes of carbon a year seems small in comparison; but it all goes into the smallest of the reservoirs, the atmosphere. The rate at which carbon leaves the atmosphere – either through photosynthesis or by dissolving in seawater – has sped up in response to this as the earthsystem seeks to restore its balance and spread the human contribution between the three pools. But the connections cannot handle net outflows from the atmosphere as large as the human input. About a quarter of the carbon humans emit each year is sucked back up through photosynthesis, and about a quarter dissolves into the oceans; the rest — what scientists call 'the airborne fraction' — stays in the atmosphere. In 1700, the atmosphere contained about 600 billion tonnes of carbon in the form of carbon dioxide. Now it contains 250 billion tonnes more — equivalent to half the carbon contained in the continents' living plants. Half of those 250 billion tonnes were emitted in just the past 40 years.

For records of how much carbon dioxide the atmosphere contained before the Industrial Revolution, scientists look to the remnants of that atmosphere frozen into the ice sheets that sit on Greenland and Antarctica. Drill down through the ice sheets and you can sample atmospheres tens or hundreds of thousands of years old. Such samples show that between the start of the Industrial Revolution and the end of the most recent ice age 12,000 years earlier — a period customarily known, before the idea of the Anthropocene started confusing things, as the Holocene — carbon-dioxide levels stayed between 250 parts per million and 280 parts per million. In the 100,000-year-long ice age that preceded the Holocene, the level reached as low as 180 parts per million. Over the previous million years, the level of carbon dioxide in the ice bounces back and forth between the higher levels seen in the Holocene, which are always associated with similarly warm and pleasant periods, known as interglacials, and the lower levels seen in ice ages. You need to look back tens of millions of years (and use proxies other than those found in ice cores) to find a time when the atmosphere's carbon-dioxide level was as high as it is today — back, that is, to a geological age where temperatures and sea levels were far higher than today's, just as today's are far higher than those of the ice ages when the carbon-dioxide levels were low.

The way in which the carbon-dioxide levels swing back and forth between ice ages and interglacials is today understood in terms of the earthsystem's response to small cyclical changes in the shape of the Earth's orbit and in the orientation of its axis. These alter the way the incoming energy from the sun is distributed according to latitude and season. Studies in the 1970s revealed that changes in the amount of summer sunshine in the northern mid-latitudes triggered the ending of interglacials – they were the answer to climate science's great enduring question, the origin of the ice ages. The change in the pattern of sunshine has effects on snow cover and ocean circulation, which together move the earthsystem into its ice-age state of low carbon dioxide (changes in circulation mean more ends up trapped in the oceans) and high albedo, brought about by what is called the ice–albedo feedback: as ice fields grow, they reflect more sunlight, cooling the planet further, thus encouraging more ice, which reflects more sunlight, and so on.

This is a story in which most scientists think humans have not, as yet, played much of a role. But there are dissenters. William Ruddiman, a climate scientist at the University of Virginia, has argued that differences between the ice core record of the Holocene and those of earlier interglacials have to be attributed to human intervention.

In previous interglacials, carbon-dioxide and methane levels had moved smoothly up and then down. In the Holocene, though, the pattern is different. Starting at about 250ppm 12,000 years ago, the carbon-dioxide level climbs to a peak just short of 270ppm around 9,000 years ago and starts to fall back down, following the pattern of previous interglacials – but then, after a few thousand years of dropping, it starts to rise again and continues to do so for most of the time that follows, hitting the pre-industrial level of 280ppm around the time the Roman Republic was founded. Methane, too, hits a peak 9,000 years ago. Again, in previous interglacials, that would have been it – a peak and then a steady fall. Again, in the Holocene things develop differently. Around 4,000 years ago the methane level starts to climb back up. It has now returned to more or less its early-Holocene peak.

Ruddiman claims that these two turnarounds are the effect of farming. He argues that early farmers cleared forests far more extensively than most people imagine, and thus that much more carbon dioxide was given off by forest clearance in the early and middle Holocene than most people expect – hundreds of billions of tonnes more, over the millennia. As to the methane – which is a more powerful greenhouse gas than carbon dioxide – he puts it down to rice paddies and cattle. Methane is produced by single-celled organisms with an aversion to oxygen. Paddies and the guts of cows, which are both low on oxygen, gave them new places to live. With their forest clearances, their cattle and their rice, Ruddiman argues, humans brought about the second-half increases in greenhouse-gas levels that distinguish the Holocene from previous interglacials. And in so doing, they staved off the beginning of an ice age. If carbon dioxide and methane levels had fallen throughout the Holocene as they did in earlier interglacials, he argues, carbon dioxide would now be down to 240ppm and there would be an ice cap forming in central Canada, with another expected in Scandinavia before too long.

Ruddiman's ideas have stimulated a lot of interesting research, but they have not yet proved correct, and many in the field think they are very unlikely to do so. Because the Earth's various orbital cycles are not precise multiples of each other, the rhythms of the climate have a syncopated swing, and the rests between ice ages vary in both warmth and length. Attending to those differences suggests that the current interglacial was always set to be longer than the previous three. The rise of methane might be the result of the northern hemisphere staying warmer longer, rather than the cause. Other modelling results account for the second rise in carbon dioxide without any recourse to deforestation by primitive farmers – and isotopic analysis of the carbon dioxide trapped in ice cores suggests that the mid-to-late-Holocene increase is more likely to have come from the oceans than from the burning of forests. But if Ruddiman's argument is far from proved, and may well be overstated, it has awakened a deeper appreciation for the amount of change that humans brought about thousands of years before the Industrial Revolution.

His ideas also have a political dimension; Ruddiman did not conceive them that way, but that's the Anthropocene for you – it makes ideas about how the earthsystem changes political more or less from the get-go. If you accept the idea that climate change brought about by the spread of agriculture constitutes an early Anthropocene, you dilute the shock of what is happening to the earthsystem today; the industrial changes of the climate become an extension of what was done unwittingly by farmers just by farming. The Anthropocene becomes an unwitting and more or less unavoidable consequence of there being people on the planet, and thus a rather friendlier concept. It is no coincidence that Erle Ellis, an ecologist at the University of Maryland, is both one of those inclined to agree with Ruddiman's idea of the early Anthropocene and someone who believes in the possibility that active human management of the earthsystem can create a 'good Anthropocene'; the two ideas sit easily together.

If, on the other hand, you keep focused on the industrial Anthropocene, you can see human interference in the earthsystem not as an unavoidable consequence of being an upright, imaginative, tool-using ape, but as the effect of a particular way of organizing the lives of such apes – industrial capitalism, a social and political contingency inextricable from the rise of fossil fuels. If you think, in the words of Naomi Klein's book *This Changes Everything*, that capitalism is at war with the climate, you will not want a good Anthropocene – you will hold out for no Anthropocene at all.

This is not a subject where people can simply agree to disagree – or impose their wills on each other through the more general mechanisms of politics. The Anthropocene derives a significant part of its rhetorical power from the idea that it can be defined with the same sort of procedures and precision as other geological periods. For centuries geologists have been setting up all sorts of committees to define the precise points in time – as represented by a particular set of rocks at a particular location – that mark the beginnings of the periods that define their world. Such a committee is currently discussing both whether there should be an Anthropocene and, if so, where it should begin.

They could conceivably just choose to set a date. Thus some argue for fixing the beginning at 1750 – a date frequently used in climate circles as a stand-in for the pre-industrial, with the steam engines designed by James Watt and Matthew Boulton just about to come on to the market. Others prefer 1950, which marks an inflection point after which the industrialization of the earthsystem picked up pace considerably, with carbon-dioxide emissions, industrial nitrogen fixation and all sorts of other processes kicking into their present high gear. It also has the advantage of being recorded in ocean sediments by the presence of moderately copious nuclear fallout from the burst of weapons testing that started in the 1940s and ended in 1963.*

There is, though, a better option. The base of the Holocene is defined not by a rock, but by a layer of ice – specifically, the layer in an ice core that was laid down 11,715 years ago, give or take a century. This layer of ice contains telltale signs of a sharp rise in temperature – specifically, signs of the rapid warming which brought to an end an episode called the 'Younger Dryas', a millennium-long spasm of frigidity which marked the last

---

* Beginning at 1950 has a nice technical appeal, too. Because it was around 1950 that carbon-14 dating started being used by earth scientists and archae-ologists, that year already has a benchmark role: when scientific publications refer to an event happening 2,500 or 5,000 years 'before present' (BP), they actually mean before 1950. If 1950 were chosen as the beginning of the An-thropocene, then the Anthropocene would, by definition, all be happening after the present – it would be in a condition of permanent futurity, hanging unsupported in the air like a Wile E. Coyote that has run over the cliff at the end of history.

gasp of the most recent ice age. There would be a pleasing technical consistency in finding a similar ice-core marker for the Anthropocene. In 2015, Simon Lewis and Mark Maslin, both at University College London, made use of recent work on the history of the carbon cycle to suggest just such a marker.

According to two ice-core records from Antarctica, the atmospheric carbon-dioxide level dropped by as much as 8ppm over the second half of the sixteenth century and the first decades of the seventeenth. The reason seems pretty clear. When Europeans arrived in the Americas at the end of the fifteenth century, they brought with them diseases to which Native Americans had no immunity. Various lines of evidence suggest that measles and smallpox reduced the population of the Americas over the subsequent decades by 90 per cent – from around 60 million to around 6 million, a far greater loss of life than that seen in Europe's Black Death. As a result, some half a million square kilometres which had been under cultivation reverted to scrub and forest, which would have sucked billions of tonnes of carbon out of the atmosphere. Though some modelling work suggests otherwise, it is a good bet that at least some of the reduction of the carbon-dioxide level seen in the Antarctic ice is a result of all that plant growth.

If that is so, these decades in the late sixteenth and early seventeenth centuries offer the only clearly defined ice-core marker of a human effect on the carbon cycle. That in itself probably makes them a plausible starting point for the Anthropocene.* The case becomes stronger when you consider what it is that is being marked. Human contact between the Americas and the rest of the world was a key ecological turning point. Plants that had been evolving in separate places since the break-up of Pangaea spread into each other's domains – sailing ships undid the work

---

* The changes Ruddiman points to, as well as being more questionable, are a lot less sharp; he himself would not advocate using them to define the base of the Anthropocene.

of hundreds of millions of years of continental drift. Maize, coffee, potatoes and chilli peppers spread from the Americas to the Old World; wheat, rice, horses and cattle travelled in the opposite direction, and these economically important transfers were far outnumbered by those due to chance, or to a yen for exotic decoration. The Earth's ecosystems have become much less distinct as a result of this 'Columbian Exchange' – in fact, the degree to which so many things now grow so widely across the Earth has led some ecologists to talk about the present not as the Anthropocene, but as the Homogenocene.

The ecological homogenization was a symptom of greater shifts that came about as Europe imposed its will on the newly emptied parts of the world. The economic growth of the seventeenth and eighteenth centuries, the growth which led to the capital investments that made fossil-fuel-powered industry a possibility, was powered to a significant extent by the acquisition of cheap land and cheap, even free, labour in the New World. England's economy was able to expand in large part because it had the power to take advantage of this situation: even more than the rest of Europe, the British had mastered the key imperial skills, in computer scientist and polymath Cosma Shalizi's phrase, of 'long-range trade backed by highly organized violence'.

In the 1960s, an environmentalist called Georg Borgstrom came up with the idea of 'ghost acres' to describe land that a nation would need to grow the food that instead it gets by trade. The subjugated lands of the New World added millions of ghost acres to England, providing calories in the form of sugar and trade goods in the form of cotton and tobacco without taking away from the stock of land and wealth devoted to other purposes. Fibre cheaper than any which could have been grown by English workers on England's pricey land – or by any other Europeans, for that matter – made British manufactures more productive, and thus more profitable, and thus more capable of sustaining investment in new technology. Some argue that without these conditions of unequal exchange, the Industrial Revolution would never have happened. In China, which was technologically similar to northwest Europe, but which lacked colonies, it did not.[*]

The path the world has followed since the harnessing of fossil fuels has been to a large extent a continuation of this process. Oil in particular has often been extracted far from where it has been used on terms that favoured the far-off users and local elites; until very recently the nations that did the using were those that came out on top in the Columbian Exchange and its aftermath through their prowess at organizing trade and violence. But where once the industrial world drew its growth just from the productivity of places far away in space, now it also draws its growth from places far away in time – from Carboniferous forests and Cretaceous seabeds whose areas far exceed those of the ghost acres which fed and fuelled Britain's Industrial Revolution. To the extent that the beginning of that world- and earthsystem-changing process, at once historical and geophysical, is recorded in Antarctic ice-core records of the late sixteenth century, its claim to mark the beginning of the Anthropocene is the best that there is.

The Greening Planet

The events of human history, and to some extent prehistory, have left their mark on the carbon cycle. The reverse is also true. Shifts in the carbon cycle were crucial to the development of human civilization.

As radioactivity revealed the true antiquity of the Earth, and the vast age over which the carbon cycle had stored up fossil fuels, similar methods pushed back the age of the oldest humans. By the late twentieth century human ancestors who in the Bible

---

[*] It is also worth noting that the most economically advanced part of China was a few thousand kilometres from the country's coalfields (a problem that still matters today), while British industry had coal on its doorstep, or beneath it.

stretched back only a few dozen generations before Christ went back for more than a million years.

A rhetorical convention similar to that which insists on the smallness of the tough and age-old Earth requires this to be seen as a short time compared with the far greater age of the Earth: if the age of the Earth is a 24-hour day, a million years is the second before midnight; if it is the span of a woman's outspread arms, a million years is a couple of millimetres of fingernail, and so on. But by the standards of the archaeological record of cities – five thousand years or so – and agriculture – twice that, at the oldest sites – the new depths of human prehistory were impressive, and a little puzzling. Why should human prehistory be so long, and human history so short?

Admittedly, for most of the past million years, what humans there were appear not to have been the intellectual equals of those around today. But archaeology strongly suggests that by 100,000 years ago in Africa, and perhaps by 50,000 years ago in most other places, they were pretty much indistinguishable in their physical form and their cognitive capabilities from you and me. They may have lived like Soddy's 'aboriginal savage, ignorant of agriculture' – but they had tools and fires and speech. They were capable of representing the world to themselves and creating beauty for each other, as their cave paintings and carvings show. They had stories and clothing and burials – they even had small millstones with which to process the grain that they gathered. What held them back from farms and towns and cities?

It seems obvious that it was the ice age. For tens of thousands of years, as the ice age grew more and more severe, there were modern humans quite as smart as those who eventually became farmers, but there was no farming. Less than ten thousand years after the glaciers started to retreat, crops were being domesticated in the eastern Mediterranean and north-eastern China. Within a few thousand years the same was going on in South America and the Sahel. It is hard to see this as a coincidence.

What, though, made the ice age inimical to agriculture? It was not just that it was cold. Many places which are warm today were quite temperate back then; average temperatures in the tropics were only about 2°C lower than they are today. And though the ice-age climate was dry, it was not so dry everywhere that there would have been no possibility of growing crops anywhere.

A bigger obstacle to early agriculture would have been the inconstancy of the ice-age climate. Whereas interglacials see slow steady rises and falls of greenhouse gases, in the 1980s it became apparent that during ice ages they shot up and down. This is part of a pattern of fluctuations which involve sudden changes in ocean circulation − slow circulation allows more carbon dioxide to be stored in the deep ocean − and which are amplified by large changes in albedo as the extent of sea-ice cover changes. It seems that all through the most recent ice age, and those before, temperatures and rainfall patterns were apt to shift dramatically over periods of decades and sometimes just years; they rarely stayed stable for more than a few millennia. Such climate craziness may have precluded the development of agriculture. The Holocene was the first period of long-term climatic stability that modern humans lived through − and, hey presto, agriculture ensued.

The discovery of the ice-age climate's choppiness had implications beyond the origins of agriculture. Until the 1980s it had been assumed that climate change was by its nature a slow thing. But the evidence in the ice cores and the sea-bottom sediments showed that it could happen much faster − cataclysmically fast. And although the conditions that allowed such abrupt shifts had apparently not pertained during the Holocene, might they not come back now that humans were pushing the climate beyond its Holocene comfort zone? The climatologist and oceanographer Wally Broecker − who, in the 1970s, was the first person to use the phrase 'global warming' in print − was at the centre of attempts to explain the ice-age flip-flops in the 1980s.

He was clear that, whatever the mechanism behind them, their existence constituted a warning that there could be much more sudden and catastrophic changes to come than the gentle word 'warming' suggested; that steady changes in radiative forcing could trigger 'non-linear' results as the earthsystem moved into new states. Because of Broecker's work, ideas about rapid climate change and 'tipping points' became central to many accounts of the risks posed by climate change. An earthsystem pushed only a little way beyond the climate norms of the Holocene might, having stumbled over some hidden threshold, suffer disproportionate and irreversible change. Thus the risks posed by extra-Holocene carbon-dioxide levels looked yet worse than they had done.*

The idea that climate stability was the key defining factor of the Holocene thus took hold, explaining both the dawn of agriculture at its beginning and the damage that might be done should it be ended. But it is not just the stability of the greenhouse-gas levels in the Holocene which distinguishes them from those of the ice age that preceded it; it is also their absolute magnitude. They were high. That is good for plants – and, quite possibly, good for proto-farmers.

In 2006 the Competitive Enterprise Institute, a right-wing American think tank with a history of shilling for the tobacco lobby, ran a campaign throwing doubt on concerns about global warming under the slogan 'Carbon dioxide: they call it pollution, we call it life'. The ads were disingenuous garbage; the slogan is an intriguing false dichotomy. Carbon dioxide can indeed be a pollutant; but it is also a necessary substrate for the

---

* It is worth noting that these ideas took hold around the time that the Antarctic ozone hole was discovered – another example of a non-linearity in which a smoothly increasing change leads to a sharply discontinuous result. Ozone depletion had been expected to manifest itself evenly, but turned out to be amplified and concentrated in a particular place. The finding fed the feeling that other unexpected non-linearities might lurk unsuspected in the earthsystem's responses to human interference.

photosynthesis on which every ecosystem depends. Plants, left to their own devices, would rather have more of it than less.

Plants are not left to their own devices; they get what they are given. And for most of the Earth's history they have been given quite a lot of carbon dioxide. When photosynthesis evolved, there was enough carbon dioxide in the atmosphere for it to be measured in percentage points, rather than in parts per million. In the Devonian, when forests were first spreading, the level was probably some five or six times that of today. The long-term decline is due to the fact that carbon dioxide dissolved in rain and seawater plays a role in the 'weathering' of silicate rocks, a process which produces rocks with carbonate ions locked into them, such as limestone and marble. Though this is a process which some life forms put to their own use – reefs are made of carbonates, as are the shells of shellfish and of tiny charmingly-armoured plankton called coccolithophores – it is completely different from photosynthesis: the carbon in carbonates is not in the reduced form from which living tissues can be formed, and energy later extracted.

Not all the carbon locked into carbonates stays there: some is brought back by the grand recycling of plate tectonics. When the ocean crust sinks back towards the underlying mantle at the edges of oceans – as it does, for example, deep below Mount Pinatubo, Krakatau and the other volcanoes of the Pacific Rim – carbonates in the crust will be heated up enough for their carbon dioxide to be returned to the atmosphere. But not all carbonates are recycled that way. And so, over the course of the Earth's history, the level of carbon dioxide in the atmosphere has fallen as the amount of carbonate rock in the crust has risen.

The molecular machinery at the heart of photosynthesis, though, has not been able to adapt to this change. Even at today's 400ppm of carbon dioxide, plants are less efficient at photosynthesizing than they would be at the higher levels of the geological past. At the low levels recorded in ice-age ice

– 180ppm was probably the lowest level of carbon dioxide the atmosphere has ever seen – they were positively gasping for it. Studies of the ice-age environment suggest that the total mass of the plants on the continents 20,000 years ago was only about half what it is today. When changes in ocean circulation at the end of the ice age started to replenish the atmosphere's stock of carbon dioxide, the plants went wild. Forests chased the ice caps back towards the poles. The amount of grain produced by wild grasses increased phenomenally.

In 1995 Rowan Sage, a plant physiologist at the University of Toronto, suggested that this carbon-dioxide fertilization might help explain the origins of agriculture. In the 3,000 years preceding the Holocene, the carbon-dioxide level rose from 200ppm to 250ppm. It was not just the productivity of cereals that would have increased in response; so too would the water retention and nitrogen fixation of soils penetrated by their more vigorous roots. This increased productivity would have meant hunter-gatherers needed to travel less and could have concentrated on just a few particularly productive species for most of their food. Researchers think that sort of sedentary life and concentrated use will bring about the selection, conscious or not, of strains that grow better under certain conditions: the beginnings of domestication.

Sage's insight is not widely appreciated. But in its outlines it is hard not to believe. Look at the spindliness of plants grown in low-carbon-dioxide conditions and it is hard to see how they could support a sedentary human population. The idea also explains why a number of different cultures domesticated a range of different species in fairly quick succession; the rising carbon-dioxide tide was a global, not regional, effect. Carbon-dioxide increases may not have caused domestication and the subsequent beginnings of agriculture, but they seem likely to have made them possible. Once they were possible, some of the many different human cultures subsisting on various different foods were bound to stumble into domesticating them; and once that happened, the advantages offered by domestication were bound to spread, through population movement, cultural contact or conquest. One group of researchers that has taken Sage's ideas seriously puts the point succinctly: agriculture in the ice age was impossible; in the Holocene it was mandatory.

It is possible that, as you read this, some shoulder-sitting pitchfork-wielding avatar of the Competitive Enterprise Institute will whisper in your ear: 'If more carbon dioxide was a good thing for not-yet-farmers at the end of the ice age, why isn't it a good thing for farmers today?' The answer is that, if you ignore the effects it is having on the climate, it is.

As noted before, only half the carbon dioxide that human industry and deforestation pump into the atmosphere stays there. A quarter dissolves into seawater; another quarter, which is to say more than a billion tonnes of carbon a year, is sucked up by plants. The evidence can be seen from space. Remote-sensing researchers have a measure, the Normalized Difference Vegetation Index, which tells them how much a given pixel in their data looks like a leaf. For the past few decades, satellite data measured this way have shown that the planet as a whole is looking more like a leaf than it used to; it is, in effect, getting greener. Estimates suggest that the 'net primary productivity' of the world's plants – the amount of the carbon they fix that ends up as biomass, rather than being metabolized by the plants themselves – has increased by somewhere between 3 per cent and 10 per cent over that time. This seems mostly down to carbon-dioxide fertilization, though as we saw in the previous chapter, greatly increased levels of reactive nitrogen matter, too.

Overall, it has been estimated that since the beginning of the Industrial Revolution, human emissions have provided enough carbon-dioxide fertilization to add about 250 billion tonnes of carbon to the planet's plants and soils, increasing the amount stored in them by about a tenth. That's a number with big caveats, based as it is on calculations dependent on a single set of models, and it is gross, not net: humans have liberated a similar amount by clearing forests for farming. But it gives a sense of the scale of the effect. And the boost to growth has not just been seen in wild ecosystems; it has helped farmers too. As at the end of the ice age (but on a much shorter time scale), higher carbon-dioxide levels have meant bigger cereal crops. David Lobell, who studies crops and climate at Stanford, estimates that from 1980 to 2008 the increase in the carbon-dioxide level – 339ppm in 1980, 394ppm in 2008 – pushed up yields for a range of crops by about 3 per cent.

A knock-on effect of reducing climate discourse to a discussion of the evils of carbon dioxide is that it makes the fact that carbon dioxide enriches life in this way rather awkward to talk about, and so advocates of climate-change action may choose, consciously or otherwise, not to do so. The World Development Report on Development and Climate Change published in 2010, to pick one example off the shelf, makes no mention of the effect of carbon-dioxide fertilization on crop yields, even though there is reason to think that it may be at its most important in marginal, semi-arid areas of developing countries. A similar caginess may explain the relative obscurity of Sage's account of the beginnings of domestication. The idea that the Holocene was the cradle of civilization because of its unprecedented climatic stability fits well with worries that the climate of the post-Holocene future may bring rapid and catastrophic change like that seen in the ice ages. The idea that the Holocene was also a period of unprecedentedly high carbon-dioxide levels, by the standards of the rest of the time modern humans have lived on the planet, and that those high carbon-dioxide levels, too, were fundamental to the growth of civilization, is a little harder to fit into a simple story.

Tough. The story is not simple; it is not even a single story. The earthsystem and its study are full of stories which head past each other in opposite directions, of stories which seem to contradict each other, of stories with different morals depending on your point of view, of stories to which no one as yet knows the end. That is the nature of the earthsystem, and, in my experience, of stories more generally. If you want to expand the imagination with which you look at and judge the possibilities in the earthsystem's future, a willingness to see that each story has its counter-stories is a necessary part of the process.

The future benefits that crops will get from higher carbon dioxide are uncertain for two sets of reasons. One is that the response to higher carbon dioxide on a real farm is hard to model. Different varieties of the same crop respond to carbon dioxide with markedly different levels of enthusiasm. Some weeds do better from the process than the crops they afflict. Other limits will come into play. As Liebig discovered, there will always be something vital that plants are not getting enough of.

The other set of reasons is that carbon-dioxide fertilization is just one factor. In many places the increased ozone levels that come with more reactive nitrogen in the environment may be taking back much, even most, of any benefit that is being gained from increased levels of carbon dioxide. And then there is the climate. Lobell and various colleagues have used the effects of weather on crops to estimate the damage that recent climate change has done to agriculture. They found that, in the 30 years in which carbon-dioxide emissions seem to have increased plant yields by 3 per cent, climate damage reduced them by a greater amount: 3.8 per cent for maize, 5.5 per cent for wheat.

Particularly hot days – the frequency of which can be expected to increase a lot even while average temperatures increase only a few degrees – can do disproportionate damage to crops. So, obviously, can extended and severe droughts. Both are widely seen as likely developments. Carbon-dioxide fertilization will still do some good to some of the crops, and net benefits may well be seen in places that avoid climate extremes. Whether, on a planetwide basis, those benefits will put a significant dent in the damage done by ozone and global warming, no one can say for sure.

But what if the planet were to enjoy a higher carbon-dioxide level without a great deal of warming – or any? That is to say, what would happen to the world's crops if a stratospheric veil were to keep the planet at today's temperatures while the carbon-dioxide level doubled?

On the basis of first principles, you might expect that the results would be pretty good for farmers. And that is indeed what the only detailed analysis of the issue so far has found. Julia Pongratz, when she was a student of Ken Caldeira, worked with him and with Lobell on models that compare agriculture on a Greenhouse Planet (one with a doubled carbon-dioxide level) with that on an Engineered Planet (one with doubled carbon dioxide and a thick enough stratospheric veil to keep the temperature at today's levels). Yields were 10 to 20 per cent higher on the Engineered Planet than on the Greenhouse Planet. Indeed, yields were higher on the Engineered Planet than they were when the same model was run under today's conditions.

The modelling involved is quite crude. It leaves out a number of the effects of stratospheric veils touched on in the first part of this book, effects which might make the bump in yields even greater. One is the diffuse-light effect, in which sunlight coming through a diffusing stratospheric veil is a bit better suited to photosynthesis than harsh direct sunlight is. This probably helps trees in forests more than crops in fields, but it might boost yields a bit. Then there is the fact that one of the ways carbon-dioxide fertilization helps plants grow is by making their use of water more efficient. One of the potential drawbacks of an Engineered Planet is that the hydrological cycle is suppressed to a certain degree. By making plants more efficient in their use of water, carbon-dioxide fertilization directly counteracts that effect. Thus the study may undersell the benefits.

The idea of an Engineered Planet making life easier for many, quite possibly most, of the world's farmers brings to mind a contribution to the geopoetry of geoengineering from the 1950s. Harrison Brown was a geochemist at Caltech who, among many other achievements, supervised the work that first showed that the Earth was 4.5 billion years old. In *The Challenge of Man's Future*, published in 1954, Brown noted that horticulturalists exploit carbon-dioxide fertilization on a daily basis, keeping the gas at much higher levels in their greenhouses than it is in the outside air, and suggested it might be worth increasing the carbon-dioxide level of the atmosphere as a whole in a similar way, either by burning coal or by breaking down carbonates. If it were tripled, crop productivity might double.

In an interesting, and perhaps instructive, irony, Brown did not mention the effects that this would have had on the climate. The reason that people did not, in the early 1950s, worry about climate change caused by carbon dioxide was not that they didn't know how much carbon dioxide contributed to the greenhouse effect. It was that they thought carbon-dioxide emissions would not stay in the atmosphere, but instead all end up in the oceans; the work by Suess and Revelle that showed that this was not the case was still a few years in the future. But Brown's thought experiment specifically required that the ocean's appetite for carbon dioxide be overcome and that the carbon dioxide stay in the atmosphere. Thus, not mentioning the climate effects is interesting. It may be that he just decided that, since he didn't really think the idea about carbon-dioxide fertilization was a serious one, the climate effect was not worth mentioning. Or it may be that even wide-ranging, polymathic scientists miss things which, in retrospect, they should have seen – a possibility everyone thinking about geoengineering should bear in mind. Either way, there is a nice irony in the fact that David Keeling, the man who would soon thereafter make the measurements which first established that a substantial fraction of the world's carbon emissions did indeed remain in the atmosphere, was laying the foundations for that work as a post-doc in Brown's Caltech lab while Brown was off at his second home in Jamaica writing the book.

Just as Brown did not in fact advocate deliberate carbon-dioxide emissions, Pongratz, Caldeira and Lobell do not suggest that the high agricultural yields on their Engineered Planet would justify creating something similar for real. They do say, though, that their work should serve as a caution to people who jump from the fact that an Engineered Planet would have a slowed-down hydrological cycle to the conclusion that it would necessarily be a place where the livelihoods of billions of people dependent on rain-fed agriculture would be put at risk. Better-fed, more water-efficient crops would make a difference in many places. The winners might greatly outnumber the losers.

Leaving aside the question of crops, such a planet would almost certainly, in terms of the total amount of plant life and its productivity, outstrip today's earthsystem – just as earthsystems with carbon-dioxide rich atmospheres often have in the geological past. Peter Cox, a climate modeller at the University of Exeter who works on interactions between plants, carbon dioxide and climate, is adamant that, if the terrestrial plants were making the earthsystem's decisions, an Engineered Planet with today's temperature but a higher carbon-dioxide level would be exactly what they would want. Such an Engineered Planet would look even more like a leaf as seen from space than today's does (though it would also look a tad hazier).

And its greenery might be not just deeper than today's, but significantly more widespread. When people study tipping points in the climate – or other non-linear systems – they often find a large role is being played by changes which feed back on themselves, their effects reinforcing the conditions that drive them in a way that cannot be stopped. Thus part of an ice sheet, in collapsing, destabilizes a far larger region of ice upstream, which, in collapsing, propagates destruction yet further. Methane, released from Arctic permafrost by warming, warms the system further, releasing yet more methane. Almost always, such self-accelerating effects look like bad news. There seems to be at least one such tipping point, though, that it might be good to cross.

In the far south of Algeria, on a plateau of wind-carved sandstone called Tassili n'Ajjer, there is a remarkable collection of rock art – thousands of pictures of elephants, crocodiles, ostriches and gazelles, and of people living among them as happy hunters, even pastoralists. They are less than ten thousand years old. In the early-to-middle Holocene, the temperature difference between summers and winters in the northern hemisphere was greater than it is today. This would have made the Saharan region hotter – but it also pulled the West African monsoons, which today bring rain only as far as the Sahel, much further north, feeding rivers that ran down the far side of the Sahara watershed and into the Mediterranean, changing the climate and landscape of all of North Africa. Some of what is desert today was also desert them – but much was savannah and woodland, studded with lakes under skies cooled by bright white cloud. In the south sat the inland sea of Lake Mega-Chad, almost as big as the Caspian Sea, covering a greater area than all North America's Great Lakes combined. The canoes used by the fishing people living on its shores were as large and sophisticated as those seen on the Mediterranean.

The relics, human and natural, of that age were part of what convinced nineteenth-century Frenchmen that the North African climate had been on the skids since the days of Rome. In fact things had got worse much earlier: around 6,000 years ago, today's desert conditions established themselves with striking speed. There was a sudden cooling event that was a faint echo of the sort of swings in temperature seen in the ice ages, and the climate never recovered. All that was left to record the plenty that had gone before was pollen, bones and those plaintive paintings.

The suddenness of the change, some scientists think, is evidence that there was some sort of amplification at play. When the Sahara, or large parts of it, were green, the plants were not just benefiting from the wetter climate – they were helping to maintain it, by holding together the soil and pumping rainwater back to the sky through their leaves, thus encouraging convection that cooled the surface and created clouds. They kept doing this even as summer temperatures fell and the monsoon shifted to the south until a dry cooling around 6,000 years ago dealt the system its death blow; the drying made life more difficult for the vegetation, which led to further drying, which killed off yet more of the plants. Patches of desert spread and merged.

In a world with a higher carbon-dioxide level, and thus more water-efficient plants, this process might be reversed; various models suggest that global warming could, in time, lead to a new greening of at least parts of the Sahara. More plants would mean more water vapour, more clouds and rain, and thus more plants – the feedbacks that brought the desert about now running in reverse. In a world with higher carbon dioxide but no concomitant increase in temperature – a world with a veil – the plants would show all the same gains in water-efficiency but suffer none of the damage of higher temperatures, and as a result, the odds of such a reversal could well be significantly better. The vast dried basins of the desert could be refilled, not by great canals like those of Verne, de Lesseps or even Lowell, but by rains from a newly sheltering sky.

9

Carbon Present, Carbon Future

'These atmospheric cleaning plants are erected in
sufficient numbers over the face of the planet in such
an ingenious manner that ALL of the air must flow
sooner or later through one of the wire cubes, here to
be purified.

'Thus does Martian intelligence safeguard the planet's
air supply. How long will it be with your coal burning
machinery till the Earth's atmosphere will need cleaning
plants?'
> Hugo Gernsback, 'The Scientific Adventures
> of Baron Munchausen' (1917)

The 'huge carbon-dioxide generators pouring the gas into the atmosphere' of Harrison Brown's imagination are an exception in the annals of geoengineering. Arrhenius saw the effect of industrial fossil-fuel burning on the climate as largely benign; Guy Callendar, a British scientist who followed up on Arrhenius's work in the 1930s, was happy to conclude that the use of fossil fuels would postpone the next ice age — 'the return of the deadly glaciers should be delayed indefinitely'. But I think only one person other than Brown (Nils Ekholm, a friend of Arrhenius) suggested increasing the atmosphere's carbon dioxide might be undertaken as an end in itself, rather than merely

243

welcomed as a side effect of industrial modernization. And since the 1970s, the emphasis has been very much on the opposite process. Returning from the previous chapters' excursions into non-climate geoengineering and deep time with, I hope, a new sense of scale, it is to these geoengineering techniques aimed at carbon-dioxide reduction that we now turn.

People wanting to suck carbon dioxide out of the atmosphere start off with clear advantages over geoengineers imagining other interventions. Unlike the would-be nitrogen geoengineers of the early twentieth century, they already know that humans can influence the carbon cycle on a global scale – there are hundreds of billions of tonnes of excess carbon in the planet's atmosphere, oceans and plant life bearing witness to the fact. And putting that carbon back where humans found it, or in some other safe store, is both ideologically more acceptable and politically more plausible than messing around with incoming sunshine. Moving carbon to safe stores feels more restorative than transformative, and sits well with common-sense notions of what to do when you have made a mess: clean it up. Reducing the amount of carbon in the atmosphere fits easily into the carbon-dioxide-as-the-thing-there-needs-to-be-less-of frame of current climate politics.

That said, the task the carbon-dioxide reducers imagine is a massive one. Only two human enterprises currently use tens of billions of tonnes of raw material a year: the supply of energy and the supply of food, and both are intimately tied into the carbon cycle. Deliberately re-routing yet another part of that cycle would mean not just adding a third enterprise to the list, but also managing its inevitable interactions with the first two.

There are three broadly distinct ways in which this could be done: by increasing the amount of carbon dioxide that plants absorb; by increasing the amount that the oceans absorb; or by creating machines like those Harrison Brown imagined, but working in reverse: huge aspirators that pull carbon dioxide out of the air. I am going to take them in reverse order, starting with the machinery of 'direct-air capture'.

As we saw in the introduction, a number of entrepreneurial physicists with big imaginations and rich sponsors are looking at this technology. The idea is to bring large amounts of air into contact with a substance eager to relieve it of its carbon dioxide. That substance needs to be greedy for carbon dioxide when in contact with the air but ready to give it up under some other circumstances. The mechanism that holds it needs to be able to bring about those other circumstances – a different temperature or pH, perhaps – as required and also to be able to scale up so as to process large amounts of air. And the whole system needs as low a capital price as possible.

A fundamental problem is that the easiest way to get leverage in the earthsystem is to intervene at pinch-points of concentration. Thus carbon dioxide is produced most readily from the concentrated carbon in fossil fuels; reactive nitrogen is fixed from the concentrated nitrogen in the air. But the concentration of carbon dioxide in the atmosphere, while high in terms of recent history, is low in absolute terms: 400ppm is 0.04 per cent. The laws of thermodynamics say that the more dilute a part of a mixture something is, the harder it is to pick it out of the mix. Imagine the molecules in the atmosphere as marbles in a bucket that need to be sorted. Four out of five are blue nitrogen marbles. One out of five is a red oxygen marble. Just one in 2,500 is a white carbon dioxide molecule. To take out an appreciable number of the carbon dioxide marbles you will have to handle a great many of the red and blue ones, which makes the process inherently inefficient.

If you could find a bucket in which one in ten of the marbles was white, things would be a great deal easier. There are, as it happens, many such buckets available. The chimneys of fossil-fuel power stations often contain gas that is 10 per cent carbon dioxide. Various systems have been designed to pull the carbon dioxide out of this flue gas, and as thermodynamics would predict, they do so with much greater ease than systems designed to pull much more dilute carbon dioxide out of the atmosphere at large. Yet even these systems have yet to make much progress in the fight to cut emissions, a failure which should – and does – give pause to direct-air-capture enthusiasts.

The idea of scrubbing the carbon out of power-station smokestacks was first mooted in the 1970s. Cesare Marchetti, an Italian physicist at the International Institute for Applied Systems Analysis (an international think tank outside Vienna; as it happens, Harrison Brown was one of the people who set it up) first used the term 'geoengineering' in the climate context to describe a system in which carbon dioxide from power stations would be stored away in deep ocean waters. In the 1990s, with carbon control the main issue in climate politics, the idea re-emerged – now mostly favouring saline aquifers and depleted gas fields, rather than the deep oceans, as possible receptacles, and no longer known, for the most part, as geoengineering, but instead as 'carbon capture and sequestration' (CCS).*

In climate-policy circles CCS is widely seen as a necessary technology. Most scenarios for producing the world's electricity without carbon-dioxide emissions rely on a significant proportion of the world's electricity coming from fossil-fuel power stations equipped with CCS technology; in scenarios that forego nuclear power, CCS is more or less obligatory. What is more, CCS is not an intrinsically challenging technology; getting carbon dioxide out of flue gas is the sort of task chemical engineers have been turning their hands to since Sir William Crookes's day.

Yet although CCS may be seen as necessary and feasible, it remains obstinately unused. It has been widely touted for more than a decade, but there are currently no full-scale fossil-fuel plants that use the technology and no plans to build more than

---

* The S has more recently come to stand for 'storage'.

a handful of modestly sized demonstration plants in the foreseeable future.

Governments would rather subsidize renewables, which are popular with some voters and can be built right now, than develop technology that might deliver in ten or twenty years' time. Any money they do spend on longer-term options is more likely to be devoted to nuclear power. Like CCS it promises constant supplies of electricity day or night and whatever the weather; unlike CCS it enjoys the benefits of an industrial establishment with decades of experience attracting government funds, and of a ready supply of engineers — Soddy's heirs — who find its Promethean challenges appealing. I do not hold out much hope for radical nuclear innovation, but some of the things on the table are undeniably exciting possibilities. CCS is, by comparison, a problem too mundane to thrill — indeed, almost too simple to solve.

Greens who would rather governments not spend money on nuclear are rarely willing to push hard for CCS as an alternative. Few greens have any real enthusiasm for a technology that is designed to allow fossil-fuel companies to stay in business. And as the S in CCS bears witness, Commoner's 'everything has to go somewhere' still applies. That means resistance from at least some of the people who live over the subterranean somewheres where the carbon dioxide might end up (NUMBY — Not Under My Back Yard — is the term of art here). Growing (if largely ill-founded) worries about fracking could well add to these concerns; in its requirement that fluids be injected into geological structures deep underground for the benefit of fossil-fuel businesses, CCS looks more than a little like fracking, and it's not a resemblance that plays well with greens.

If companies thought they could make money from storing carbon underground they would probably find a way to do so in the face of opposition, just as in many territories they have found ways to frack. But CCS on a grand scale could only promise its practitioners profits in a world where climate policy put a significant price on carbon – at least $50 a tonne, quite possibly double that – that would be paid as readily for the billionth tonne as for the first, and that was guaranteed either to stay stable or to rise for decades. This is not that world.

There are, though, a number of industrial markets for carbon dioxide, some of which might make CCS viable in particular circumstances. The biggest such use is for enhanced oil recovery: in some oil fields, squirting carbon dioxide down one well can increase the flow of oil up an adjoining one. This technique is already widely used in Texas; one of the reasons that Summit Power plans to build a coal-fired power station with CCS in Texas is that the company thinks the money it can earn selling the carbon dioxide to the oil industry should, along with some government incentives, put it in the black. Yet while tying CCS to the enhanced-oil-recovery market this way may well help get a few pilot plants built, it makes it even less likely to be accepted by greens.

The oil-recovery market is not the whole story; some governments have plans for CCS pilot plants in their climate-change-technology portfolio, too, on the basis that it will one day be needed. But the lack of ambition and progress seen in these efforts suggests that none of these governments really foresees a future with hundreds of thousands of kilometres of pipeline taking carbon dioxide from thousands of power stations to tens of thousands of wells.

If there is no real intention to use CCS, why research it? One answer is that research funding can function, like hypocrisy, as the tribute vice pays to virtue. Clive Hamilton makes this point with vigour, arguing that the implicit promise of CCS research – low-emission fossil-fuel power available in the not-too-distant future – is being kept alive purely to provide cover for the continuing use of fossil fuels in the infinitely extensible meantime. It is like St Augustine's 'make me chaste, Lord – but not yet', but imbued with an uncanonizable cynicism. I think it is a strong argument, and one that underscores the concern that talk of climate geoengineering could serve as a delaying tactic in just the same way.

As an aspiration, direct–air capture has advantages over CCS. It suffers neither from being too mundane to thrill nor from being too simple to solve; it has Promethean world–changing promise, and finding a way to do it cheaply looks really hard. Still, in practical terms, as Robert Socolow pointed out at that snowy Calgary meeting, deploying a technology dedicated to the truly hard task of capturing carbon dioxide from the atmosphere on a global scale before mounting a comprehensive attempt to capture carbon dioxide from the much more concentrated sources that industrial emissions offer seems both daft and unlikely. Those working on direct–air capture at the moment almost all agree; most see their work fitting into niches where tight regulatory frameworks and a need for carbon dioxide that cannot easily be met by current suppliers make their product peculiarly valuable. Such niches are not large – measure them in tens of thousands of tonnes a year, not in billions. Against that, one must remember that small things grow; the oil industry was small once. But one must also remember Hamilton's criticism of CCS.

What, then, of the two channels by which the earthsystem is already taking a significant proportion of humankind's emissions out of the atmosphere – photosynthesis by plants and absorption in the ocean? At the moment these sinks between them soak up about half of the annual 10 gigatonnes of carbon emitted by human activity without any encouragement. If one or both of them could be made just a little more capacious, could not the whole problem be solved, or at least pushed off into the future?

The ocean looks like the obvious place to start. After all, it represents a reservoir of dissolved carbon that dwarfs any conceivable amount of fossil-fuel emissions; the amount of carbon needed to double the pre-industrial level in the atmosphere would increase the stock of inorganic carbon in the ocean by just under 2 per cent. Finding a way to engineer such an increase in the ocean's stock using the atmosphere as a source would seem to solve today's climate problems more or less completely.

The degree to which the oceans absorb carbon dioxide depends not just on how much carbon they have already absorbed, but also on what state that absorbed carbon is in. Inorganic carbon in the oceans comes in three forms: carbonic acid, bicarbonate ions and carbonate ions. You can imagine them as three compartments in a row, with carbon moving back and forth between them up and down the line; when the system is in equilibrium, the flow in and out of any of the compartments is equal. If you want to get the sea to suck up more carbon dioxide, you have to upset that equilibrium in some of the surface waters, creating conditions in which carbon is happy to keep flowing from the carbonic acid compartment to the bicarbonate compartment to the carbonate compartment.

Chemistry offers a straightforward option. The eagerness of carbon to move from one compartment to the next depends on the water's alkalinity. Increase the alkalinity and carbon moves more readily towards the carbonate compartment; reduce it and it moves the other way. Make some of the surface waters of the ocean more alkaline and they will increase their carbon dioxide consumption. The problem is that this requires a great deal of work. Imagine using lime – calcium oxide – to add alkalinity. For every five molecules of calcium oxide thus added to the seas, you could expect to draw down eight or nine carbon atoms from the atmosphere. Unfortunately, because calcium is a heavier atom, this means that pulling a billion tonnes of carbon out of the atmosphere would require pumping two billion tonnes of lime into the ocean.

There is also the problem that lime is made by heating limestone, a process that both takes up energy and produces carbon dioxide. If the three billion tonnes of lime needed to take up a billion tonnes of carbon were made over a year, the energy requirement would be a continuous 50 gigawatts or so – a power requirement similar to that of the United Kingdom. Unless you have CCS of some sort to get rid of the carbon dioxide produced in the liming process the whole idea is more or less futile.

This does not make liming a non-starter at small scales; there are ways of taking carbon dioxide out of flue gas by bubbling it through alkalinized seawater that seem to me to be under-appreciated alternatives to existing CCS technologies (though because they produce bicarbonates and carbonates rather than carbon dioxide they can't service markets such as that for enhanced oil recovery, which may tend to marginalize them). It's also possible that in some places – such as around a reef you valued very much – you might want to change the alkalinity of the seawater anyway to combat acidification.

There are other approaches to making parts of the surface ocean more alkaline and to creating more carbonates by encouraging the weathering of rocks, some as simple as crushing suitable rocks up into fine powders, sprinkling them over beaches and enjoying the speeded-up chemistry you get by increasing something's surface area. It is not clear how quickly these things could be made to work. It is clear that they would all require large net inputs of energy – crushing rock is hard work – and a capacity to move billions of tonnes of some substance or other around for every billion tonnes of carbon you pulled out of the atmosphere. It's the sort of thing that makes one wish for a lever. And in the 1990s and 2000s, it seemed as if biology might provide one.

Ocean Anaemia

Again, we return to Liebig. When something isn't growing, it is often because it lacks an essential nutrient. For Crookes, look-ing at the British Empire's future wheat supply, the lack was of nitrogen. For the scrawny plants of the ice ages, it was carbon dioxide. For a great many microscopic photosynthetic organisms in the ocean it is iron, which is vital for photosynthesis.

There are large patches of the ocean rich in nitrogen and phosphorus (the element which, after carbon and nitrogen, living tissues need the most of) but comparatively bereft of life – marine biologists sometimes call them deserts. In the 1930s the Norwegian oceanographer Haaken Hasberg Gran suggested that this was because there was not enough iron in those areas for the plankton at the bottom of the food chain. Proving this is hard; measuring very low concentrations of iron in seawater is a challenge when you are sitting in a ship made of steel. But in the 1980s an American oceanographer, John Martin, was able to use new 'ultra-clean' techniques to show that the plankton in the deserts were indeed starved of iron.

The fact that the concentrations of iron in question were so low as to need ultra-cleanliness was, in itself, evidence of quite how powerful the iron fertilization was. A theoretical calculation in the 1990s suggested that by adding one tonne of iron to the surface ocean you could pull 100,000 tonnes of carbon out of the atmosphere. That's leverage on a par with that offered by stratospheric veils, where a few million tonnes of sulphur can offset the warming from 100 billion tonnes of carbon.

Martin saw his discovery as a way of answering the great enduring question of the ice ages. By the 1980s it was known that there was a lot of carbon tucked away in the deep oceans during the ice ages – it wasn't in the atmosphere, it wasn't in those puny terrestrial plants or their impoverished soils, so where else could it be? Martin suggested that iron fertilization put it there. The ice ages were windy, dry and dusty, and mineral dust contains iron. Dust from the plains of South America, which are a lot more extensive in the ice ages thanks to the lower sea level, would blow into the Southern Ocean, which is particularly short of the stuff. That would increase the amount of photosynthesis going on there. In parts of the Southern Ocean, cold, dense surface water sinks into the abyss, and more photosynthesis would mean that this sinking water had more carbon in it. By allowing life to pump carbon out of the atmosphere and into the ocean depths this way, Martin argued, iron fertilization kept the ice ages cold.

Something very similar, he went on, might serve to temper greenhouse warming. Create an iron-rich, ice-age-like Southern Ocean and its plankton would squirrel away a lot more carbon dioxide. It would not stay in the depths indefinitely – but it would be out of the atmosphere for centuries, providing a breathing space during which, like Michael Caine in *The Italian Job*, humankind would undoubtedly think of something. And this could be done with comparative ease. At the 100,000:1 leverage that theory suggested, the billion tonnes of carbon that would require three billion tonnes of lime could be had for 10,000 tonnes of iron.

A lot of Martin's colleagues saw the ice-age suggestion as an exquisite piece of geopoetry, and more than a few were intrigued by the geoengineering, too. A number of groups arranged for seagoing experimental voyages to release iron in the ocean's deserts and see what happened. Various entrepreneurs took note, too. The idea came to prominence in the period leading up to and immediately following the Kyoto protocol, when carbon markets were all the rage. If ocean fertilization could cheaply dispose of carbon from the atmosphere, it could be used to generate carbon credits which could be sold to people putting carbon dioxide into the atmosphere; one could, as one of the proponents put it, 'save the world and get rich on the side' – all while messing around in boats.

The oceanographers did pretty well out of this enthusiasm. The possibility that iron fertilization might have an application in fighting climate change helped to get expensive research cruises funded. The experimental fertilization carried out on those cruises produced blooms clearly visible from space – which was impressive – and lots of data on how marine ecosystems work that could not have been gathered in any other way. This research has gone some way to substantiating Martin's ideas about the ice age: the biological pump was not the only factor in keeping the deep oceans carbon-rich in the ice ages, and probably not the biggest either, but there is now broad agreement that it certainly played a role. Pulling the changed geography of South America, the molecular biology of photosynthesis, the dynamics of ocean circulation, the carbon-dioxide level and the global climate into a single story this way was a wonderfully rich piece of earthsystem explanation, and very satisfying for those involved.

But the findings were not all that encouraging for would-be geoengineers. The experiments found an iron-to-carbon ratio of at best thousands to one, not hundreds of thousands to one. The plankton that benefited most – there is always something that benefits from fertilization most – were not those best suited to getting carbon dioxide into the depths. More carbon was turned back to carbon dioxide by the creatures that grazed on the plankton bloom than had been hoped; less sank into the depths in the form of faeces and undigested fragments. To make things worse, the low-oxygen conditions underneath the bloom (a little like those seen in the dead zones in coastal waters caused by the run-off of fertilizer) encouraged the production of nitrous oxide. Given the high global-warming potential of nitrous oxide, the most moderate emission of it could be enough to scupper the whole idea. The complexity uncovered by the experiments also demonstrated things that should have been obvious from the start: tracking the fate of the bloom's carbon reliably enough to certify the process as a way of producing carbon credits that could be traded on exchanges would be immensely difficult.

If the experiments did little to make geoengineering more plausible, though, they certainly raised its profile. The grandiose claims of one of the ocean-iron proponents, Russ George, an American entrepreneur also involved in trying to sell 'cold fusion' systems, caught the attention of the ETC group, a Canadian environmental organization which had previously been concerned mostly with agricultural issues. ETC denounced George's plans vociferously, thus raising both its own profile and that of the topic; the group has taken a leading role in environmentalist opposition to geoengineering ever since. For those keen on geoengineering research this involvement can be a pain, as ETC's opposition to almost all such work is more or less implacable, and it has been effective in fostering, mobilizing and focusing a widespread distrust of geoengineering among environmentalists. For there not to be strident debate on such a subject, though, would be wrong. And by and large no one gets to pick their adversaries.

Protest and lobbying by ETC and others meant that in the mid-2000s iron fertilization became the first type of climate geoengineering to be discussed in various interlocking intergovernmental forums. The parties to the 1972 London Convention on dumping at sea, which regulates the disposal of waste in international waters, started to take an interest, and decided that iron fertilization came under the convention's purview. The parties concluded that, for the time being, fertilization in the pursuit of scientific knowledge could be allowed, even in the Southern Ocean, much of which enjoys special protection under the Madrid protocol of the Antarctic Treaty. Iron fertilization as a means to non-research ends, such as selling carbon credits, is not allowed at the moment, though the convention has not been amended in such a way as to rule it out for good, and it does not apply to coastal waters.

So the prospects of practical geoengineering using iron fertilization are currently slim. That seems to suit most oceanographers – many who were interested in the original research want no part in any application that it might have, and some have been active in opposing it. Environmentalists who dislike the idea have cause for satisfaction, too, as do those who care about making use of international institutions. Though it is often said that geoengineering raises particularly vexing problems in terms of governance, and this may well in future turn out to be true, in the only case where any sort of climate geoengineering activity has come close to real-world operation, relevant parts of the existing web of international organizations and agreements were pressed into service fairly effectively. The issue was raised; expert advice was sought; the relevant jurisdiction was decided on; decisions were made: what more could one want? Not only was the process good, the result seemed fairly unexceptionable, too. Anyone offended by the scam-like feel of some of the iron-fertilization schemes promulgated by George and his ilk – many of which sought to fertilize surface waters where there was little, if any, possibility of long-term carbon storage in the deep water – couldn't help but feel a certain satisfaction at seeing them come out the losers.

But there are oceanographers who still wonder whether more shouldn't be done. One of the results of the iron-fertilization experiments has been better modelling of how the ecosystem in the oceans actually works when prodded this way; another has been a better understanding of what the models do not yet cover. Natural fertilization by dust seems to be more effective by orders of magnitude than the human interventions have been to date, demonstrating room for improvement. Might not more experimentation, and more modelling, provide the basis upon which a better form of intervention could be designed – one in which the iron was better used, the bloom more likely to export carbon to the deep oceans, the production of nitrous oxide and other unwanted side products truly negligible? And might well-designed fertilization of this sort be worthwhile for governments even if today's dysfunctional carbon markets are not capable of making them profitable for entrepreneurs?

To most of the people who have taken part in these debates to date, I'd guess the answer to both questions will remain no.

In all serious discussions of geoengineering people worry about the side effects on ecosystems and biodiversity. Indeed, in part as a result of effective campaigning by ETC, the UN's Convention on Biological Diversity has some tough-sounding (but hardly binding) language on limits to geoengineering research supposedly enacted to preserve biodiversity. I, and others, find this odd; the damage done to biodiversity by global warming is expected both to be extensive and to be largely driven by changes in temperature, which threaten to shift the habitats of various species at speeds that test or surpass the species' ability to keep up. Shifts in patterns of precipitation and the increased possibility of war, which I think could pose real problems for people in some geoengineering scenarios, matter a lot less to wildlife than temperature does. If your only concern was preserving biodiversity, the choice of an Engineered Planet where temperatures stay stable over a Greenhouse Planet where they climb would seems a pretty obvious one.

That said, people worry, with at least some reason, about the unintended effects on ecosystems of any geoengineering scheme. And with iron-fertilization geoengineering the ecosystem effects aren't even unintended. They are the whole point; changing the composition of the ecosystem from its primary producers on up is how the iron-fertilization geoengineering does its thing. For such a scheme to come close to sequestering a billion tonnes of carbon a year, models suggest every iron-deprived part of the Southern Ocean would have to be targeted. Wholesale and deliberate interference in the web of life will always run the risk of unintended consequences, given the complexities of the food chains involved. When the process is spread out over millions of square kilometres, those unintended consequences could be similarly huge. The risks seem even worse when you take into account the various sorts of damage humans are doing to ocean ecosystems all around the planet – by over-fishing, by creating dead zones, by reducing alkalinity, by increasing surface temperatures. It is hard not to feel that, even if a healthy ocean might be able to deal with the affront of fertilization on a massive scale, today's stressed and sickly waters must surely be insufficiently resilient.

Even if one could somehow feel that the models were reliable enough to deserve confidence, and that increasing ocean productivity would not combine with the other insults being inflicted on the ocean in some cataclysmic synergy, the prospect might still seem a step too far. Deliberately recreating something akin to an ice-age ocean some 20,000 years past the peak of the ice age would mark a huge new step into a cobbled-together Anthropocene. Humans have already given the earthsystem an atmosphere that looks like that of twenty million years ago — importing the carbon needed for the job in large part from the Carboniferous. They have created an everything-everywhere distribution of species that brings to mind Pangaea. Add on a carbon-hungry ocean from the ice ages and the earthsystem looks ever more like a Frankenstein planet stitched together by geological resurrection men.

And yet: when I mentioned the possibility of reviving the green Sahara of the early Holocene — of streams and savannah where now there is barren sand, of animals grazing where today they would die, of rock paintings that might once again reflect the reality of life — was your response one of straightforward disgust? Did you not at least entertain the thought that more life, restored life, could be a boon to the desert sands? If you did, are you sure that reviving desert waters is necessarily a sin? There are undoubtedly ways that it could go wrong. But there are ways that it could go right, too. If there were a team of engineers genuinely looking for ways in which it might be a boon, one that was willing to give up if no good option were discernible, one that accepted that the decision as to what a 'good' option might be was not its own to make, and which accepted the supervision of a transparent and equitable international governance — if there were such a team, should it not be allowed to look at the question? You may think that there can be no such scheme, that such attempts will always be perverted by some sort of technological inertia, by scientific hubris, by commercial interests or power politics. That may be so. If it is, though, you have to accept that the same is true of more or less any big international attempt to deal with the climate, not just those that employ geoengineering.

Cultivating One's Garden

The oceans are not the only place where life fixes carbon. Most of the world's biomass, which includes the plants that are mopping up about a quarter of humankind's current emissions, is on land. How much more could be added to their stash?

This was the question behind a 1977 paper by Freeman Dyson, a mathematical physicist famously broad ranging in his thinking, called simply 'Can We Control the Carbon Dioxide in the Atmosphere?' In part as a response to Marchetti's ideas about CCS, Dyson made some back-of-the-envelope calculations on how much carbon dioxide could be absorbed by deliberately growing more plants. Aiming to find a way to store five billion tonnes of carbon a year (which he correctly estimated would be the size of the airborne fraction around the turn of the century), he found that sycamores – a fast-growing and well-studied species – grown on all the land in the United States that was suitable for their growth and not already devoted to growing crops or covered in forests would be good for about a tenth of that. Similar projects in other countries might be able to provide the other 90 per cent.

Dyson's idea of what a really ambitious global effort might achieve is, as it happens, similar in scope to what humankind has done to the planet's forests in the past few centuries – except done faster and in reverse. The total amount of carbon put into the atmosphere in the past few centuries through land clearances and the like is about 250 billion tonnes – the amount that would be stored up by Dyson's five-billion-tonnes-a-year forest over the fifty years he imagined that it might be necessary to suck up carbon. Only fifty years were needed because Dyson's scenario was a breathing-space one; he imagined that half a century would be time enough for people to replace fossil fuels with some mix of biofuels, solar cells, geothermal energy and nuclear power.

The problem with such a scheme is that it would need a hard-to-come-by amount of land. The 700,000 square kilometres of new forest he imagined in the United States would be an area equivalent to that of Texas. The seven million square kilometres needed for the whole world is an area twice the size of India and equivalent to half of the planet's arable land. By comparison, the world's largest current afforestation effort, in China, aims to cover 50,000 square kilometres a year. Overall, despite progress on halting deforestation in recent years, the Earth's forests are still shrinking. Simply stopping deliberate deforestation is a challenging enough target.

And there are a couple of wrinkles that make such efforts less effective than you might at first imagine. One is an albedo effect. Forests reflect less sunlight back to space than cropland and pasture. If you plant forests in places that have snowy winters, the warming produced by this albedo effect may well outstrip the cooling offered by the carbon-dioxide reduction that comes with the forests' growth. This is not in itself a reason not to plant forests – the global benefits of lowering the carbon-dioxide level could well be worth the winter warming at high latitudes, which might itself be a regional benefit. But it is a complicating factor.

Another complication is the counterbalancing effect of the ocean. If you push carbon dioxide into the atmosphere the seas suck some of it up; if you pump carbon dioxide out of the atmosphere the seas give some up, reducing the effectiveness of your pumping. This means that to get a net reduction of a billion tonnes of carbon in the atmosphere, you need to pull out well over a billion tonnes. All told, if you grew enough forest to suck up the amount of carbon given off by the past two centuries of deforestation, according to work by Jo House of the University of Bristol and her colleagues, you might reduce the carbon-dioxide level by as little as 40ppm – that is, by two decades of emissions at the current rate of 2ppm a year. You might do almost twice that well – the models are not precise – but either way, it's pretty small beer. Replanting the forests of the nineteenth and twentieth centuries can't do all that much while the ghost acres of the Carboniferous burn on.

If huge afforestation projects would have only a relatively small effect, the same will be true for other schemes people have suggested for increasing the amount of carbon in the biosphere. You can bury unwanted biomass – for example, the residues left in fields after harvest – in places where oxygen is scarce and respiration thus suppressed, such as the sediments of the Gulf of Mexico; you can create wetlands (though watch out for the methane); you can use much more wood, and much less steel and cement and glass, in your buildings; you can farm and manage pastures in ways which increase the amount of organic carbon stored in the soil.

Some of these ideas might have great benefits beyond what they could do for the climate, and some of them are much closer to the sort of action that most environmentalists value than anything else which can even remotely plausibly be called geoengineering. Environmental action is by preference small and local, close to the soil and the practices of daily life – action which pulls back from some of the contradictions of the present to build up a different future. Climate geoengineering offers few such possibilities; it is for the most part a 'think global, act global' deal. For many greens, the difference speaks not just to a matter of style or approach, but to a question of political credibility and acceptability. They see geoengineering as requiring top-down solutions which can only be imposed by the systems of power already in place, rather than by an alternative built up from personal practice, local innovation and dissent. They do not believe that such worldly systems can be changed other than from the bottom up, by the spread of beliefs which start off as marginal and end up as universal. And they have a point. The idea of creating both a new political world and a changed earthsystem – a hand and a thermostat, a fulcrum and a lever – in a single process, the needs of each driving changes in the other, is largely unprecedented. It certainly requires action far beyond the grass-roots. There is no guarantee that such things can be done.

But that leaves the problem of what things can be done that meet the challenge. In this specific instance, there are undoubtedly ways to encourage the storage of carbon in the biosphere through soil management, agronomy and forestry. But such actions do not store carbon on the scale needed to put a serious dent in the fossil-fuel-driven trajectory of atmospheric carbon dioxide, because the reservoirs into which they put the carbon are quite constrained. There is only so much woodland you can plant, only so much soil you can enrich, only so much farming you can do better. The biosphere is not that big.

This constraint has led some – notably a New Zealand economist called Peter Read – to imagine separating the photosynthetic removal of carbon dioxide at which the biosphere excels from the long-term storage it does less well. The fact that plants use carbon dioxide to make more leaves and stems works well for them, and for the animals that eat them, but it is not the greatest way to store the carbon away. Why not make photosynthesis a route into some other storage system – such as the aquifers, depleted hydrocarbon reservoirs and other geological systems that are favoured by people who work on CCS? Such stores should have a much higher capacity than soils and standing wood can offer. To create a pipe that takes carbon more or less straight from the atmosphere to the lithosphere, Reed argued, all you needed to do was feed biomass into a power plant fitted with a CCS system.

Biomass energy with CCS – BECCS to its devotees – has real attractions. The things that are so hard for direct-air capture – filtering the air, recycling the substances that absorb carbon so they are ready to do so again, and so on – are as natural to plants as growing leaves is, because that's how they do it. And whereas direct-air capture uses energy, BECCS produces it. That means that as well as taking carbon out of the atmosphere, BECCS also reduces the need for power plants which might otherwise be putting carbon back into it.

These notional charms, though, come with practical and political drawbacks. For one thing, as we have seen, there is not yet any large-scale CCS technology being used. For another, burning biomass has an increasingly bad reputation. Biomass and biofuel programmes have often used unsustainable sources, and the harvesting and transport of the fuel can generate a lot of carbon dioxide. Even if everything is done responsibly, biomass systems run into the problem that, for all its excellent cheapness, photosynthesis is pretty inefficient. Plant growth typically turns sunlight into fuel at a rate of less than one watt per square metre in temperate places, a bit more in the tropics. Power stations fitted with CCS are unlikely to burn that fuel at an efficiency of more than 30 per cent. This means that a one-gigawatt BECCS facility in New England would need an area of well-tended woodland a fifth the area of Massachusetts to keep it fuelled. That's a lot of land not to be using for other things. And Massachusetts uses more than five gigawatts of electricity; use its whole surface for BECCS and it still wouldn't be self-sufficient.

Yet despite these drawbacks, there is a lot of interest in BECCS. As mentioned in Chapter Six, for a good chance – two-to-one, say – of keeping warming below 2°C, the world may well need to do more, at this stage, than simply cut emissions to zero; it may well need to go into 'negative emissions'. And BECCS, relying as it does on something proven – the burning of biomass – and something more or less available – CCS – is the only technology for large-scale negative emissions that currently looks even remotely plausible.* It was because its authors felt that they had to talk about negative emissions that the most recent IPCC report gave geoengineering a prominence in its conclusions that greatly perturbed critics such as the ETC group. Like many people and institutions that care about sustainable agriculture, food security and the rights of indigenous people, ETC hates industrial-scale biomass energy even when it has no geoengineering attached.

That negative emissions are likely to be necessary if the chances of two degrees of warming are to be kept low surprises many people. There is a widespread belief that simply stabilizing emissions – emitting the same amount every year – solves the warming problem. But if the world stabilizes its emissions, it is still adding to the stock of carbon in the atmosphere, and it's the stock that sets the temperature. The goal of cutting emissions by half, which sounds dramatic, is insufficient for the same reason: the world would still be adding to the stock, year by year; it would just be doing it a bit slower. (It is tempting to think that cutting emissions by half would be enough, because only half the emissions stay in the atmosphere. If people cut back, and the plants and oceans kept their giant sucking sound going, surely things would be OK? But to do so ignores the fact that the carbon uptake by the oceans and land depends on the rate of change of carbon in the atmosphere. If people cut back on emissions, the ocean will cut back on its absorption.)

If you are not going to indulge in some form of sunshine geo-engineering, the only way to stop the warming is to bring emissions to zero. What would happen then? There would be a little

---

* There is another way of taking energy out of biomass that leaves you with something storable: burn the wood into a form of charcoal that can then be used as a soil additive. This 'biochar' approach may work well in some places as a way of producing energy, storing carbon and improving the soil, but like other forms of soil management and local enhancement of the biosphere, it is very hard to see it being used for hundreds of millions of tonnes of carbon a year, let alone for billions.

helpful sucking up in the decades that followed as the biosphere and the oceans settled into a new equilibrium. But once that equilibrium was reached the carbon-dioxide level would, in the absence of negative emissions, sit more or less where it was for a long, long time. It would only be brought down by the slow processes of mineral weathering and carbonate deposition in the oceans – the processes that regulate the Earth's carbon-dioxide levels over the geological long term. Models of the carbon cycle suggest that such a decline would take thousands of years – quite possibly tens of thousands.

Scientific papers dealing with these changes to the carbon cycle often use the word 'irreversible'. And if you rely on natural processes to get the carbon out of the atmosphere but think on human timescales, then they sort of are. The lesson of this chapter, though, is that if you don't feel constrained to stick with natural processes, they aren't.

There is no cheap direct-air capture around today. Iron fertilization of the oceans has serious problems; adding alkalinity would require a huge effort. The BECCS capacity which could be fielded in a responsible, sustainable manner might not be huge. And – what should always, really, be the first fact in such a list – emissions are going up, not down. Though they will start falling at some point, there is no political or economic process in place to force that fall to start soon, or to ensure that it continues all the way down to zero. And none of this looks likely to change in the short term.

But there is a lot of room for manoeuvre between today's political, economic and technological short term and the millennia it takes for the inorganic carbon cycle to get things done. Does the status quo on direct-air capture and other forms of negative-emission technology look likely to persist through to the end of this century? And the end of the next? And the one after that?

Once, perhaps it did. Before the seventeenth century the technical and organizational basis of human life changed pretty slowly. It may be that the huge growth in wealth, population and energy use that has marked the subsequent era of fossil fuels, capitalism and their discontents, since then cannot go on, and that civilization will at some point step off the whirligig of progress and return to something more staid, even stagnant. It may also be that civilization collapses – or tears itself apart – so thoroughly as to never rebound. If either were to be the case, then high carbon-dioxide levels really would be here to stay. But if neither of those things happens, then the option of changing the carbon-dioxide level deliberately seems almost sure to arise. It may never be feasible or desirable to engineer fast changes. But just drawing the level down by a few parts per million every decade would restore things to the way they once were in a matter of centuries, were people to choose to do so.

They might not. To have a technology is not to have to use it. In *The Fountains of Paradise*, a 1979 novel by Arthur C. Clarke, the current interglacial draws to a close in around a thousand years. From the cabin ecologies of their orbital cities the humans of that future age watch the strange new beauty of the glacial world develop. They could stop it. They have the technology to warm the planet. But they prefer to allow the earthsystem to retain its own rhythms than to impose a lasting summer of their own design.

When my father bought me that book in 1980 the idea of technology that could stall an ice age seemed remarkable. It still seems remarkable; but it does not seem a remotely unlikely technology a millennium from now. It does not seem outlandish to think that well before then humans will be capable of pulling greenhouse gases out of the atmosphere or putting them back, or of thinning or indeed thickening the flow of sunshine into the earthsystem. It does not seem outlandish to think that they might choose to use such powers. But nor does it seem outlandish to think that, as Clarke suggested, they might choose not to. What can be said of such powers for sure?

What does feel outlandish is that humans more or less have such power already – if only in rough and ready form. In 3,000 years the Earth's orbital cycles will hit the sort of alignment that would seem likely to trigger an ice age. David Archer of the University of Chicago has calculated that, if humans go on emitting at something like the current rate for just a few more decades and do not subsequently choose to get into the carbon-removal business, there will still be enough industrial carbon dioxide left in the atmosphere three millennia hence to stop that ice age before it starts. Guy Callendar would appear to have been right when he wrote in 1938 that industrial carbon dioxide might postpone the return of the glaciers indefinitely.

The Holocene was an interglacial. The Anthropocene could be post-glacial. Or it could not be. It is the fact that there are two possibilities which makes this long-term speculation so rich, and so unnerving. That humans act on a scale previously reserved for forces of nature is a remarkable enough development. That the ends of such action might come to be a matter of some sort of choice is more extraordinary still.

10

Sulphur and Soggy Mirrors

*If you have built castles in the air, your work need
not be lost; that is where they should be. Now put
the foundations under them.*

Henry David Thoreau, *Walden* (1854)

In the early 1970s eight-year-old Mike Latham looked out at
the Irish Sea from a hilltop in Wales and asked his father John
why the sunset looked so wonderful. Latham senior explained
that the clouds were made up of countless tiny drops of water,
and that the surfaces of these drops reflected the sunlight in
various ways, playing with its colours. Mike got the gist: 'They're
soggy mirrors!'

His father could have gone into a lot more detail. John
Latham claims to have taken the university course that got him
interested in clouds in the late 1950s under the misapprehension
that meteorology was about meteors, but he found it fascinating
and stuck with it; by the 1970s he was an established expert
on cloud physics. Forty years later, he is trying to apply that
expertise to a form of geoengineering.

He is not the first person to think the study of clouds should
be an applied science. In 1946 researchers at the General Electric
research laboratory in Schenectady, New York discovered that
both dry ice (frozen carbon dioxide) and silver iodide could
make tiny water droplets freeze, something which the droplets

268

were unwilling to do on their own even at temperatures well below 0°C. Because ice crystals tend to grow much faster in clouds than water droplets do, they reasoned that these new tricks might have dramatic effects outside the lab, and this proved true. When one of the researchers, Vincent Schaefer, dropped three kilograms of dry ice out of a plane in 1946, he said it seemed as though the cloud into which the dry ice fell 'almost exploded' as its droplets turned to ice en masse.

The Schenectady lab was run by Irving Langmuir, who knew an opportunity when he saw one. He had no background in cloud research, but he had connections, ambition and the stature conferred by a Nobel Prize. And now he had a newly discovered lever with which to move the earthsystem: the work by Schaefer and his colleague Bernie Vonnegut seemed to show that you could get tonnes of rain with just grams of dry ice or silver iodide. Langmuir was on the phone to the *New York Times* before Schaefer's first cloud-seeding flight had landed.

The idea of using science to bring rain was hardly a new one; it has often enjoyed a particular place in America's imagination. In the 1840s James Espy became a national figure – 'the Storm King' – by applying his state-of-the-art ideas about the atmosphere as a massive heat engine to the problem of precipitation and attempting to bring rain by setting fires in woodlands. In a history of these ideas, Horace Byers wrote of how, in the late nineteenth century, 'a rash of rainmakers . . . armed with crude pyrotechnics and a convincing sales pitch, became a conspicuous part of the American rural scene'. Langmuir's sales pitch was that he would be the man to finally make such dreams come true. He set up a number of field tests and talked grandly to the press, to his peers, to GE's shareholders and to Congress about the possibilities offered by the new techniques of cloud seeding. Hurricanes would be re-routed, dust bowls damped down, enemy states subdued; it was a power to rival or surpass that of the atomic bomb, the new yardstick for both promise and catastrophe.

He did not deliver on these claims. Seeding could not be made to produce rain with anything approaching reliability. There is a particularity to clouds. Because you never get to work with the same one twice it is impossible to know what, left to itself, a cloud would have done – which makes controlled experiments hard – or what the next one will do if seeded. Seeding could, however, be relied on to generate litigation: any and all subsequent weather-related damage in the broad vicinity of an experiment offered such opportunities. In this it fell squarely into Simon Schaffer's category of 'Promethean science', with its distinguishing mix of vaulting ambition and troubling hazards.

To the great relief of GE's lawyers, responsibility for the hazards of Langmuir's research was soon in the hands of the government, which sought to develop the work done by the GE team in a programme called Project Cirrus. Military interest continued for decades, and cloud-seeding techniques were covertly deployed in attempts to render Viet Cong supply routes in Laos too muddy to use during the Vietnam War. When these actions were revealed they provoked an international outcry; along with the defoliation campaigns conducted at the same time, they led to the UN Convention on the Prohibition of Military or Any Other Hostile Use of Environmental Modification Techniques, known as ENMOD, which outlawed all such activities when it came into force in 1978.

But the military did not dominate cloud seeding and weather modification in the United States; most of the research spending in the 1960s and 1970s was not by the Pentagon but by the Department of Commerce and the Bureau of Reclamation, a part of the Department of the Interior concerned with water resources. The focus was on moistening the western states, and in particular on increasing snowfall in the winter so as to bolster the region's rivers in summer. Though no longer federally funded, dozens of cloud-seeding programmes have continued in the western states to this day. In 2014 a long-term research effort in Wyoming reported that seeding increased snowfall over a season by 5–15 per cent (in the nearest thing to a controlled trial you can get, the researchers looked at data from two mountain ranges, without knowing which one had been seeded at any given time). The California Department of Water Resources has estimated that cloud-seeding efforts over the Sierra Nevada, which are mostly undertaken by companies with hydroelectric plants in the area, increase precipitation over the state by about 4 per cent.

America is not alone. According to the World Meteorological Organization, in 2013 there were 42 countries using cloud seeding for hail suppression, precipitation enhancement or both; some of these programmes were more than fifty years old.* India and Thailand are making significant investments in the technology, and there is a history of both research and application in Australia and South Africa, but by far the largest effort is China's. While many people know that the Chinese used cloud-busting technologies to try and keep the weather fair in Beijing during the 2008 Olympics, few appreciate the scope of the country's efforts. There is an active weather-modification programme in every province, and thousands of batteries of cloud-piercing missiles. The scope of this effort, though, should not be taken as evidence for effectiveness; like nineteenth-century American hucksters, Chinese government programmes are quite capable of keeping themselves going without a proven track record of success.

As well as setting off interest in the applied science of cloud seeding, the work by Langmuir and his team also led to an increase in general scientific interest in clouds, in part by raising interesting questions, in part by providing a context in which such studies

---

* Bruno Latour, a French social scientist of huge importance to those concerned with the coming into being and passing away of the boundaries between society and nature, tells me that his father used regularly to commission hail-busting services to protect the family's vineyards.

might some day have application. Schaefer and Vonnegut were leading lights in the field. When John Latham started to make research trips to the west of America in the late 1960s, he met and befriended them both: he wrote a number of papers with Vonnegut; he enjoyed winter expeditions to Yellowstone with Schaefer. He also got to know Bernie's kid brother, Kurt, who had been Langmuir's public-relations man and who drew on his experience of GE and his brother's work in his fiction. 'Ice-nine', the McGuffin in *Cat's Cradle*, Vonnegut's black comedy of the apocalypse, brings cloud seeding and its military applications down to the planet's surface. A catalytic speck of Ice-nine will cause water to freeze even at room temperature in much the same way that a speck of silver iodide catalyzes the freezing of liquid water in clouds, though to more world-ending effect.*

Latham's own interests, though, were not in precipitation or weather modification. He was a lightning man. The idea that something as insubstantial as a cloud – a set of the tiniest particles, sparsely arrayed through otherwise empty space – could produce bolts of celestial fire provided him with a source of sublime wonder. He flew through thunderclouds to try and understand their workings, in search not so much of specific data as of a general sense of the processes going on. (The general sense he got was that the processes were scary; he made such flights only rarely.)

It was not until much later that Latham started to think about cloud seeding himself, and to take the idea in a whole new direction. In the 1980s he took early retirement from UMIST, the university in Manchester at which he had spent all his career: he was afraid that he would again be made Head of Department, a responsibility he had hated when it was first pressed upon him,

---

* The writer said that he got the idea of room-temperature ice from Langmuir who, when H. G. Wells had visited the laboratory, had tried to convince him that room-temperature ice might be a nifty idea for a story. Wells did nothing with the idea, so when the time came, Vonnegut seized on it.

and he was eager to devote more time to his a second career as a poet. The independent life, though, was more challenging than he had anticipated, and by 1990 he was looking for a new position. Britain was setting up a new climate-change-research outfit in its Met Office, the Hadley Centre. Thinking it might offer opportunities for him, Latham read some papers by one of the researchers who would be leading it, Tony Slingo.

One of those papers talked about the effect that different concentrations of cloud-condensation nuclei – the specks on which water droplets form – have on cloud brightness. It put Latham in mind of staring out at the sunset sea with Mike twenty years before and appreciating together the glories of the soggy mirrors. And thinking about it, he realized that brightening the mirrors might give you a lever with which to actually do something about the climate change that the Hadley Centre was being set up to study.

The insight that a change in the supply of cloud-condensation nuclei could increase the amount of sunlight that clouds reflected dates back to work done by an Irish meteorologist, Sean Twomey, in the 1970s. Other things being equal, Twomey argued, more nuclei would make a brighter cloud, because they would mean there would be a greater number of smaller droplets – more surface area for a given volume of water, which means more scattering of light by reflection. The smaller nuclei might also tend to make clouds longer-lived. The smaller the droplets in a cloud start out, the longer it takes for them to grow into drops heavy enough to fall as rain.

Clouds over the sea tend to have much lower numbers of nuclei than those over land. Latham reckoned that you might rectify that with some sort of system for making very little droplets of water near the sea surface; the water would evaporate, leaving behind tiny salt crystals which, when sucked up into the clouds above by convection, would become condensation nuclei. These would brighten the clouds through the Twomey effect.

Back-of-the-envelope calculations suggested that an idealized system capable of producing condensation nuclei from seawater droplets would need to process just 30 tonnes of seawater a second, worldwide, to offset the warming caused by a doubling of carbon dioxide.

Thirty tonnes a second is the rate at which water flows over just a single metre of the lip of Horseshoe Falls. But if Latham was right, such a trickle could make a small but measurable difference to the power of the Worldfalls as a whole.

After the geological perspectives of the carbon cycle, Latham's ideas bring us back to today's debates around sunshine geoengineering. But though planetwide possibilities still play a role in this chapter, as do the earthsystem's biogeochemical cycles, so does something we have not considered much so far – the possibility that climate geoengineering techniques might not be global at all, and that far from being a radical alternative to what is going on today, they might instead be a continuation of some current trends.

Global Cooling

Latham was the first person to think about deliberately cooling the Earth with clouds. The idea that such cooling was going on by accident, though, was by that stage old hat.

Unlike the side effects of industrial nitrogen fixation or massive carbon-dioxide emissions, there has never been anything subtle or see-you-next-decade about the effects of the sulphur given off by burning coal. It was the sharpest-stinging part of the air pollution that increasingly blighted city life in the first part of the twentieth century. The soot of the Victorian era had been at least somewhat abated by better furnaces, turbines and the like: if you burn coal more efficiently, then by and large you will produce less soot. A greater use of oil had helped too. But however well you burn either coal or oil, if it starts out with some sulphur in it – and it mostly does – you will burn that sulphur, too. That will end up producing acidic aerosols very much like those that volcanoes or veilmakers might create in the stratosphere, but in much greater profusion and much closer to people whom they can hurt.

The natural sources of sulphur in the atmosphere are volcanoes and plankton. The role of the plankton, which is larger than that of the volcanoes though rather less flashy, was for a long time unappreciated. It only really came to light when James Lovelock of Gaia fame measured the emissions from plankton in the open ocean when on board *RRS Shackleton* in 1971. His discovery that living things in the ocean pumped sulphur into the atmosphere played a part in his growing conviction that life changes the environment to its own benefit. If it was not for the plankton, the sulphur that dissolves into river water and washes out to sea would stay there, and the plants and animals on the continents would run short of the stuff. By recycling sulphur, plankton in the oceans make the continents more habitable.

By the time that Lovelock had elucidated this crucial part of the natural sulphur cycle in the 1970s, though, humankind had easily outstripped it. In terms of the proportion of the flow provided by industry, the human role in the sulphur cycle was already then, and still remains, greater than it is in the nitrogen cycle, and far greater than it is in the carbon cycle, though in absolute tonnage the numbers are smaller. This is all the more remarkable when you consider that humans aren't trying to move sulphur around through the atmosphere at all – it is an utterly unwanted side effect of other activities.

In the 1960s scientists began worrying that sulphates and other anthropogenic aerosols might not just be harming human health, they might be cooling the planet, too. Reid Bryson, a climatologist from Wyoming who was one of the people already convinced that there was a link between volcanic eruption and cooler climates, started to talk about a 'human volcano' of industrial aerosols and dust released by poor farming practices that could have similar cooling effects. The idea caught on, at least in some environmental circles. In 1970, on the first Earth Day, Stephen Schneider, then a young graduate student just getting interested in climate modelling, heard Barry Commoner talking about the future of the climate. Commoner told the audience about Roger Revelle and the greenhouse effect; he also talked about Bryson and the human volcano. Which of the two human influences on the climate would win out, he said, was up in the air.

Schneider, already eager to tackle big questions of real human importance, thought the ice-or-fire question sounded like the sort of thing one might want to sort out. He and his mentor in climate science, S. Ichtiaque Rasool of NASA's Goddard Institute for Space Studies, borrowed some code that was good for modelling aerosol effects from a young colleague, James Hansen, and set to work.[*] Rather than try to provide absolute values for the amount of greenhouse warming and aerosol cooling, they looked at how much the two effects would increase as concentrations of the relevant pollutants built up. A then-recent assessment had suggested that aerosol pollution could increase by as much as eightfold over the following fifty years. And even a fourfold increase in its effects, Schneider calculated, would produce a cooling of 3.5°C – much greater than the warming their model predicted as a result of carbon-dioxide increases over the same period. Their paper concluded with the observation that simple models of the planet's energy balance, one of which had recently been published by Mikhail Budyko, suggested that sustaining such a cooling for a few years might trigger an ice age by encouraging longer, snowier winters, creating an ice–albedo feedback that would tip the planet into full-on glaciation.

Their results became public during a conference on the 'Study of Man's Impact on the Climate' held in Stockholm in 1971 – a

---

[*] Hansen, as we saw in Chapter Three, went on to apply his code to the question of stratospheric aerosols from non-human-volcanoes.

warm-up event, as it were, for the first great UN environmental summit, which was to be held in the same city the following year. Other researchers who had reached similar conclusions were also at the conference, as were Budyko and Twomey, whose ideas about the effects of aerosols on clouds were just taking shape, and who thought the cooling effects predicted by Schneider, Rasool and the other researchers severely underestimated the power of aerosols.

Since the 1990s some of those who want to diminish the standing of climate science and cast doubt on its broadly agreed and well-supported conclusions have promulgated with glee the claim that climate scientists in the 1970s were as concerned about global cooling and incipient ice ages as climate scientists today are about global warming. Defenders of climate science reply hotly that this isn't true, and they are right. The worries voiced about global cooling in the early 1970s were fewer, far more tentative and never treated to anything like a community-wide IPCC-style assessment. Even at that time, warnings about warming were as common or more so.

That said, though it may in large part have been little more than idle speculation, there was some talk about ice ages in the 1960s and the 1970s – the Stockholm meeting and the Schneider and Rasool paper, which Schneider later realized had under-estimated the strength of the greenhouse effect considerably, probably marked its high point. That this would have gone on is hardly surprising. After all, ice ages were the biggest problem in climate science; no one knew quite how they came about; it had been a while since the last one: that was all good fodder for geopoetic speculation. What was more, the idea that humans could have damaging environmental impacts on a global scale had, through the work of Commoner, Rachel Carson and others, come into its own, which raised the possibility of a self-inflicted ice age; what could deliver a more sublime shiver? When the possibility that aviation might thin the ozone layer was raised in the 1960s, a Canadian scientist, John Hampson, suggested that it could bring forward the next ice age by flipping the stratosphere from one way of working to another. Another ice-age theory of the 1960s suggested, with seeming perversity, that an ice age might be triggered by surface warming of the sort a strengthened greenhouse effect could lead to; if the Arctic got warm enough for open water, evaporation and therefore also precipitation would increase in the region, and thus so would snow cover; cue the ice–albedo feedback.*

If any more encouragement for such speculation were needed, over the 1950s and 1960s there was in fact a small dip in global temperatures. It was nothing like severe enough to cause an ice age. But it is now widely believed by climate scientists that a significant part of it was due to Bryson's human volcano. Anthropogenic aerosols were contributing enough cooling to counteract some of the effects of greenhouse warming, though the natural variability of the climate system undoubtedly had a role to play, too.

From the mid-1970s on, the global temperature began to increase consistently. The cooling ended in part because the greenhouse effect, depending as it does on the stock of carbon

---

* In the spirit of the times, just as there were fears about ice ages, there was also speculation about how geoengineering might avert them. Lovelock was one such speculator. When in 1966 he was commissioned by Shell, the oil company, to write a short report on the world as it might be in the year 2000, he drew to his conclusion by talking about the 'good chance of an unpleasant surprise, such as a brush with an ice age, during the next decade or so'. He went on to suggest that in such an eventuality there might be call for dark tars to be spread over the deserts to reduce the planet's albedo, or perhaps for specially formulated greenhouse gases to be released into the atmosphere. In a later speculation along similar lines, the astrophysicist Fred Hoyle suggested pumping warm water from the surface of the ocean into its depths to provide a store of heat with which the planet could ride out an ice age that would be caused, he thought, by naturally occurring stratospheric aerosols. And as we saw in Chapter Six, in their 1997 paper on geoengineering with stratospheric aerosols and other scatterers, Edward Teller and Lowell Wood included stopping ice ages as one of the benefits on offer.

dioxide, was getting stronger. But the aerosol effects diminished, too. From the 1960s onwards, aerosol emissions from fossil-fuel plants in the countries that were at that time industrialized – the main culprits – were reduced quite dramatically, first because of direct concerns over human health, and then as a result of worries about acid rain. The use of high-sulphur fuels was restricted; the sulphur dioxide in flue gases was scrubbed out of chimneys.

How much aerosols have masked the effects of global warming in the past, and how much they continue to do so in the present, is a question of great uncertainty. Estimating the effects of anthropogenic aerosols is a harder project than quantifying anthropogenic greenhouse warming. Because they are almost entirely confined to the churning troposphere the effects of aerosols are patchy, which makes them hard to model. Their heterogeneous nature makes things harder still. There are shiny sulphates, heat-sucking soot, organic compounds with all sorts of properties – and they all come in a wide variety of sizes. And then there are their effects on clouds. As Twomey argued, some aerosols amplify the cooling effects of shiny clouds. Other aerosols, though, can destroy clouds; aerosols which absorb sunshine and radiate in the infrared, like soot, can make the air that holds them too warm for cloud droplets to form. The complex physics of clouds means that capturing their behaviour would be a big problem for climate modellers under any circumstances. The presence of all sorts of aerosols makes matters worse.

The latest IPCC assessment, taking all this into account, reports that the net effects of anthropogenic aerosols, including the warming caused by sooty ones, the cooling caused by shiny ones and the various interactions with clouds, add up to about one watt per square metre of cooling – roughly half the forcing due to carbon dioxide. The error bars on that number are similar in size to the number itself, meaning that the possible net effect could be as little as 0.1 watts per square metre or as much as 1.9 watts per square metre. This means that it is possible that more than half of the warming due to carbon dioxide is currently being masked by cooling due to aerosols, though it is also possible that only a little is.

What is fairly certain is that the aerosol cooling will weaken yet further as time goes by. Take China. Its economic boom, beginning in the mid-1990s, was accompanied by a ghastly worsening in air quality. The copious aerosols created by Chinese pollution in the 2000s may indeed have played some role in the slowdown in warming seen in that decade, rather as aerosols from Europe and America seem to have masked the effect of carbon dioxide in the 1950s and 1960s. But this effect is set to diminish as China cleans up its act. It is both moving away from coal, to some extent, by building more renewable and nuclear generating capacity, and also cleaning up the emissions from the coal-fired plants which remain the mainstay of its electricity generation. Similar stories will be told elsewhere. The world will go on building fossil-fuel capacity for decades to come, but that capacity is likely to get cleaner because increasingly affluent citizens will demand a healthier environment and the technology needed to keep sulphur emissions down will get cheaper. More warming is thus set to be unmasked – perhaps a lot, perhaps a little.

What, though, if China and the rest of the world could banish the damage aerosols do but keep their cooling effects? That was one of the questions that Paul Crutzen used to shape the 2006 paper with which he rekindled and partially legitimized interest in geoengineering. The paper suggests a stratospheric veil not as a source of new cooling, but as a replacement for the cooling that tropospheric aerosols due to fossil-fuel burning already provide.

The cleverness of Crutzen's suggestion rests on the fact that stratospheric aerosols stick around for years while tropospheric ones are sluiced out of the air by rain within a week or so; this means that a tonne of sulphur in the atmosphere gets you 25 to 50 times more cooling than a tonne in the troposphere simply by staying around longer. So if you put just 4 per cent of today's total sulphur emissions into the stratosphere rather than the troposphere, you could get rid of all the other 96 per cent, along with all the damage that they do to human health, while preserving the cooling they had until then provided.*

It was a typically cunning bit of rhetoric. Rather than making climate geoengineering a radical novelty, it made it a sanitized version of Anthropocene business-as-usual. The cooling emissions were going on already; the proposed geoengineering was a way to shape them so that their beneficial effects were maintained even as the harm they did was eliminated. No new level of interference in the earthsystem was needed – merely a new level of thoughtfulness.

The elegance of the idea does not, however, make it a practical piece of politics. Cutting tropospheric aerosols is a pressing problem; stratospheric veilmaking is still just this side of geopoetry, and hugely divisive to boot. Countries are willing to endanger their citizens through poorly regulated air pollution because of the economic importance they put on generating energy for industry and having cars and vans to move people around. They may well also endanger them because of the political muscle of

---

* Actually, not quite all the damage. The stratospheric aerosols do eventually fall back to the surface, and so some will be breathed in. But not only would stratospheric aerosols be emitted at a far lower rate than the tropospheric aerosols they 'replaced', on a tonne-for-tonne basis their health effects would be much less grave. Tropospheric sulphates are emitted near where people live; stratospheric aerosols returning to the surface would be much more evenly spread (though, like fallout in the 1950s, they would favour the jet-stream zones). Also, as they fall down through the whole of the troposphere, the stratospheric aerosols are more likely to be washed out of the sky by rain than those which come up from the surface. Compared with the fatalities due to current sulphur aerosols, which are seen as numbering well over a million, the lives shortened by the direct effects of the stratospheric aerosols that would be needed to make a moderately thick veil would probably be in the fairly low thousands. The net effects would be more complicated, as the harm done by some tropospheric pollution depends on temperature and climate geoengineering seeks to change that temperature; this is an area where a great deal more work needs to be done.

the people who do the generating and who make use of the cars. But they will not leave them suffering as they try to develop and agree on some pie-in-the-sky alternative, the benefits of which are speculative and global. Sulphur emissions will be cut regardless of whether their cooling can be replaced.

The most striking example of this, I think, comes not from one country but from 171 of them acting in concert as members of the International Maritime Organization (IMO), the UN agency that deals with shipping. One of its recent initiatives has been to clean up the emissions that come from those ships, setting progressively stricter standards for the amount of soot and sulphur they can emit when out in the heart of the oceans. Taking on this task had given the IMO regulatory influence over the albedo of more than half of the Earth's surface – and thus a far greater ability to play with the parameters of the climate than that currently enjoyed by the UNFCCC.

The effects of sulphur emissions at sea can be particularly strong because of the Twomey effect. Over land, where everything from dust to dandruff is constantly being blown into the air, clouds are rarely constrained by a lack of condensation nuclei. Over the seas, where the air is cleaner, they are often limited in this way. Most of the condensation nuclei that are available come from plankton-produced sulphur compounds; in the 1980s Lovelock worked with the aerosol expert Robert Charlson and others to show that sulphur emissions by plankton are probably keeping the planet significantly cooler than it would otherwise be by encouraging cloud formation in this way. Changes in ocean circulation or fertilization patterns that discouraged them from this work could lead to bad effects on the climate, a non-linearity to bear in mind.

But some of the mid-ocean sulphur comes from ships, and thus so does some planetary cooling. It has been calculated that the new emission standards the IMO is bringing into force this decade will reduce the cooling effects of global pollution by something like a third of a watt per square metre – a considerably greater effect, models would suggest, than that of all the carbon dioxide emitted by every generator and engine in the world over the same ten years. Those new standards will also, according to a companion analysis, save something in the region of 40,000 lives a year, because what is emitted over the mid-oceans does not stay over the mid-oceans; it is blown to shore, where it increases the damage done by pollution to susceptible lungs.

Do I think it realistic to imagine that the IMO might, as the result of a far-reaching envelope-stretching boundary-breaking debate, have come to a Crutzen-like grand bargain in which it sought to make good the cooling it was taking away by implementing a replacement brightening? Not really; but it remains striking – no, shocking – that as far as I can ascertain no one even mentioned the matter, even though the IMO's own technical advisers used the term 'geoengineering' in some of their analyses. Conversations have to start somewhere, and that would have been a good place to start one.

Cloudships

A much smaller example of the lack of interest in non–carbon-dioxide approaches to climate change can be seen in the response to John Latham's soggy-mirrors idea. Published in *Nature* in 1990 under the eye-catching headline 'Control of Global Warming?', it attracted more or less no interest at all. Latham succeeded in getting more scientific employment, ending up at NCAR in Boulder, Colorado, home of Kevin Trenberth and many others. But for more than a decade his soggy-mirrors paper might as well not have been published. In 2003, though, Latham's paper came to the attention of Stephen Salter.

Salter is one of the very few people working in geoengineering who is actually, by training and avocation, an engineer*. He spent

---

* Indicatively, he may be the first engineer mentioned by name in this book since Matthew Sankey and Archimedes, back in Chapter Two.

his career – like Latham, he is now strictly speaking retired – designing various sorts of marine hardware and teaching students to do the same. Over the course of it he has built up a phenomenally rich knowledge of ways of building things, of the properties of materials, of techniques for articulating and controlling mechanisms, of the workings of joints and bearings and gears and circuits. It reminds me of the knowledge jazz musicians have of scales – exhaustive and necessary, but hard for an outsider to connect to the creativity of the uses to which they are put.

Last time I saw Salter, in the hotel restaurant at the Berlin geoengineering conference in 2014, he was outlining his ideas for something that, if I understood it correctly, would be a cross between a flight of very long thin kites and a Venetian blind. The idea was to fly it over Singapore and obviate the need for air conditioning. By the end of the conversation something similar had been set flying over the Grand Canyon too, for reasons I forget (they may have had something to do with wind power). On an earlier occasion the subject was a way of making big, efficient, cheap pumps for taking warm water off the surface of the ocean in the hurricane season, thus depriving the hurricanes of the power they needed. Though he delights in the detail he brings to the imagining of the hardware necessary for these fancies, the hardware itself is never the point. The fancies start – they always start, Salter says with great sincerity – with an idea for helping people, an idea for doing good.

In the early 2000s, Salter wanted to do good by bringing peace to the Middle East. It was a dry place, he noted in his big-picture way, and some of its conflicts were exacerbated by that lack of water. So why not give it moisture? He started to think about wind-powered systems that could sit in the region's seas lifting fine mists of water on to the winds, there to drift to where such moisture was needed. Unfortunately, he could not see a way to avoid lifting salt too, and sowing salt on people's lands tends not to go well. But the mist-making-at-sea technologies were established in his mind when a colleague mentioned Latham's idea to him — an idea in which lifting salt was not a bug, but a feature. Riffing off the idea, he came up with the cloudship.

Salter's cloudships would be moved with 'Flettner rotors', peculiar spinning-chimney-like contraptions that look nothing like sails but interact with the wind in somewhat sail-like ways, and can thus move a ship forward (the forward motion can be used to generate the electricity that spins the rotors, which sort of feels like a free lunch but really isn't — the wind is doing the work). Flettner rotors had enjoyed a following on the fringes of marine engineering for decades, but had never proved a workable solution to a real problem. Salter saw that the rotors could be used both to drive the cloudships, obviating any need for fuelling, and to provide funnels through which the extremely fine water droplets Latham's idea required could reach the sky. A fleet of a few hundred such ships, he reckoned, could produce cooling on a global scale.

Salter's interest gave new life to Latham's ideas in various ways. It showed Latham that he was not alone, and encouraged him to start working on the soggy mirrors again. And the cloudships gave other people a way of thinking about cloud brightening, and climate geoengineering more generally, in terms not of abstract models and ethical conundrums but of pieces of hardware. Cannily, Salter commissioned some illustrations of a cloudship that showed a white trimaran topped by three weirdly ridged Flettner rotors — the sort of thing the Tracy brothers, who rushed around the world in finned and winged rockets in *Thunderbirds*, might have favoured if their international-rescue remit had extended to geoengineering. It was a very good piece of PR. Faced with the difficulties of illustrating an article on geoengineering, picture editors jumped at the chance to put this neat but weird-looking piece of science-fictional kit on to the page.

In the upturn of interest that followed Crutzen's 2006 paper Latham and Salter attracted new collaborators. One who proved particularly helpful was Phil Rasch, a widely respected climate modeller who was a colleague of Latham's at NCAR. Rasch, encouraged to work on geoengineering by his friendship with Crutzen, ran climate models for simulated centuries to see where they most reliably produced clouds of the sort which lent themselves to Latham's sort of brightening, Then he ran simulations in which fleets of cloudships patrolled the 25 per cent of the ocean where such clouds were most likely to be found.

With fairly reasonable assumptions about the efficiency of the process, the models suggested that such an intervention could produce enough cooling to more or less offset a doubling of carbon dioxide. The fact that the technique could produce quite large coolings mattered to its proponents – and Latham and Salter, much more than any of the natural scientists working on stratospheric veils, did give the impression of being proponents. They saw themselves, I felt at the time, as the Avis to veilmaking's Hertz: number two, and thus required to try harder. They didn't feel a need to make a case for geoengineering so much as to make a case that, if you were considering geoengineering, you should be sure to consider their version of it. They were always keen to find opportunities to promote the advantages of cloud brightening: it used nothing more toxic than seawater; if it were turned off, the clouds would return to normal in a day – maybe two, tops.

In 2009, Ken Caldeira and David Keith were thinking of putting some Gates money into this cloud-brightening research. Kelly Wanser, a Bay Area entrepreneur with a yen to save the world, had taken a shine to Latham's ideas. She put together a small meeting of the geoclique and various cloud experts in Edinburgh to discuss cloud brightening. It showed that the idea had some promise and a lot of unanswered questions. The flat decks of cloud that the technique seeks to brighten may look simple, but they are not; they have their own dynamics, which the added nuclei could disrupt in various ways. It was possible that the brightening might be self-limiting, with the brightened clouds less apt to gather up more condensation nuclei from the air below, or more apt to dry out. Some models suggested that aerosols added in the way Latham envisaged would actually darken clouds more than they would brighten them. The biggest hurdle, though, it seemed, was the problem of making the nuclei in the first place.

One of the people at the Edinburgh meeting was Armand Neukermans, a retired Silicon Valley engineer, who has worked on the spray-making problem ever since with a posse of collaborators supported and chivvied by Wanser. Neukermans spent some of his career at Hewlett Packard, where he worked on very early ink-jet printers; he thus had experience with little droplets. And he had become curious about engineering the earthsystem, possibly encouraged by the fact that one of the people he met at Hewlett Packard was Jim Lovelock; the company produced commercial versions of Lovelock's electron-capture detector, the very subtle instrument with which he had first shown that CFCs were building up in the atmosphere and by means of which he first came to appreciate the importance of the sulphur emitted by plankton.

Neukermans and his colleagues explored various recondite ways of making a spray that would meet Latham's specifications without relying on tiny physical nozzles that were all but certain to get gummed up. In the end they found one that seemed to have real promise: if the pressurized water coming through a nozzle has some gas in it, it will spontaneously break up into droplets a good bit smaller than the diameter of the nozzle just after passing through it. So you can use nozzles large enough not to be cloggable and still get droplets of the appropriate size. An array of a few hundred such nozzles could provide the required 1,000 trillion particles a second; commercial snowblowers could blow them high enough into the air to do their thing.

'It seems feasible,' Latham wrote in his 1990 paper, 'to conduct

an experiment in which [cloud–condensation nuclei] are introduced in a controlled manner into marine stratus, and in-
cloud microphysical measurements and above-cloud radiative
observations made.' A quarter of a century on, the equipment
to mount such a test programme is ready for prototyping;
Neukermans, Latham, Salter, Wanser and others are looking for
people willing to pay for them to build it and start experiments,
probably somewhere on or near the California coast.

Bright Patchwork Planet

Such tests would not only show the degree of local cooling
that cloud brightening might be capable of. They would also,
properly instrumented, produce a great deal of knowledge about
the mechanisms at work within clouds. This, their proponents
hope, could win them support from a broader community
of scientists than just those interested in geoengineering – as
happened with iron-fertilization experiments, and could happen
with the experiments in the stratosphere that David Keith and
others plan to propose. Experiments aimed at understanding
earthsystem processes as well as possible geoengineering
approaches seem much more likely to be acceptable to scientists
– and to civil-society organizations – than those geared towards
simply looking at the feasibility of a geoengineering project.

And if cloud-brightening proved its mettle, it would in
principle be possible to start using it fairly quickly for regional
cooling projects. When it comes to climate geoengineering, this
book has so far looked at the planet as a whole. And when considering the carbon cycle, or long-term interventions in the
stratosphere, that is the only plausible perspective. But the sort of
adaptation offered by sunshine geoengineering does not have to
be global. Brightening things to cool them down is something
you can do on scales from that of an individual house to a lake to
a sea, using everything from micro-sprayed cloud-condensation
nuclei to plain old paint.

People in hot climates have been whitewashing their houses for millennia. A number of experts now argue for making cities reflect more sunshine with whitened roofs and road surfaces as an adaptation to warming. On the planetary scale, such efforts make only a tiny difference, but they might provide a bit of summer cooling to the cities involved – and since the world is increasingly given to living in cities, and cities tend to be warmer than the surrounding countryside, that could be a real advantage.

Another regional brightening option, suggested by Andy Ridgwell of Bristol University, is to genetically engineer crops so as to give them more reflective leaves. The idea of crops that throw away sunshine sounds mad – it is, after all, what they eat – but Ridgwell's arguments are intriguing. As we saw when discussing the way plant growth can increase under a volcanic veil, direct sunshine can be too bright for the photosynthetic apparatus; in strong sunlight a bit more reflection could help. And sunlight reflected from a leaf lower down on a plant can be used for photosynthesis by another leaf higher up: most leaves can use light from below pretty much as well as light from above. You can imagine a crop canopy that reflects more sunshine back into space, uses more or less the same amount of sunshine as a normal crop uses for photosynthesis, but, thanks to big reflective leaves towards the bottom of the plant, lets rather less light fall to the ground.

The fact that you can imagine such a crop doesn't mean you can make it. But the biotech equivalents of Neukermans and his buddies might find it interesting to have a go. Even if they succeeded, the brighter crops would at best provide only a little cooling, and then only in summer, since that would be when they would be in leaf. But it is in summer that farmland most needs cooling. If crops could be cooled just a little it could make a worthwhile difference, particularly on the hottest days. If there were a prize for the form of geoengineering with the highest ratio of promise to research effort so far invested, I think

Ridgwell's – which has been explored in just a couple of papers – would be on the shortlist.

By employing genetic engineering in the cause of geo-engineering it might also find itself in the running for a green-baiting prize of some sort, which is in itself rather interesting. Ridgwell still has the dress sense and dreadlocks he sported when he was a protester trying to shut down road-building projects through civil disobedience in the 1990s; eco-activism turned him on to earthsystem science. He still cares about the planet's prospects a lot, and doesn't much like extraneous roads – but earthsystem thinking has encouraged him not to confuse caring with letting-be.

Ridgwell studied with Andrew Watson, an oceanographer with the distinction of being the only person Jim Lovelock ever supervised as a PhD student; Watson is one of the group of oceanographers who remain interested in exploring iron fertilization as a form of geoengineering. Another of Watson's former PhD students, Tim Lenton, is perhaps the closest thing to an intellectual heir that Lovelock has, adapting and defending his ideas about the regulation of the earthsystem in various ways. He, too, takes geoengineering seriously, and has published widely cited papers comparing the merits and risks of different proposed techniques.

There is a pattern here. Gaian thought, which stresses the degree to which feedbacks can, in some circumstances, bring stability to the earthsystem, has often been looked at askance by mainstream science on the basis that it has vaguely religious overtones. These overtones are in fact much less marked today than in Lovelock's original conception in the 1970s – Gaia theory has been largely stripped of the teleological tendencies which made it look a little like a religion. Besides that, though, the peculiar interest its adepts take in imagining the disassembling and re-engineering of their god would mark it out as a very odd sort of a faith. The truth is that the everything-depends-on-everything-else aspects of a complex system strongly reward a tinkerer's sense of how changes ripple through systems and how systems respond; it is a sense that Lovelock has in natural profusion and which his followers either shared from the outset or picked up as they went along. To be a Gaian in this way thus predisposes you to a geoengineering-friendly mindset – and it is possible that, to some extent, the reverse is true too. As I noted in the introduction, there is a particular appreciation of the earthsystem that can be gained only by imagining how it could be changed. To think like a geoengineer can enrich one's wonder at the earthsystem's workings.

Back to brightening, and to local and regional approaches. With cities and crops there may be a good case for local brightening. Something similar could apply to lakes and reservoirs. Russell Seitz, an independent researcher, has suggested schemes in which evanescent foams and bubbles could be used to whiten standing water, thus cooling the surrounding area and reducing evaporation from the water itself. You could imagine doing something similar for selected patches of sea and ocean with something like Salter's cloudships encouraging flecks of foam like those seen on choppy waves. The idea would be to have an effect along the lines of the ping-pong balls that Roger Revelle proposed to President Johnson – except without any permanent litter.

It might also prove possible to use veilmaking techniques in the stratosphere or upper troposphere on a regional basis, as long as you only wanted temporary effects – relief during a particularly dreadful heatwave, for example. Because the veil would not stay put the cooling would spread downwind pretty quickly; the uncalled-for effects on other places would thus be similar in scope to the effects in the region intended, which doesn't sound very desirable, and very little research has so far been carried out into how such a regional veil could be made to work. But that does not mean such a scheme might not appeal to some governments at some times, or as an extra but occasional use for a global veilmaking infrastructure. Although this idea doesn't fit with previous thinking about veils, or with GeoMIP models, it might be one that ends up pursued.

How might Latham's cloud-brightening ideas, if experiments show them to work, fit into this range of local and regional approaches? The flat cloud-decks on which the technique is designed to work are only found at sea, and because people tend not to live or work on specific bits of the sea, there would seem to be rather less call for local cooling there than on land. But there are some bits of the sea that might merit special cooling. People are very keen on coral reefs, and though some reefs can survive high temperatures, others cannot; in water that gets too hot, the reef-building polyps may 'bleach' themselves, expelling the photosynthetic symbionts they depend on for energy. At best this means the reef goes quiescent, awaiting recolonization by other symbionts; at worst, it dies. If cloud brightening works, I could imagine its use to provide tropical reefs with parasols being both helpful and not terribly controversial.

A grander – and doubtless more controversial – use for a technology which cools the surface of the sea but not the surface of adjacent land would be to boost the power of monsoons. A modelling study by Caldeira's former colleague Govindasamy Bala and various colleagues has found that an Engineered Planet in which greenhouse warming is offset by marine-cloud brightening sees more precipitation over land than a model that uses veils to produce a similar cooling – the encouragement of monsoons seems to be part of the reason why. Similar approaches on a smaller scale might provide welcome onshore winds during heatwaves in coastal areas.

You can thus imagine a patchwork, or a hierarchy, of brightening, one that might come in time to underline the idea that geoengineering is simply a form of adaptation writ large. The white roof adapts the house, brightened cities and lightened crops adapt the country, foams cool the lakes and clouds the seas, a veil brightens the planet as a whole. Thus local brightening might make regional and planetary efforts seem more approachable – or, perhaps, if the local and regional effects deliver, less necessary.

Unfortunately, when you start getting close to global scales, brightening just a part of the world can have rather unfortunate consequences. A veilmaking study by Jim Haywood, a modeller at Britain's Met Office, makes the point clearly. Ringing the changes on one of the GeoMIP scenarios described in Chapter Four, Haywood and some colleagues compared a Greenhouse Planet with a pair of Engineered Planets; in one of them all the cooling took place in the northern hemisphere, and in the other it was all in the southern hemisphere, a regionalization which would probably be achievable if the cooling veil was emplaced only by aircraft in the appropriate hemisphere and a fair way from the equator. He found that in the northern-cooling-only scenario there were horrendous droughts in the Sahel, because the Intertropical Convergence Zone, a rainbelt around the equator, moves away from the cooler hemisphere, taking its rain with it. This effect does not just appear in models. Of the Sahel's four worst years of drought during the twentieth century, three took place after volcanic eruptions sufficiently far north of the equator to cool only the northern hemisphere – the year after the Katmai eruption in the Aleutians (1913), and the year of and the year after the El Chichón eruption in Mexico (1982 and 1983). There is also a substantial body of evidence suggesting that the cooling effects of sulphate aerosols from the industrial north played a role in the Sahel droughts of the 1970s and 1980s.

The most obvious lesson from Haywood's simulation is that no one should try to cool the planet just by cooling the northern hemisphere. The more general lesson is that cooling some parts of the planet can have large effects in others. The climate system works in such a way that if you perturb one bit of it you would expect to see responses in other bits a long way away; there are patterns of atmospheric pressure and circulation that can shift from one state to another on a global scale. It is due to just such 'teleconnections' that the effects of a warm body of water moving across the Pacific – the essence of an El Niño event – can be reliably expected to produce a broadly predictable pattern of change from Cape Town to California.

Stephen Salter imagines that it might be possible to use such teleconnections to fine-tune a geoengineering scheme's effects on the climate, avoiding some areas and doubling down on others in order to get a pre-selected pattern of beneficial effects across the planet. Few people with a background in climate science believe this. Rather, they worry that by concentrating cooling in particular places you could do damage to others without fully understanding why – that teleconnections are a bug, not a feature. Some modelling work at the Met Office, for example, has suggested that brightening the marine clouds off Namibia (one of the best prospects for the technology, according to Phil Rasch's model-based search for the most susceptible clouds) can lead to drought in northern Brazil.* It might be possible to do local cloud brightening without too much concern for anyone other than those in the immediate region. But the larger the scale, the more likely it seems that teleconnections would start to matter.

This leaves various ways in which cloud brightening could be used for local and temporary relief. In a 2014 paper Latham and colleagues looked at the possibility of focusing cloud-brightening efforts on protecting reefs from bleaching, saving polar ice and damping down sea-surface temperatures in order to forestall hurricanes. This doesn't necessarily require a sophisticated knowledge of teleconnections. It might simply mean

---

* That said, Met Office models are prone to predicting drought in Brazil in many circumstances where other models do not. The finding really only shows that models are not yet remotely good enough to provide the sort of detailed assessments that would make it possible to design a large-scale cloud brightening programme with confidence as to its teleconnected effects.

using your knowledge of oceanography to do the cooling 'upstream' of the places it is needed, and letting the ocean currents do the rest. Such approaches, they wrote, might make it possible to use cloud brightening 'in an almost "surgical" manner at local and regional scales'.

It is hard not to hear an echo of Langmuir's boasts of cancelling New England ice storms and stopping hurricanes in their tracks in such claims; what started as an approach to geoengineering the global climate starts to look like case-by-case weather modification. This has some advantages. In offering short-term benefits, it has the potential for prompt improvements in human welfare. But like all Promethean science, weather modification has its own troubling hazards. As Langmuir and his successors found, showing that what the weather does after an attempt at modification is different from what it would have done had no such attempt been made is very hard. All clouds have particularities; controls for experiments are hard to come by. Proof of success is mostly statistical – showing an increase in the likelihood of rain, or in the likely intensity of rain, rather than any true control over whether it rains or not. You cannot guarantee the benefits you seek to offer.

And just as it is hard to take credit for good weather, so it is hard to avoid the blame when it is bad. In 1947 Project Cirrus tried to weaken a hurricane by seeding some of its clouds; the particular hurricane in question, which went by the name of King, was chosen because it was heading northeast away from Florida out into the empty Atlantic, and thus the experiment would not have any effects on the public at large. Unfortunately, once seeded, the hurricane promptly and sharply changed course, hitting the coast near Savannah, Georgia. It seems fairly unlikely that the seeding was responsible – earlier hurricanes following similar courses had on occasion done similar things in similar circumstances, and the scale of the seeding was pretty small – but the people of Savannah were understandably unimpressed, and hurricane modification was struck from Project Cirrus's roster of interests. If something similar were to happen today, in a more litigious culture often less accepting of reasons of state, the aggrieved voices would be raised much louder.

As we have seen, though, for all the issues that they bring up, weather-modification programmes based on Langmuir's continue around the world. As climate change raises worries about water supplies, it seems quite likely that they could be deployed more widely; it is also possible that the techniques they make use of could be improved. And it is quite conceivable that countries and companies which increasingly try to manipulate clouds for one set of purposes might more readily take on the possibilities of manipulating them to other ends – perhaps to protect coral reefs, or increase onshore winds during a heatwave. If it were to prove possible to use Latham-like techniques to make clouds from scratch under some circumstances, as Daniel Rosenfeld, a professor at the Hebrew University of Jerusalem who specialises in aerosols and clouds, has suggested it might, such targeted interventions could be all the easier, and thus more attractive.

If those in power in Beijing, to take an example not at all at random, were to perceive some national benefit in deploying cloud-brightening technology in the South China Sea, why would they not act on that perception, given that China already seeds clouds from Xinjiang to Guangdong? The fact that others dispute China's dominion over that sea might simply add to the appeal; how better to express your dominion than to commandeer the sky? And further justification could be found, as we have seen, in the fact that China is already cooling the South China Sea, along with much else, by means of the power-station aerosols that drift east from the mainland. It would be Crutzen's stratosphere–troposphere trade-off achieved by other means: the unintentional effects of pollution replaced by intentional engineering that preserves perceived benefits while causing less harm.

That sort of argument could also be applied more widely. As well as China's sulphates, there are huge clouds of soot, organic compounds and sulphates hanging over large parts of south Asia, southeast Asia and Africa. These aerosols are undoubtedly modifying monsoons and shifting ocean currents. Given the radiative forcing they exert, their effects are probably quite as appreciable as those you might expect from a modestly ambitious climate geoengineering scheme, with effects that may be felt well beyond the areas where the aerosols are emitted. What exactly the effects are, it is hard to say – but it seems a priori unlikely that they are all damaging.

You do not need to imagine a grand global bargain like that sketched by Crutzen to think that researchers will be far more curious about the climatic effects of pollution control in Asia during the 2010s and 2020s than they were when Europe and America cleaned up their acts from the 1950s to the 1980s. That will surely increase the chance that, as inadvertent modifications are reduced, countries will look at ways in which seeding and brightening clouds might possibly iron out any unwelcome side effects from that reduction, not on a global basis, but regionally or locally.

The chance may remain small; the ability of policy makers not to think about such things even when researchers are aware of them is, as the IMO has demonstrated, quite well developed. And that could be a good thing. Such ironing-out obviously offers the possibility of conflict (especially when it comes to the South China Sea, where more or less everything offers the possibility of conflict). As with global approaches, changes that are to one country's advantage might be to another's detriment.

At the same time – indeed, as a direct result of those risks of conflict – there is a possibility that national desires to make intentional regional changes might bring about serious strategic discussions about what could be acceptable. One reason that the IMO did not discuss the climate effects of stripping shipping's sulphur from the sea air was that no one thought that they had anything particular to win or lose from it. When people feel that something more is at stake, they might negotiate.

What the Thunder Didn't Say

We could leave this part of the story here, in the confusion of the regional and the global, the accidental and the deliberate, the worthwhile and the dangerous. Instead, though, a personal postscript. John Latham lives in Gold Hill, a small community about 20 kilometres outside, and 1 kilometre above, Boulder, Colorado. Five years ago I spent some time on the deck of his small house — basically a log cabin — on a wooded hillside, looking west at a spectacular view of the peaks of the continental divide, listening to stories and telling a few. There were tales of the colourful Gold Hill community (the one about the mayor, the mule, the marijuana, the magnum and the monument is a corker) and there was the discovery of mutual acquaintances (Latham has some in common with my wife, also a poet). There was a sense of being part of the same thing which went beyond shared membership of the geoclique; perhaps we were part of the same *karass* – the hidden fellowships that bind together followers of Bokononism, the religion of utopian lies with which Kurt Vonnegut underlines the absurdity of the world and its end in *Cat's Cradle*.

As the sun prepared itself to sink we both became aware of a cloud a little way off to the southwest which was slightly larger than the others dotting the sky. It was beginning to stretch itself up in an arched-back-of-a-cat sort of way that hinted at the possibility of a late-afternoon thunder shower. I like cloudscapes, but rarely take the time to just observe the development of a particular cloud. In conversation with someone who has studied them for decades, though, it seemed too good an opportunity to miss. We watched, sometimes chatting, sometimes in companionable silence, as something like a thousand tonnes of water divided among ten billion trillion droplets ceaselessly rearranged itself according to the flow of energy through the air that held it.[*] The flow of energy shaped the cloud; the cloud shaped the flow of energy, soaking up heat and releasing it as droplets evaporated and condensed around its edges. As new lobes bubbled up and out from its sides, turning over and subsiding, the whole thing took on a sort of twist, like someone swivelling from the hips to ease their back and shoulders after too long at a desk. We watched over the course of three quarters of an hour as it morphed from one shape to another entirely different one while always being, from moment to moment, clearly and entirely the same thing. The shape changes; the process endures.

Some of its upper edges started to lose the distinct surfaces they had previously shown – a sign that some of the liquid droplets, which evaporate very quickly if they stray into the drier air outside the vapour-saturated body of a cloud, were being replaced by ice, which can persist in dry air. The difference explains why wet clouds have much more distinct surfaces than icy ones. The change was a good omen for lightning, which depends on ice particles rubbing against each other. Part of Latham's contribution to cloud theory, back when he was more interested in understanding lightning than he was in cooling the planet, was a model that showed the key role the interactions of very small and much larger ice particles play in the process, allowing positive and negative charges to build up in a way that creates the strong electric fields needed to tear the air apart.

John recalled flying into the heart of bigger, much scarier clouds with his friend Paul MacCready, who, utterly intrepid, would head straight for whatever looked most exciting to him. It strikes me that Paul, an imaginative inventor of unusual aircraft

---

[*] To calculate the number of droplets while writing this passage I checked Google Scholar for a reasonable set of measurements; I came up with an NCAR study of a cloud in Montana in the summer of 1981 which was, of course, chock full of references to work by Latham.

whom I had the pleasure of meeting myself, might have been interested in designing veilmakers.

That day, though, we were to be denied the sort of aerial scariness MacCready delighted in – at least as far as our particular cloud went. Its extension towards the tropopause petered out, its flanks cinched in, its leeward trail of ice never developed into anything like the anvil head of a thundercloud. By the time the sun had set it was much reduced.

We sat on in the dark, drinking good Scotch, talking of joys and sadnesses. John told me about the death of his Mike, the boy who had talked of soggy mirrors who became a man whose life never really worked out, unable to balance the forces within him and without, dead of an overdose at the age of 37. I told him of my friend Dominic, with whom I had last spent time that spring, looking out over a different English sea. He had walked drunk off a cliff on his birthday a couple of weeks before I had flown to Boulder.

As it happened – as it was supposed to happen, Bokonon would say – the lightning, when it finally came, was far off, farther off than I would have thought possible. It lit up the silhouettes of mountains I had barely seen, flashing between and beneath clouds the best part of a hundred kilometres down the mountain range. Indeed, it was so distant that the accompanying thunder could not be heard. I had, until then, always thought that a storm's thunder and lightning had a sort of synergy, deriving their drama from overwhelming two senses stretched apart in their response by the speed of sound. And they certainly do bring out that tension – not to mention, sometimes, the right-here-right-now jolt of the sublime which comes from lightning drawn to something very close, like the lightning conductors of the tall hall that stands next to my home in Greenwich, striking at almost the instant that the thunder reaches your ears. But lightning without a hint of thunder – that has a grandeur too, a sense of something godlike playing in the distance, too far beyond the human to even bother announcing itself.

It is a strange world than has lost children and mute lightning in it, and where two men can sit overlooking an abandoned mining town and talk in sadness and hope of building ships to whitewash far-off ocean skies.

The next morning, before we drove sober back to Boulder, we talked a little more about the Vonneguts. John liked Kurt, with his anger and humour and fierce sense of the absurd; they played practical jokes on each other through the post. But it is Bernie, who had learned how to turn one sort of cloud into another, who brings real life to Latham's voice. Not because he gave humans a new lever with which to prod the earthsystem. Because 'he was the nicest man I ever knew'.

PART THREE
Possibilities

11

The Ends of the World

I can guide a missile by satellite
By satellite
By satellite

And I can hit a target through a telescope
Through a telescope
Through a telescope

And I can end the planet in a holocaust
In a holocaust
In a holocaust
In a holocaust
In a holocaust
In a holocaust

I can ride my bike with no handlebars

The Flobots, 'Handlebars'
(2005)

Global warming was not the first potential climate catastrophe to trouble the politics of the 1980s. There was something else that came before, stirring up the worlds of science and politics. That this scene-setting furore was triggered by Paul Crutzen should come, by this point, as no surprise.

In the 1970s some physicists had realized that, like grotesquely distended lightning bolts, the fireballs produced by nuclear weapons would create a great deal of NOx, a lot of which would rise up into the stratosphere. Thus it was calculated that a nuclear war would thin the ozone layer by up to two thirds over the northern hemisphere. In 1982 Crutzen, the go-to man on ozone worries, was commissioned to review and update this work. The basic science still seemed sound, but because military fashion had shifted towards smaller, better-targeted nuclear warheads over the course of the 1970s, the threat seemed less fearsome than it had a decade before: the smaller fireballs created by the new weapons would lift less NOx into the stratosphere than those of their forebears. Unwilling to send the message that nuclear warfare was less environmentally destructive than it had previously appeared, Crutzen and his co-author, John Birks, widened the range of their study to other atmospheric changes that a war would bring about.

Crutzen, who had studied fires in tropical forests, was struck by the scale of the burning that might be set off in cities, at oil refineries and around rural missile silos — all prime targets. The conflagration, he and Birks calculated, would cover an area equal to that of Scandinavia, and would create more NOx than the blasts themselves, not to mention lots of other pollutants. What was more, its smoke would darken the sky; the amount of sunlight that reached the surface would drop by at least 50 per cent, maybe by more than 99 per cent. They called their paper 'The Atmosphere After a Nuclear War: Twilight at Noon'.

Crutzen and Birks did not look at the climate effects of this darkening in any detail; they made no mention of surface cooling. But some of the scientists who in the 1970s had nailed down the link between volcanoes, stratospheric sulphur and climate were already at work combining models of aerosol behavior and the climate to investigate the climate effects of a nuclear war — a remarkably unstudied area at the time. After reading Crutzen and Birks they added soot and smoke to their calculations, and the fires became a crucial part of their prognosis. In late 1993 those researchers – Richard Turco, Brian Toon, Thomas Ackerman, James Pollack and Carl Sagan, later abbreviated as TTAPS – published their results in a paper in *Science*: 'Nuclear Winter: Global Consequences of Multiple Nuclear Explosions'.*

The TTAPS results were staggering. The team calculated that a 5,000-megaton exchange between America and the Soviet Union – an all-out war, but not the largest imaginable – would ignite more than a billion tonnes of flammable material, producing 225 million tonnes of smoke particles; hundreds of millions of tonnes of dust would also be thrown up, some of it into the stratosphere. This would reduce the amount of sunlight getting to the surface by about 98 per cent: not so much a veil as a shroud. The upper levels of the shroud would warm as sunshine was absorbed by soot and smoke; the lower atmosphere would cool because the sunshine was being absorbed higher up. Thus the warmer-on-top-of-cooler layering normal to the stratosphere but rare near the surface would extend right down to the ground and stretch across whole continents.

This stable layering would curtail the convection that normally produces clouds and thus rainfall; the stalled hydrological cycle would stop the aerosols from getting washed out, perpetuating their cooling effect. After three weeks of this, according to the TTAPS calculation, surface temperatures in the northern hemisphere would be down as far as −23°C. Crops would die; rivers and lakes would freeze; the resulting death toll might easily be higher than that of the war itself. Stanford

---

* Like 'solar radiation management', 'nuclear winter' is a term created as a result of official nervousness at NASA Ames Research Center, where a number of the TTAPS researchers worked. NASA decided the paper could not be published if it had the words 'nuclear war' or 'nuclear weapons' in its title. Turco came up with 'nuclear winter' as an alternative, thus giving the paper a much more memorable and evocative title.

biologist Paul Ehrlich, reporting on a meeting held to discuss the biological consequences of such an event just before the TTAPS paper was released, talked about the possible extinction of the human race.

When people treat the climate challenge as completely un-precedented, or say – as Johan Rockström, the director of the Stockholm Resilience Centre and an influential proponent of the 'planetary boundaries' approach, has said – 'We are the first generation to know we are truly putting the future of civilization at risk,' they are flattering the importance of the present in a way that demeans the past. Civilization has been at risk – and yet not lost – since the 1950s.

That nuclear risk did not merely predate today's concerns about climate change; it shaped them politically, scientifically and emotionally. One can worry about climate change without appreciating this prefiguring. But if you want a full appreciation of the origins of climate geoengineering and how it is imagined, such an appreciation is vital.

The development and use of nuclear weapons has had a far smaller direct effect on the earthsystem than has the Haber–Bosch process. But the spectacle and implication of their possible effects make them the technology which, more than any other, evokes power beyond the human. The prospect of nuclear war feels different not just in scale, but in kind. It is no longer remotely comprehensible as a manifestation of popular will, as the hugely increased powers of the national armies of the nineteenth and twentieth centuries were; it is so much smaller – the creation of a scientific, military and technical elite, its triggers in the hands of just a few politicians – and so much vaster. Look at it afresh and it can but feel absurd. And yet submarines dedicated to this absurdity patrol the oceans day in and day out, machines more terrible than any others ever created, the power to eradicate whole nations poised beneath the hatches of their missile tubes.

Those deathboats are designed to keep out of sight, and that which they make possible is, most of the time, successfully kept out of mind. But the fragility such systems have introduced into the world resurfaces in other contexts. Nuclear weapons encouraged people to imagine other phenomena that might do damage on similar scales. From the early days of the cold war people sought to draw on the newly conceivable apocalypse to burnish their own agendas: thus the parallels drawn between the nuclear bomb and the 'Population Bomb' by the Malthusians of the 1950s. Others, such as those taking part in the 'environmental warfare' programmes at NATO and elsewhere in the 1950s and 1960s, looked at ways that nature might be subverted on scales similar to that offered by the bomb, and to similar ends. The historian Jacob Darwin Hamblin argues that this military speculation about other modes of destruction nuclear-bomb-like in scale played a role in smearing the specific fears associated with nuclear war into a more generalized fear of environmental catastrophe. As mentioned in the discussion of this idea in Chapter Five, it was in these military speculations that the idea of the ozone layer as something fragile enough to have holes torn in it first appeared. Similarly, images of a flooded New York were first created to illustrate fears that the Russians might melt the Arctic's ice.

Whether it was by that specific route or through some subtler imaginative metastasis, new catastrophisms began to spread in the 1960s, dark doubles of the mushroom cloud. Science fiction started to deal with all sorts of environmental crises, likely and unlikely: the world was drowned, burned and transformed into crystal; it was stripped of all grass, shattered by earthquakes and infested by triffids. Sometimes – for example in Cormac McCarthy's *The Road* – what exactly happened is never specified, nor is it felt necessary that it should be. In his excellent bibliographical work, *Nuclear Holocausts: Atomic War in Fiction, 1895–1984*, Paul Brians notes, 'In a surprising number of cases, it is uncertain whether a nuclear war has occurred or not. There are many vague holocausts to which no cause is ascribed.' The war and its dark doubles could not be distinguished.

When the literary critic Fredric Jameson remarked in the 1990s that it was now easier to imagine the end of the world than the end of capitalism, the truth of what he said was surely in part due to the remarkable amount of practice people had been getting in imagining ends to the world – practice which stood them in good stead when worries about greenhouse warming began to gain traction. Yet at the same time as the future integrity of the natural environment felt ever more at risk, society felt ever harder to change, especially if you were on the left. The idea that the political world and its economic underpinnings might be open to radical revolutionary change, a threat or promise central to European thought from 1789 onwards, went into retreat after 1945. There are a number of features of late capitalism which might be called on to account for this, most notably the shift from identities built on work to identities built on consumption, but there is a role for nuclear fear, too, both as a political fact – a world of superpower conflict was one in which revolution and war might be synonymous – and as the driver of a deeper change in the imagination.

The anthropologically attuned historian Joe Masco has pointed out that the nuclear age brought with it two incompatible futures: amazing plenty at some distant time and unparalleled devastation always only a few minutes away. Fending off that prompt apocalypse meant perpetually preserving the present – freezing it as if in Ice-nine. The increasing ease of imagining the end of the world and the difficulty in imagining change in the way it is run are not independent of each other; each is to some extent the cause of the other.

If fears about the consequences of nuclear war helped give rise to a more generalized concern about environmental catastrophe, fears about the causes of such war fed a new distrust in technology. In the 1940s and 1950s the fear of nuclear weapons is the fear of being attacked by an enemy using them. As time goes by, though, the nuclear holocaust's context within some imagined war starts to become irrelevant – the sting in the tail of Pat Frank's widely read 1959 novel *Alas, Babylon* is the discovery by the impoverished, diseased survivors of a nuclear war that they are in fact its winners. Increasingly, the holocaust no longer requires a war: it can come from a general's madness, a circuit tripped by accident, a programmer's error, even an emergent malevolence in the computers themselves.

The effect of the nuclear age on ideas about the human relationship to nature, extreme technological risk and environmental catastrophe conditions the ways people think about geoengineering. Fear of nuclear war brought with it a generalized fear of catastrophe which changed the way people thought about the environment. One of the problems faced by those who see nuclear energy as a vital part of the response to climate change is that many of those most worried about climate catastrophe have a similar fear – in some way, possibly, the same fear – of nuclear technology. Geoengineering taps into the same well of concern. Climate geoengineering does not require nuclear technology, much less nuclear bombs. But it cries out for being imagined as just the same sort of assumption of godlike power. It is easily read as embodying a Dr Strangelove sensibility and an I-am-become-death-the-destroyer-of-worlds hubris; as being bewitched by a silver-arrows-through-the-stratosphere technocratic ambition in just the way that the Strategic Air Command was; as being deaf to – or scornful of – the voices of the masses it claims to protect, and whose lives it holds at risk. It too feels like the creation of a scientific, military and technical elite, its control in the hands of just a few politicians – if it can be controlled at all. It too feels like the outcome of some Faustian pact in which the power that comes from knowledge is seen as needing no other moral justification. Like the faulty transistor which launches the missile, like the madman at Burpelson who sends the bombers past their fail-safe, like the game theorist who loses the game, it feels like something that does not care.

If I thought this was a necessary truth at the heart of climate geoengineering I would not have written this book. But it would be foolish to ignore the power of geoengineering's nuclear resonance, not least because it has considerable historical foundation. The idea of deliberate climate change was bound up with thinking about nuclear weaponry from the very beginning of the cold war.

Control and Catastrophe

In 1945 ENIAC, the first fully programmable electronic computer, was given its first problem: a simulation of the hydrogen-bomb design then being touted by Edward Teller. The task was set by John von Neumann, a mathematician at the Princeton Institute for Advanced Study who was attached to the Manhattan Project. Von Neumann was determined that the computer's potential for solving problems should be applied to the most pressing issues of the day. And that was why, even as the million-punch-card H-bomb explosion was trundling through ENIAC's circuits, he was already working out how the machine and its successors could be programmed to predict the weather, and produce new insights into controlling it.

Over the following years von Neumann built up a team, eventually led by a brilliant young meteorologist called Jule Charney, to turn the question of how weather patterns would develop over time into a set of calculations that the machine could handle. In 1950 the calculations began. Even after the bugs were ironed out, it was hardly an operational success − a prediction of the weather 24 hours ahead took 36 hours to calculate. But the principle that computers could do such things had been established. Later that year the machine even produced what could be taken as a forecast with some real value.

The computer hardware for modelling the atmosphere was the same as that needed for bomb design. The software approaches were similar, too – the textbook that climate modellers came to rely on for programming differential equations in the 1950s was the same one the weaponeers used when building models of imploding plutonium cores. And there was a deeper complementarity. The science of the atom held out the promise of an ultimate power; the science of the climate held out a promise of the ultimate control.

Back in 1906, rhapsodizing about radium, Frederick Soddy had imagined the power of the atom being so great that its wielders might 'transform a desert continent, thaw the frozen poles, and make the whole world one smiling Garden of Eden'. Within months of the destruction of Hiroshima and Nagasaki, Julian Huxley, a British biologist who had become the first secretary general of UNESCO, told an audience at Madison Square Gardens that the new era of atomic power promised to do just that. When it came to the transformation of desert continents, atomic power would make it possible to 'pump fresh water into numerous depressions in [the Sahara's] vast area until they again blossom into the fertility they had in the ice age'. As for frozen poles, 'How many people realize that we could alter the entire climate of the North Temperate Zones by exploding . . . at most a few hundred atomic bombs at an appropriate height above the polar regions? As a result of the immense heat produced, the floating polar ice sheet would be melted *and stay melted*.'*

Huxley was not just extolling power; he was extolling control, both technical and political. As H. G. Wells argued in *The World*

* The emphasis on permanent change came from Huxley's tipping-point model of the Arctic ice; he believed it to be a relic of the ice age which, if removed, would not re-form. This might not have been true in his 300-ppm-of-carbon-dioxide world; it may well be the case in today's world of 400ppm or more.

Set Free, written under Soddy's heady influence and required reading for British scientific intellectuals of Huxley's generation, the power of the atom required the new sorts of control that only world government could offer. Unsurprisingly for a high official in the newly formed UN, Huxley supported Bernard Baruch's idea of a super-governmental body to take control of all nuclear weapons, seeing it as a route to such world government. And once the world was thus controlled, so might the climate be. Waldemar Kaempffert explained to the readers of the *New York Times* that the context for understanding Huxley's ideas about flooding the Sahara and nuking the Arctic was that they would be part of such a government's 'social planning on a world-wide basis'. This control would stretch from the Gulf Stream to the genome; Huxley was a convinced eugenicist, believing in the scientific betterment of human stock just as he believed in the scientific betterment of the climate.

Von Neumann's politics were quite different. He had no time for the Baruch plan; while Albert Einstein, whose Princeton office was next door to von Neumann's, was contributing to 'One World or None', the world-government manifesto of the Federation of American Scientists, von Neumann was advocating a pre-emptive nuclear war against the not-yet-nuclear Soviet Union which was occupying his native Hungary. He still saw a need for control, but he looked for it, at least in part, in the computer: control over industrial processes and industrial workers; control over the global economy; control over the climate.

In late 1955 researchers gathered in Princeton to talk about the sort of modelling which would be needed to underpin that control. This was what von Neumann called 'the infinite forecast' – the use of computers not just to predict tomorrow's weather but to model the persistent features of the climate as a whole in the way of the GCMs now central to climate science. In his introductory remarks, Robert Oppenheimer compared the gathering directly to the meeting at Los Alamos that had set forth the first designs for atomic bombs, the only technology of comparable ambition and importance – both in its overarching scope and, perhaps, in its military application.

A few months before that conference, in June 1955, von Neumann wrote an essay for *Fortune* in which his thoughts about climate control, war and the future were expressed more fully than they had been before, at least in public – along with ideas about nuclear power and the end of the world. It is called 'Can We Survive Technology?'

Referring to 'the great globe itself' in its first line, the essay seems to draw inspiration from the soliloquy at the end of *The Tempest* which ends with Prospero, the mage, breaking his staff and quitting the island world where his power and control have been total, choosing in its place the larger world where he will be a lesser but better man. However, in von Neumann's version the vision is inverted. The staff cannot be broken. There is no larger world to retreat to.

The growth in technology's powers, von Neumann wrote, was once offset by the ever-expanding realm over which the progress-oriented, capitalist, technological civilization that housed that technology held sway. Previously there had always been new places for technological powers to spread, and ever-larger institutions within which they could operate: 'geographical and political Lebensraum'. No more. 'Literally and figuratively, we are running out of room. At long last we begin to feel the effects of the finite, actual size of the Earth in a critical way.' The world was coffin-cornered; Prospero's cell shrank even as his powers increased.

Computer models of the climate were both symbols of the world's shrinking – the workings of the Earth could now be represented in something that could be fitted into a room – and tools for its furtherance. When von Neumann wrote about powers that could be 'projected from any one to any other point on the Earth', he was surely thinking about intercontinental ballistic missiles, in the development of which he was playing a key part as a member of the Atomic Energy Commission. But he was also thinking about the climate models that would reveal ways in which 'the Arctic may control the weather in temperate regions, or measures in one temperate region critically affect another, one quarter around the globe'. Getting a handle on those teleconnections 'will merge each nation's affairs with those of every other more thoroughly than the threat of a nuclear or any other war may already have done'.

But while promising control of the climate, he offers no possibility for the control of technology itself, nor any way of prioritizing technologies that might help over those that might harm; to von Neumann, technology is always becoming more powerful, and is always indifferent to the uses to which it can be put.* At the essay's end he retreats to the reassuring idea that 'the human species has been subjected to similar tests before and seems to have a congenital ability to come through' – an idea that the body of the essay, with its claims about the unprecedented nature of the small-world big-technology trap, directly contradicts – and the bromide that the human qualities that will be required will be patience, flexibility and intelligence. It feels like an optimism of duty, not of belief; it does not succeed in masking the undercurrent of dread that I read as running through the piece. Ten years earlier, shortly after the bombing of Hiroshima and Nagasaki, von Neumann speculated in a letter to Lewis Strauss, the first head of the AEC, as to whether supernovae were not an unavoidable end-point of stellar evolution, as astronomers thought, but rather the culmination of technological evolution – 'Evidence that

---

* One of the things for which von Neumann is remembered today is his theoretical treatment of 'von Neumann machines' that could reproduce themselves, growing in number exponentially – the *nec plus ultra* of technology as its own reward.

sentients in other planetary systems had reached the point in their scientific knowledge where we now stand, and, having failed to solve the problem of living together, had at least succeeded in achieving togetherness by cosmic suicide.' The same grim thought underlies the later essay. One cannot help but feel that his private answer to the question of survival was 'Probably not'.

Doom and Denial

Shortly after 'Can We Survive Technology?' was published, von Neumann was diagnosed with metastasized cancer. The 1955 colloquium on the infinite forecast that Oppenheimer addressed was the last scientific meeting he attended, and after a painful decline he died in early 1957. His legacy can be seen in all the computers of the world, but perhaps nowhere more so than in those housed at the laboratories that design nuclear weapons.

From the first simulation onwards, Teller was convinced that the computer would be crucial to all such work, and when Livermore was set up as his fief in the 1950s he made it a priority that the lab should have the best computers. That commitment continues: at the end of 2014 Livermore boasted the third and ninth most powerful supercomputers in the world. Teller also remained interested in von Neumann's ideas of climate control, and made sure that the lab's computers were used for GCMs, too (as they still are today). Chuck Leith, who created Livermore's first GCM, remembered Teller explaining the vision in the 1960s in just the way von Neumann had put it in the 1940s: the stable aspects of the physics of the atmosphere would allow its behaviour to be predicted; the unstable parts would allow it to be controlled.

It did not come to pass. The climate science that grew up around the new techniques of computer modelling did not share its progenitors' interest in control, becoming instead a 'pure' science concerned with basic facts about the Earth and pursued at America's elite academic institutions. There was still some interest in weather modification by means of cloud seeding and the like; but after the heady years of Project Cirrus it became increasingly marginalized, the province of less prestigious universities in the continental interior, deprived of any great theoretical or computational underpinning. As concern about the climate effects of pollution grew, their study and that of cloud-seeding increasingly came to be seen as part of the same sub-discipline – deliberate and inadvertent forms of weather modification. The idea of operational climate control did not entirely vanish, but it was deferred *sine die*. What was needed for the time being, the academic scientists said – as they can be relied on to say in almost any area where they get to set their own agenda – was basic research.

The idea that nuclear weapons, von Neumann's other project, might impact on the climate described in these models was one in which neither climate scientists nor weapons scientists took much interest. The public was widely convinced that they did; throughout the period of atmospheric testing there were claims that strange new weather patterns had been brought about by the explosions. Some scientists lent their imprimatur to the idea – the elderly Soddy argued implausibly that the release of radioactive material was changing the ionization of the atmosphere – but most did not; nor did they explore in any great detail the climatic effects of the far more numerous explosions that a heated-up war would bring. At a time when suggesting causes for ice ages was something between a cottage industry and a parlour game, the possibility that nuclear wars would chill the world to its marrow was hardly mentioned.

This lack of curiosity served the cold-war nuclear establishment well. The possibility that nuclear weapons might be used to fight wars required that the effects of the blasts be limited. Thus the possibility that a nuclear exchange might change the climate was one that those closest to the question had an interest it downplaying. Studies of the indirect effects of nuclear war repeatedly found that they were far smaller than the direct destruction. As the historian Paul Edwards puts it, '[A]mazingly, the nuclear damage estimates . . . took virtually no account of the dust, fire, and smoke effects of nuclear blasts. Instead, in an astonishing case of self-inflicted organizational blindness, researchers focused only on shock waves and fallout.'

The irony is palpable. The power of the atom underwrote the sweeping new technocratic ambition that imagined melting ice caps and blooming deserts – earthsystem-changing potential was part of the atom's Promethean appeal. At the same time, the politics of maintaining a nuclear-weapons program required people to believe that real applications of nuclear energy would not in fact change the climate even when they were used to more-or-less end the world. People at Livermore and in the Soviet Union imagined various 'peaceful' uses for nuclear explosives – such as the building of canals and harbours – but never came up with a way of applying them to the climate.

Mike MacCracken, who worked on climate models at Livermore from the 1960s to the 1990s and went on to run America's Global Change Research Program, tells an anecdote which illustrates the point. In 1976–77 California suffered a severe drought. When, one hot summer afternoon, MacCracken got a call summoning him to Teller's office in an hours' time he knew – 'I just knew' – that Teller was going to ask how this problem could be solved with nuclear explosives. And so he did a set of calculations aimed at getting a sense of the energies involved in the problem. He calculated the energy needed to evaporate the amount of water California had gone without, and thus needed. He calculated the energy needed to warm up by a couple of degrees the cold area in the Pacific that some meteorologists blamed for the drought. He calculated the energy needed to melt the extensive winter snow cover that other meteorologists blamed. Each calculation came up with an answer of a similar magnitude: 10^{21} calories, in the energy units

American scientists used at the time, which in Teller terms was equivalent to the yield of a million one-megaton nukes.

Three different approaches, each yielding the same sort of outlandish figure, served to stop Teller from pushing on with the question. And it is true that, compared with the energies of the earthsystem as a whole, nuclear weapons are pretty puny. But as we have seen in previous chapters, punier powers than these can make a difference on a global scale. It is just that in the nuclear world no one looked for the levers. Leverage such as that which might be available if soot were to prove 'self-lofting' – that is, if, by warming the air in which it sat, it kept itself airborne long after you might have expected it to drift back down to the surface – went undiscovered until the days of Crutzen and TTAPS.

It is easy to look back on this self-interested curtailing of curiosity as evidence of an impoverished intellectual approach, and of a community lessened by its relations with the military. It is harder to accept the implication that such restrictions on vision can surely not have been unique to that particular group and period. If people today think their vision unblinkered, may that not simply be because the blinkers are doing their job? What effects is today's climate research failing to appreciate in face-doesn't-fit ways similar to those which hid the possible climatic effects of nuclear weapons? Unexpected upsides? Unexpected downsides?

Returning to the willful blindness as to the climate effects of nuclear war, though, how was it that, eventually, the possibility of nuclear winter came to be discovered? The answer piles irony on irony. The catastrophic turn in the imagination that had been shaped by fears of a nuclear apocalypse was not restricted to novels, disaster movies and generalized dread. It moulded the course of science as well. Specific insights derived from studies of nuclear explosions led to a widened scientific appreciation of other ways in which the world might end. It was the search for possible examples of such natural catastrophes that eventually led to predictions of nuclear winter.

In the first part of the twentieth century geology was ideologically and methodologically committed to 'uniformitarianism' – the belief that although the Earth's features change, the processes by which those features come into being and pass away are constant. From the 1950s on, though, the discarded idea that the Earth's past had been shaped by sudden catastrophes of sorts unseen in modern times began to return to favour. When Crutzen's work on the effects of NOx on stratospheric ozone was published in the early 1970s, Mal Ruderman, a physicist at Columbia (and a member of the JASON group of advisers to America's military), did what von Neumann had done three decades earlier, but in reverse: where von Neumann had looked at supernovae and seen nuclear wars, Ruderman looked at nuclear wars and saw supernovae. The radiation released in a nearby stellar explosion, he calculated, might create stratospheric NOx in much the same way that bomb blasts did, and on a sufficient scale to badly damage the ozone layer. He went on to speculate that such astronomically induced damage might explain spates of sudden extinction seen in the fossil record.

As yet there is no evidence that this has indeed been the case.[*] But back in the 1970s the idea of a link between supernovae and extinctions caught the imagination of Walter Alvarez, a palaeontologist, and through him, that of his father Luis, who had been one of the bomb-makers of the Manhattan Project. Looking for isotopic evidence of a supernova that might have finished off the dinosaurs, they eventually found, instead, a layer of iridium-rich clay spread all around the world. Iridium is rare in the Earth's crust, but it is common in asteroids, and the Alvarezes decided that that was where it had come from. An

---

[*] In a rather remarkable piece of scientific consilience, though, a supernova observed by Chinese astronomers in 1054 CE has been correlated with a NOx-rich layer in an Antarctic ice core that dates to the same year.

asteroid, or possibly a comet, had hit the Earth 66 million years ago, throwing up so much dust that it had blocked out the sun and stopped plant growth for years, crashing the biosphere.

Giving asteroids a role in the Earth's history was controversial. The *New York Times* was reflecting an initially widespread view among scientists when it editorialized against the idea on the basis that looking for the causes of earthly events in the skies was the business only of astrologers. But the idea of an impact did not come, as it were, out of the blue. In the 1960s Gene Shoemaker, a geologist who did more than anyone to foster the application of his science's principles to planets beyond the Earth, used studies of craters excavated by nuclear blasts to show that some natural craters had been created by asteroid impacts, a possibility which astronomers and geologists had for the most part ignored or denied. Those studies, coupled with an Apollo-era understanding that the craters of the moon – and the newly discovered cratered highlands of Mars – were also created by impacts, led some scientists to start taking the idea seriously. That prepared the way for the Alvarezes' work.

Shoemaker's work was pure and respectable science; that it was of a piece with the broader catastrophic turn in popular culture can be seen in *Lucifer's Hammer*, a bestselling 1977 novel by the Californian writers Larry Niven and Jerry Pournelle. *Lucifer's Hammer* uses the blockbuster format of then-popular authors like Arthur Hailey and the visual imagination of disaster films like *The Poseidon Adventure* and *The Towering Inferno* to tell the story of a comet's impact with the Earth. Its account of what a comet strike might bring about – a superb piece of science-fictional extrapolation which, tellingly, includes hints of an ice age – was produced at a time when serious science had given almost no thought to such things, and it is for this that the book is remembered by scientists who went on to study similar phenomena; as Luis Alvarez put it in one public lecture, 'Lucifer's Hammer killed the dinosaurs.' Less readily remembered, perhaps because less easily celebrated, is the way the science is deployed in the novel as part of a conservative counter-narrative to the counterculture of America's west coast. The end of the world is a relief as well as a scourge; environmentalist hippies and urban African-Americans are incipient cannibals. A paranoid millenarian miasma is seeping up the canyons of Los Angeles long before the comet-induced tsunami sweeps in from the sea, carrying Santa Monica's surfers to their doom.

The Traditions of Titans

This is more or less where we came in at the top of the chapter. The Alvarez hypothesis led the climate modellers who would form the core of the TTAPS team to look at the effects of putting huge amounts of dust into the atmosphere. Because the vast explosive force of the Alvarez impact made it a dark double to nuclear war that was beyond any denial, their work on that dinosaur-killing dust cloud led to studies of what the debris cloud that would follow on from a nuclear war might do. It was to those studies that the insights into smoke and soot in the 1982 'Twilight at Noon' paper by Crutzen and Birks about smoke added their crucial, darkening touch. A way in which the power of the atom could be used to change the climate had finally been found – and it was horrific.*

At the same time, though, the work that led to the TTAPS

---

* It is hardly surprising that, having looked at this unintended climate cooling, the TTAPS scientists went on later to look at climate geoengineering. In 1993 Sagan and Pollack, both now dead, looked at the issue in the context of 'terraforming' proposals for other planets, and while noting that more work was needed, concluded that 'comparatively inexpensive and environmentally prudent methods of mitigating greenhouse warming on Earth [by planetary engineering] may be within reach in the next few decades'. Turco and Toon have both published papers on veilmaking schemes (Turco is very critical of them) and Ackerman is one of the leaders of the attempt to test John Latham's ideas using the spraying technology developed by Armand Neukermans and his colleagues.

paper engendered not one but two attempts to save the world.

The first of these was Carl Sagan's crusade to use the results to change America's nuclear strategies. It is hard to read it as a success, for all the effort he put into it. In 1983, when the nuclear-winter paper came out, Sagan was famous in a way few scientists ever get to be. He was a bestselling author; his TV show *Cosmos* had been a hit; for more than a decade he had averaged two guest slots a year on Johnny Carson's *Tonight Show*, which for anyone who wasn't Bob Hope was pretty remarkable. And he had long worried about nuclear weapons – one of the reasons he gave for his interest in the search for extraterrestrial intelligence was that success would show that technological civilizations did not have to head down the path to supernova suicide. But until the early 1980s he had not used his charisma and nous to campaign on that or any other political issue, even though friends on the left, such as Stephen Schneider, had long encouraged him to do so.*

The TTAPS results might not have been enough to turn Sagan into a campaigner on their own – a brush with death on the operating table and a new marriage seem also to have had a lot to do with it – but they gave him ammunition. At a time when officials in the Reagan administration were happy to talk of winnable nuclear wars, and new weapons in America's arsenal seemed to have been designed to launch a first strike, Sagan argued that the science of nuclear winter revealed such ideas to be suicidal as well as genocidal. In the words of the writer Tom Levenson, who covered the issue for WGBH-TV in Boston, Sagan set out to generate 'fear, then belief, then response' – pretty much the modus operandi for subsequent climate activism. Accepting the TTAPS results showed that policies had to be changed; instead of the build-up of new nuclear systems

---

* Schneider was also a guest on the Carson show for a time in the 1970s – the only climate scientist, I think, ever to be so. But after making the mistake of upstaging the host he never got invited back.

then under way, there had to be a freeze followed by a steep reduction in nuclear arsenals.

I saw him make this case when I was an undergraduate at Cambridge in 1984: it was a heady experience. But it is worth noting that while, for Sagan, the line from the TTAPS results to his preferred policy was compelling and direct, there were other responses available. TTAPS could have been taken as making a case for new weapons and strategies: for weapons that exploded underground rather than in the air, thus setting fewer fires; for a 'Star Wars' missile shield; for large stockpiles of food. Nuclear winter could thus be used to argue for a new arms race, rather than for an end to arms races. There is only very rarely one possible policy response to new science.

When someone claims there is, the obvious rejoinder for those who dislike that policy response is to attack the science. The TTAPS results undercut the logic of America's (and Russia's) strategic stance – they were always going to be controversial. But Sagan's campaigning made them more so, and probably served to intensify the criticism the results received; the cocksure charm and lyrical certainty that made Sagan a draw on television did not endear him to his scientific peers. The TTAPS team had missed out some effects that others thought might ameliorate matters (or, perhaps, worsen them). Research that made different assumptions and used other models tended to show environmental effects that, while often pretty catastrophic, were less dramatic than those seen in the TTAPS paper, sometimes markedly so. What would normally have been quite technical arguments about things like the possibility of self-lofting, the different behaviours of wet and dry soot and the importance of properly capturing the day/night cycle in your model became more heated than would usually have been the case. Modelling by Stephen Schneider and colleagues at NCAR which led them to talk of 'nuclear fall' rather than 'nuclear winter' damaged Schneider's friendship with Sagan badly; Schneider, to whom it was important that climate models keep their credibility because of their importance in other debates, felt that Sagan was sacrificing science in favour of impact. The irony that he had in the 1970s pressed Sagan to become more politically engaged was not lost on him.

Criticism founded on genuine differences of opinion blended into attacks based on an unwillingness to accept Sagan's policy prescriptions. The idea that climate science was a rat's nest of left-wingers stretching their computer models to unreasonable lengths in order to try and subvert America's national interests flourished in the minds of the *Wall Street Journal*'s editorial board, among other places; a decade later it would be reused more or less unchanged in attacks on attempts to curb carbon-dioxide emissions.

Sagan and Turco later claimed that fears of a nuclear winter were instrumental in ending the cold war. There is no strong evidence for this, and it is not a common view – indeed, it is quite easy to find histories of the cold war that make no mention of nuclear winter at all. In 1983, after a pre-publication briefing on the TTAPS paper, an arms-control expert told Sagan that 'if you think that the mere prospect of the end of the world is sufficient to change thinking in Washington and Moscow you clearly haven't spent much time in either of those places'. He may well have been right. That said, the idea may have contributed to a longer-term shift in attitudes. In the past few years the idea of aiming for a world with no nuclear weapons has gained increasing currency in policy circles, and it is possible to see the shadow of nuclear winter, and the metamilitary threat it poses to combatant and non-combatant countries alike, behind the impulse to change the basic terms of the nuclear debate. But that may be projection; the idea is rarely evoked specifically. When President Obama spoke about moving towards a post-nuclear age in a 2009 speech given in Prague he made no mention of nuclear winter; he did, though, make mention of global warming as one of the new threats to which the world must turn its attention.

What of the second campaign to save the world which I earlier claimed had flowed from the same science as TTAPS? It was much less high-profile, but arguably considerably more successful. In the three decades since the Alvarez paper made stratospheric shrouds a subject for concern, this second campaign made the Earth a significantly – though not noticeably – safer place. The anthropologist Margaret Mead famously said that one should 'never doubt that a small group of thoughtful, committed citizens can change the world; indeed, it's the only thing that ever has'. I don't actually believe this, though I can see the sentiment's attraction. But if I did, it would be to the clique at the heart of this second campaign that I would point as evidence.

Mike MacCracken told me his illustrative anecdote about Edward Teller and the California drought while we were sharing some wine in a courtyard in Erice, a Sicilian hill village which, thanks to the schemes of an Italian physicist called Nino Zichichi, has become one of the scientific world's most agreeable conference venues. The topic that had brought us there was neither nuclear winter nor geoengineering – though as a climate modeller at Livermore MacCracken had worked a lot on nuclear winter in the 1980s, and he has also been a proponent of increasing research and policy discussions on geoengineering. We were there to think about the danger posed by asteroids that might hit the Earth.

Before the Alvarez paper, most astronomers gave little thought to this danger. They lacked both the evidence with which to calculate the frequency of such impacts and a sense of what damage they might do. And the whole subject risked making them look more like astrologers prophesying doom than they would wish (witness that silly *New York Times* editorial on the Alvarezes). When in the 1970s Gene Shoemaker and a colleague, Glo Helin, started using an old telescope to look for asteroids in orbits that crossed that of the Earth, and which thus might one day hit it, they were seen as indulging an eccentricity more than as doing science, let alone contributing to the planet's safety.

The Alvarez paper started to change that. There were questions at the time – and there have been, on and off, ever since – as to whether the impact the paper described was the single decisive cause of the dinosaurs' demise or just one thing among many. But there is now no serious doubt that one day a little more than 66 million years ago an asteroid about 12 kilometres across hit what is now the Yucatan with an energy equivalent to that of 20,000 simultaneous all-out nuclear wars.

In the wake of the Alvarez paper, a NASA workshop put together arguments for a systematic programme aimed at finding asteroids in orbits which crossed the Earth's and which thus might at some point pose some similar sort of hazard. Shoemaker, charged with writing up the workshop's proceedings and recommendations, never got round to doing so; but some of the arguments were later published in *Cosmic Catastrophes*, a book by Clark Chapman and David Morrison. The two space scientists argued, counterintuitively but convincingly, that the average person's chances of dying in an asteroid impact were as high as one in 30,000 – similar to the chances of dying in a plane crash. The largest contribution to the risk came neither from the most common impacts capable of killing a lot of people – objects 100 metres or so across, which explode at or near the Earth's surface with the force of a pretty big nuclear bomb a few times a millennium – nor from giant impacts such as those described by the Alvarezes – which would wipe out almost all the human race as well as half the other species on the planet, but which come by only once every 100 million years or so. The biggest risk came from asteroids a couple of kilometres across, the sort of thing that might be expected once or twice in a million years.

Again, it's a matter of leverage. Small impacts have regional effects because their effects are direct: heat and shock waves. If such an impact happened to a city it could kill millions. But most regions on the Earth are pretty empty, and so most such impacts would be expected to kill relatively few people. Once the asteroids get up to a kilometre or more in size, though, they kick up enough dust to have global effects more or less wherever they hit. If the impact were big enough, the shroud would act like a severe nuclear winter, shutting down agriculture for a year or so.* Such a climate emergency would not be enough to drive humans, or any other species, to extinction, but it would kill a substantial fraction of the world's population. Even if the odds of an impact in a given year were a million to one, over a lifetime that would add up to a small but appreciable risk.

But even after the catastrophic turn in the world's imagination, the threat from asteroids was hard to take seriously. Most people hearing such calculations for the first time laughed or threw up their hands. The media seemed all but obliged to mention Chicken Little and her fear of the falling sky whenever the subject came up. The statistical nature of the threat was hard to communicate. Impacts did not fit into any easily understood narrative; unlike environmental or technological disasters, they were hard to understand in terms of nemesis or hubris, or the wages of sin, or the revenge of an angered nature. And when it

---

* 'Big enough', here, means releasing a lot more energy than a nuclear war, as an impact which deposits all its energy in one place is a lot less good at kicking up aerosols and triggering fires than a war which spreads its *Schrecklichkeit* across thousands of targets. Indeed, by an interesting coincidence which may also owe something to physicists' fondness for round numbers, the energy in one of these globally important impacts is the same as that which Mike Mac-Cracken calculated as being needed to do something about the California drought of the late 1970s: 1 million megatons.

came down to it, were there not enough things to worry about without adding an unknown asteroid that was, on average, 5,000 centuries from doing any harm?*

Yes, there were. But the fact that the threat was almost certainly not immediate did not make it uninteresting or utterly negligible. Some people took the question seriously. I was one; much more importantly, a Capitol Hill staffer called Terry Dawson was another. In 1991, largely due to Dawson, George Brown, chair of the science committee in America's House of Representatives, put into law a requirement that NASA should look into how it might detect all the asteroids that posed a planetary danger and how, if the need arose, such threats might be dealt with.

David Morrison – once, as it happens, a student of Sagan's – was charged with studying the problem of detection, and a team he led produced a report saying that with a fairly modest investment in Earth-based telescopes – $300 million over a decade – it would be possible to discover almost all of the asteroids a kilometre or more in size in Earth-crossing orbits. Once discovered and characterized, their trajectories over the coming centuries could be calculated. If none of these trajectories intersected with the Earth's then the impact problem was effectively solved. The risk of a globally disastrous impact by an unknown asteroid would have been greatly reduced, simply by increasing knowledge; the risk of an impact by a known asteroid would in all likelihood remain zero.

And that is what has happened. There was never a systematic programme of the sort that Morrison envisaged, but new sensors developed for the Air Force made it possible do more with small existing telescopes; soon there were a number of programmes discovering nearby asteroids at a gathering pace. The population

---

* A few people do find the prospect of asteroid impacts a genuine concern in terms of their own lives, rather than just an abstract statistical worry. As it happens, a dear friend of mine who has various anxiety issues is one of them.

of potential impactors came to be better understood at a statistical level – well enough understood that it was possible to say with some assurance how many of the big ones were left to discover. When Morrison wrote his report, less than 10 per cent of them had been discovered. As of today, well over 90 per cent have been found, and none is on a collision course. An apocalyptic threat which really could have been there has proved not to be, which I think makes it fair to say that the world is a safer place as a result – thanks to the work of a group of scientists which, as Morrison liked to point out, never numbered more than the staff at a medium-sized McDonald's.

What, though, if things had turned out differently? If a large asteroid had been discovered in an orbit that would have led to a collision the obvious response was to nuke it – how, in the late twentieth century, could it have been anything else? Teller, who in the early 1990s was taking a predictable interest in the subject, suggested with a certain enthusiasm that destroying such an asteroid might require hydrogen bombs hundreds or even thousands of times more powerful than any in the military arsenal.

A more subtle analysis suggested that, rather than try to destroy the incoming asteroid, one might instead nudge it aside – an option which, it was thought, would probably also require nuclear explosives, but on a more modest scale. Such nukes could, in effect, turn an asteroid into a dust-powered rocket. A nuclear explosion a little way off would heat up the dust on one side of the asteroid enough for it to stream out into space. Just like a rocket exhaust, this would push the asteroid itself in the opposite direction. A sequence of such nudges, with the results from the first feeding into the decisions about when and where to let off the next, could make a difference to the orbit which would, over time, grow large enough for the asteroid to go from being a hit to being a miss. And time would in all likelihood be on humanity's side – a threatening asteroid was much more likely to be spotted a few decades out than a few months out.

Duncan Steel, an astronomer who took a keen interest in the asteroid hazard, found a quotation from Lord Byron, as set down by Byron's friend Thomas Medwin, which wonderfully encapsulated the grandeur of such thinking. Like Soddy a century later, Byron compared the growing powers of humankind to those of imagined ancient civilizations, and saw them as making possible tasks as magnificent as those that Steel, Chapman, Morrison and their colleagues feared might, just possibly, be necessary:

> Where shall we set bounds to the power of steam? Who shall say, 'Thus far shalt thou go, and no farther?' We are at present in the infancy of science. Do you imagine that, in former stages of this planet, wiser creatures than ourselves did not exist? . . . Might not the fable of Prometheus, and his stealing the fire, and of Briareus and his earth-born brothers, be but traditions of steam and its machinery? Who knows whether, when a comet shall approach this globe to destroy it, as it often has been and will be destroyed, men will not tear rocks from their foundations by means of steam, and hurl mountains, as the giants are said to have done, against the flaming mass? – and then we shall have traditions of Titans again, and of wars with Heaven.

A Tale of Two Cliques

If there is a project that evokes the traditions of titans today, it is surely solar geoengineering. And there are instructive and intriguing parallels between what is happening in that field and the attempts, over the past few decades, to understand, and if necessary avert, the climatic emergencies that impacts might bring about.

Both efforts identified technological interventions which might offset massive risks to the earthsystem. Both have pointed to natural spectacles that illuminate and demonstrate their analysis of the risk and the potential for action – the Pinatubo eruption, in the case of geoengineering; the 1994 impact of comet Shoemaker-Levy 9 on Jupiter, in the case of impact mitigation. The sight of a fragmented comet leaving visible scars on the swirling cloudscape of another planet did a great deal to make the impact hazard seem real.

The size and character of the scientific community involved in the two areas is another clear similarity. An 'asteroclique' formed in the 1980s and 1990s in much the same way as the geoclique formed in the 2000s – a set of people who would turn up at conferences to swap results, reiterate positions and argue. Like the geoclique, its common interest was spiced up by scientific and political differences, as well as supported by friendships. It had dominant personalities, but after the death of Gene Shoemaker no clear leaders. Like the geoclique, it was essentially interdisciplinary, made up of practitioners of observational astronomy and government policy; experts in the design of space missions, in orbital mechanics and in the physical properties of asteroids; and students of the impact history of the Earth and other planets and of the effects of nuclear weapons. As in the geoclique, almost all these people had other research interests that took up most of their time; later on in the process some people were funded to work specifically on the problem, but up to the late 1990s it was hardly anyone's primary job.

Scientists outside the asteroclique often saw the impact problem as unimportant and faintly disreputable. Astronomers are covetous of telescope time, and many didn't want to see it wasted on hunting asteroids of little scientific interest. Nor did they want to share in the media panics the field seemed prone to. Every few years a story would hit the media about some object or other that might hit the Earth, and on each occasion the asteroclique would be riven by arguments about who had irresponsibly abetted the story, or failed to explain the errors, or whatever. The geoclique has been similarly marginalized by the climate science community, though, as we have seen, for rather deeper reasons. It has not experienced anything quite like the flaps over newly discovered asteroids, but it has from time to time had its internal divisions thrown into focus by bursts of media attention – not, in itself, at all a bad thing.

In both cases a particular source of worry to many outsiders – and to some insiders, too – was the closeness of the clique to the military–industrial complex, and in particular to the part of that complex centred on Livermore. It was hard to look at analyses of the impact risk from the lab (some of which greatly overstated it) without thinking that this was some sort of an attempt to find a partial replacement for the lab's original cold-war mission. Livermore was, of all America's national laboratories, the one most interested in strategic missile defence – the 'Star Wars' programme. And if impacts were a dark double to nuclear war, impact prevention was a dark double to Star Wars.

The Pentagon's Star Wars office – formally, the Strategic Defence Initiative Organization – took a keen interest in the subject. Part of the rhetoric surrounding the *Clementine* mission to the moon that the star warriors mounted in 1994 – and which would, but for an accident, have flown on to the Earth-orbit-crossing asteroid Geographos – was that it served to demonstrate the ability to scout out such asteroids. A mooted follow-on mission would have tried firing (non-nuclear) interceptors at one. The Erice meeting at which MacCracken and I chatted about how one might, or might not, nuke rain out of a clear California sky was paid for by the Pentagon.

The level of military–industrial involvement was in fact considerably higher in the case of the impact hazard than it has been with recent discussion of climate engineering; with the exception of the crucial early paper by Teller and Lowell

Wood, and the willingness of General Pete Worden, the begetter of the *Clementine* mission and until recently the head of NASAs Ames Research Center, to host meetings on geoengineering, there has been little close involvement. But the deep links between the military, weather modification, climate control and nuclear-power Prometheanism are undoubtedly worth keeping in mind.

The key difference between the efforts was that the astero-clique's observations have so far given it no reason seriously to consider any action beyond observing. It is worth bearing in mind that the geoclique's work could have a similar outcome; it might resolve the questions about geoengineering firmly on the won't-work side. If that were to happen, it would be an achievement. Just as it matters to how you think about the world's future that the odds of being hit by a large asteroid can reliably be put much lower than they could twenty years ago, so it would also matter if through further research and discussion people became convinced that veilmaking could never work. If there is truth to the moral-hazard argument, knowing for sure that there is no technological cooling on offer ought to matter in terms of increased mitigation effort.

But the fact that the asteroclique has not seen action does not mean its story is over. In an obvious parallel with the possible course of geoengineering, the aims of the asteroid-detection effort have shifted as it has gone on, and the shift has increased the chances that, eventually, attempts will be made to change an asteroid's orbit.

Once the asteroid hunt got under way it built up its own internal logic – why would you stop looking for things that might hit the Earth just because the ones that had originally worried you had been found? At the same time new technology allowed the detection of ever-smaller bodies. The risks posed by such collisions are much smaller than those attendant on big climate-changing impacts. But if one is actually found it is easy to imagine that the risk calculus might start to look a bit different – especially to the people living on the small part of the Earth that was going to take the hit. There is something here which seems to me quite similar to the way that discussion of marine-cloud brightening has moved from a stress on global averages to looking at coral reefs, glaciers and hurricanes, and to the sometimes-raised possibility of temporary regional veils for heatwaves. It's not what people started off thinking about, and it's rather smaller scale, but it might be attractive nonetheless.

In the case of increased demand for the deflection of small asteroids, the good news is that asteroclique thinking on these matters has moved away from the more Tellerian options. With a lot of advance warning – always the most likely scenario – it seems plausible that a small asteroid could be nudged off course without recourse to nukes. There are various other ways it could be pushed and pulled around, perhaps using nothing more troubling than the pressure of sunlight. Such gentle pushing is plausible not only because the asteroids in question are small ones, but also because a better understanding of some of the subtleties of orbital mechanics has opened up strategies where the Earth's own gravity amplifies almost trifling changes into deflections worth making – a very Archimedean lever indeed.

The bad news is that this brings up a 'winners and losers' problem which provides another echo of geoengineering debates. If an asteroid is going to hit the Earth, it is going to hit a particular spot on the Earth. If you change its orbit enough, it will miss the Earth completely. If you change its orbit insufficiently, though, you will just move the point of impact.

The impact points for different levels of insufficient nudging, like the track of a solar eclipse, form a well-defined path across the face of the planet, a bit like a tram track. And there will be people living at various points along that path. The question of deflection thus becomes not just one of can-it-be-done, but also one of what–if–it–fails. Is it worth deflecting an impact that would leave a substantial hole in Nebraska if, in so doing, you run the risk of dropping it on Mexico City? Or deflecting an impact from a fairly empty bit of Russia if the tram tracks pass over Beijing? Opinions might differ – just as they might if models were able to project the risks and benefits of geoengineering with anything like the same accuracy.

Such complexities are part of the price of making what was once an act of God a human responsibility. It is a price that, I have come to realize, some think not worth paying.

When I first wrote about asteroid impacts in the 1990s I was frustrated by the difficulty people had getting past the 'giggle factor'. It took me a while to understand that the laughter was not uncomprehending but, at some level, anxious – anxious not so much about the asteroid threat as about the human powers that countering it implied.

It was not the idea of impacts per se which made people laugh and change the subject. It was the thought that humans might be able to act on a scale necessary to avert such things. That they might move the stars – albeit very small ones – in their courses. That is a more uncomfortable feeling to some people, I think, than the idea that they are at risk of a cosmic accident. In the abstract, a natural disaster can feel more acceptable than the idea of tampering with nature to avert it. This is the same uneasiness that sits at the heart of the deepest concern about geoengineering, concern not about its possible evil consequences, but about the sheer scope of the idea itself. To act in ways that avert Armageddon, whether by means of asteroid wrangling or by means of aerosols, seems immodestly to overstep limits that should be fundamental. It changes what it is for humans to be humans and what it is for nature to be nature – it takes human empire over the border of blasphemy.

I believe imagining human action on inhuman scales can reveal something sublime, a distant lightning of the mind. It offers the humble awe of being part of something so much greater than oneself; of knowing that people like you built the curved cliff of a dam that holds back a trillion tonnes of water, the space station seen climbing through the night sky, the far-off city lights brightening half the night horizon. But I realize that not all people have the privilege or inclination to sympathize with the builders of such things. And I can see that their scale can be deeply troubling.

It may be a matter of aesthetics. The conflict between the fear and power chambered together in the heart of the sublime becomes much more uncomfortable if no beauty is experienced, and the beauty to be found among the abstractions of radiative forcing and orbital dynamics, or in the unnatural – counternatural – machinations of technology, is not for everyone.

To say this is a matter of aesthetics is not to say it is not of importance. Quite the reverse. These feelings are not trivial, or simple questions of taste. Aesthetics matter because beauty matters. I can see arguments for geoengineering based on compassion, on duty and on virtue. In the end, though, should there not – must there not – be an argument that stems from beauty? If there is no beauty, merely power, can the technological traditions of titans be anything other than monstrous?

After Such Knowledge

Carl Sagan must have had his own thoughts on the uneasiness of the technological sublime. In 1958, fresh out of graduate school, he worked on an Air Force project that looked into the possibility of setting off a nuclear device on the moon. It had a scientific rationale – to provide insights into cratering – but also a more important political purpose: to invite other nations to consider the similar things which might be done to them. It's hard to imagine that that work did not feed his later interest in making sure that humans didn't set off one of von Neumann's supernovae.

Sagan's main contribution to the asteroid-deflection debate, though, was rooted not in aesthetics or the impropriety of playing God, but in a later-developed distrust of the military–industrial complex. There are, he argued in the 1990s, many more asteroids on courses that just miss the Earth than there are on courses that hit it – the outer ring of a target is necessarily bigger than the bull's-eye. If people could nudge away the rare asteroids that would otherwise hit the Earth, they could just as easily nudge the more common asteroids that miss it on to trajectories that would hit. To develop the capacity to deflect asteroids, Sagan warned, was to develop the capacity to wage asteroid war. The cold-war tendency to imagine natural catastrophes as the dark doubles of nuclear weapons, and vice versa, returns.

Like most such ideas about weaponizing geophysics, though, the thing that Sagan was afraid of was far-fetched. It is true that there were a few people at Livermore arguing that there should be an experimental programme aimed at demonstrating that asteroid orbits should be changed, and that an operational capacity to do so should be set up on the off chance it might prove necessary one of these millennia. I remember being shown (though I cannot now find) a copy of a presentation in which the possibility that the Russians might weaponize asteroids was raised: 'Mr President,' you can imagine some Strangelove saying, 'we must not allow an asteroid gap.'

But as with many of the more baroque ideas of the cold war, a standing anti-asteroid force was never going to happen. The fringes of cold-war thinking were full of wild schemes that were kept from amounting to anything by some mixture of clear impracticality, mind-numbing expense and wiser counsel. That scheme to nuke the moon is one example. Vast space colonies were another, the excavation of canals and harbours by means of nuclear explosives another. You can find reports in the archives of work on such ideas, but the enthusiasm of impractical scientists, the willingness of researchers to get paid for writing up analyses of the most untethered fantasies of under-employed colonels, and even the ability of Edward Teller to win funding for a pilot programme should not be mistaken for evidence that such things would be taken seriously at the level of government where they could be turned into national policy. The dark geopoetry of the cold war repeatedly came up against strikingly conservative attitudes in the higher echelons when it came to trying such things out. In 1963 President Kennedy gave explicit instructions that all geophysical experiments with large impacts should be reviewed at the highest level.

Most plans for weather and climate modification and environmental warfare fell into this category of quasi-realistic impracticality. And they were a bit pointless, too. It wasn't just that there was no very clear idea about how to go from the first part of von Neumann's programme − using computers to forecast the weather and model the climate, both endeavours which have worked out pretty well − to the second part − using the understanding the computers provide to exert precise control. It was also because the ideas didn't actually seem to offer much by way of military advantage. You could knock the Van Allen belts out of kilter − indeed, this idea got through the net and was tried out with a few nuclear tests. You could look at ways to make tsunamis by means of vast bombs exploded under water. But when it came down to it, why bother? The grim capability that had started all the speculation about environmental warfare also made it somewhat otiose. The bomb gave the major powers the wherewithal to reduce each other to ruins. How useful was it, really, to find additional ruinations?

Environmental warfare was not all bluster. But though disturbing, it never became crucial to military planning. The capability was easily and readily given up in the negotiations which led to the UN ENMOD treaty just a couple of years after America's armed forces were found to have been using cloud seeding and defoliation in Indochina. No one in the military or outside it much mourned its passing.

What, then, of using solar geoengineering as a weapon? Its effects could conceivably cause wars, but as a way of fighting them it seems as impractical as most forms of environmental warfare. If it is done locally it can be undone locally – a fleet of Salter cloudships could be sunk by anti-ship missiles as easily as any other fleet. If it is done globally it would be so untargeted as to be hardly a weapon at all. Even if you came up with a reason to seek to damage some adversary by cooling their climate, rather than by more normal means, it is very hard to imagine how you could use veilmaking to do so without cooling a great deal of the rest of the planet too.*

There are, though, a few possibilities for weaponization that should at least be mentioned. One is sometimes called 'counter-geoengineering'. If a strong solar geoengineering programme were enacted unilaterally, it would be possible for other powers to counteract its effects by temporarily strengthening the greenhouse effect. Difluoromethane is a compound rather like the CFCs banned by the Montreal protocol, in that it has a very high global warming potential – almost a thousand times greater, molecule for molecule, than that of carbon dioxide. Unlike CFCs, though, it has no effect on ozone and an atmospheric lifetime of only half a decade or so. If you really didn't like someone else's solar geoengineering, emitting millions of tonnes of difluoromethane would allow you to strengthen the greenhouse effect in response. When the other party blinks, and stops its veilmaking, you stop emitting and the status quo is back within a few years.

---

* That said, in a world in which geoengineering technologies exist, or are rumoured, strange weather events will often be ascribed by some to their use whatever the evidence or lack of it. Similarly, even though there are no current asteroid-diversion technologies, a damaging asteroid strike would, in many parts of the world, be blamed on America.

This seems to me an incredibly unlikely bit of statecraft. If you don't like someone else's geoengineering it makes much more sense to round up a strong coalition to oppose it, perhaps by mounting a credible military threat against the geoengineering facilities.* I would imagine that counter-geoengineering would make that a lot harder, and maybe lay you more open to attack yourself. But I feel it would be remiss not to mention it.

The other thing that geoengineering can offer is a doomsday device. David Keith has noted some subtle physical effects that would seem to make it possible to create particles which self-lofted much more effectively and reliably than soot does (if soot does — it remains an open question). Such particles could lift themselves up above the stratosphere and stay up for decades rather than years. Unlike stratospheric geoengineering, which needs constant replenishment that could, if desired, be interrupted, geoengineering using these designer aerosols could be semi-permanent and unstoppable. Deployed in sufficient number, they might be able to decrease the incoming sunlight enough to cool the planet by a number of degrees. If they did so for long enough — a century, say — they could conceivably start an ice age.

Indeed, in principle they could go further than that. In some circumstances — it has happened a few times over the Earth's history — the ice–albedo effect can drive glaciation a lot further

---

* A third option — if they geoengineer, you should geoengineer even more — is explored in a remarkable 1947 novelette by Gerald Heard, a friend of Aldous Huxley, whom we encountered championing the sublime power of the realms beyond the stratosphere in Chapter One. In 'The President of the United States, Detective', published to some acclaim in *Ellery Queen's Mystery Magazine*, a rather *übermenschlich* president responds to a Chinese plot to flood America by melting the Arctic ice with a programme of his own which melts even more of the ice. Thus China, too, is flooded, while America is presented with the magnificent new territory of an ice-free Greenland where it can enjoy its manifest destiny beneath the transcendental energies of the aurora. It is quite remarkably bonkers; I would love to know what Julian 'Nuke-the-Arctic' Huxley made of it.

than it has gone in the recent ice ages. In these 'Snowball Earth' events it appears that almost all of the continents and most or all of the oceans become ice bound. Triggering such a snowball event would rearrange the earthsystem to a degree far beyond anything a nuclear winter, or even an Alvarez–scale dinosaur-killing asteroid, could manage.

At this point, something like difluoromethane looks quite attractive.

For the most part, such speculation says more about how easy it is to imagine the world ending than it does about practical threats from geoengineering. It is sobering to understand that such power is at least a part of what we are talking about — that the levers under discussion could, in the extreme, push the Earth into more or less total inhabitability. But the mundane horror of an old-fashioned nuclear war seems to me far greater than the chances that the rest of the world might stand by as someone tried to push it into an ice age; nuclear winter worries me more than this dark double. Though it is good to have the outside edges of possibility defined, it would be a mistake to let them dominate the way you see the world. Better to try and understand what might actually come to pass in a world where solar geoengineering gets under way — a process that provides ample room for fears of its own, but is not without its offerings of hope.

12

The Deliberate Planet

Far more can be mended than you know
Francis Spufford, *Unapologetic*
(2012)

In 2003, I wrote an opinion piece for the *New York Times* about a little sort-of-geoengineering project. My friend Euan Nisbet, a Zimbabwean geologist, was exercised by the fact that the glaciers on top of Kilimanjaro were melting, and that if they continued to do so then the record they contained of earlier African climates would be lost for good – as would a tourist attraction which brings a significant amount of foreign money into the Tanzanian economy, and as would a thing of ancient beauty. Simply covering the cliff-like sides of the glaciers with white tarpaulins would, he thought, be enough to bring the melting almost to a halt by shielding the ice from sun and wind. The tarps of Kilimanjaro could buy at least a few decades of breathing space, maybe a century of it. In that time the forests on the flanks of the mountain could be replanted, moistening the air rising to the summit in a way which might go some way to restoring the ice.

My contribution to the idea was to suggest that such an exercise in care might appeal to Christo, the artist who, with his partner Jeanne-Claude, has become famous for wrapping up things as various as the Reichstag and the islands in Miami's

Biscayne bay. The preservation of beauty could in itself become a work of art, an incarnation of the technological sublime which would offer inspiration to those who worked on it and witnessed it alike: 'The white tarps would float above the clouds a tentative hope: the hope that human will and ingenuity just might be able to meet the challenges of a century in which more change will be faced, and more protection needed, than at any other time in human history.'

I still think it's rather a good idea, and at the time I nursed a private hope that it might actually develop a little traction, though I now know that Christo absolutely rejects all suggestions from others. What I did not expect was that it would draw a riposte from the very newspaper that published it. Op-ed articles often take a line that the paper publishing them does not, but it is rare for a paper to take time to rebut an idea that has only ever been floated on its own pages. The editorial that ran a few days later was just such a rarity. An artwork of beauty, care and hope, it suggested, would not be welcome. Better to get the data that the glaciers have to offer as quickly as possible and then abandon the ice to its fate, in the process leaving 'Africa with a new icon – a bare mountaintop underscoring the folly of the reckless destruction of the forests.'

It was an interesting lesson. It seemed to me at the time to show that there were people who would pass up an opportunity to care for beauty, to respect the Earth's history and to nurture the livelihoods of others if they could score points about human folly instead. In retrospect, though, calling it point-scoring seems perhaps unfair. The heart of the rebuke was a sense that the folly would not be recognized as folly unless it did all the damage it could. And that full recognition of the folly was the only route to avoid further reckless destruction.

I can't say that that there is no wisdom in that stance. I can only say I do not share it. I refuse to accept a world in which nothing can be protected and only pain and loss instruct. It is sometimes a bad idea to shield people from the consequences of their actions, and when those consequences damage other people's lives such shielding poses particular moral problems. But I do not think that such protection is always, in all circumstances, a bad idea. I think that saving people from some of the harm their actions do can help them and others. Look, for example, at needle exchanges. They let people go on doing things that harm them – but they take away part of the harm to the people making the exchange and to others. What's more, they provide avenues that may make it possible to talk about that harm and the behaviours that underlie it.

I understand why people are afraid of climate geoengineering. Its various techniques pose risks that are not as well understood as they need to be, and pursuing some of them with any seriousness may weaken other efforts to limit the risks to which the build-up of carbon dioxide exposes the world. It tampers with what people understand to be natural, which rouses feelings from uneasiness to disgust. When such tampering seems to be an essential part of the problem, many find it unimaginable as the solution, treating the mere suggestion as stupid.

I would never say that geoengineering was the solution, or even a solution. But then I think that it is a mistake to treat climate change as a problem to be solved. Something as complex as the relationship of industrial civilization to the earthsystem that it shapes and is shaped by isn't the sort of thing that is simply solved, once and for all, and it's a snare to think that it is.

There are worthwhile things to do, though, which are not solutions. I can imagine ways for geoengineering to reduce harm to people at risk, and to preserve things that would otherwise be lost. What is more, I can see it helping in a renegotiation of what it is for things to count as natural that may ease some of the contradictions of the Anthropocene. I can even see it as an inspiration.

As the previous chapter showed, it is easy to imagine

catastrophes. It is much harder to imagine unabashed utopias – or at least, with the knack out of fashion, it seems so. Instead, drawing on something somewhere between the planning of wars and the scripting of movies, people tend to deal with scenarios. Here is one to explore.

The Concert

It is not a large nation that does it – indeed, it is not a single nation's action at all. There is a small group of them, two of which are in a position to host the runways. They call themselves the Concert; once they go public, others call them the Affront. None of them is a rich nation, but nor are they among the least developed. All of them already have low carbon-dioxide emissions, and all of them are on pathways to no emissions at all. Not too hard for small developing nations with sun, sea breezes and no aptitude for, or ambition towards, heavy industry, but still impressive: in climate terms, they look like the good guys. But their low emissions and the esteem of the environmentally conscious part of the international community are doing nothing to reduce the climate-related risks their citizens face. Indeed, the fact that their commitment to mitigation does nothing to make their people safer is a point to which they return on numerous occasions in the years following the group's establishment.

The hardware is mostly sourced from the Brazilian and Indonesian aerospace industries. The only bits of the planes' design that are really new are the wings and the spraying kit – the engines and the fuselage are adapted from other craft. The tanker planes that extend their lifting capacity and their range are utterly standard. The aircraft's development had been overseen and paid for by one of the world's increasingly numerous billionaires, who had made her money from high-density data-storage systems. Her cover story was the development of a space-tourism follow-on to Virgin Galactic; that project's name was Espedair, a name that stuck even when the cover was blown. The project cost more than the most extravagant of her peers had ever spent on a yacht – but not all that much more.

The Concert has two sites for operations, one in Central America, one in the South Pacific. With a few flights a day from each site, they deliver a few tens of thousands of tonnes of aerosol to the stratosphere over the first year. Sprayed out comfortably above the tropical and subtropical tropopause in both hemispheres, this forms a tolerably even, remarkably tenuous veil. There had at one time been a satellite devoted to measuring stratospheric aerosol density which might have allowed researchers to notice the veil's creation, but after that satellite's life was over no one replaced it – there's always some group with an even more interesting set of measurements to make, and there are only so many missions you can launch.

After 18 months of operations, the Concert announces what it has been up to at the UNFCCC climate summit of 202–.

The Concert presents its programme as an act of civil disobedience. Not, the countries say, that they are actually breaking international law. If their actions were hostile they would have been in breach of the ENMOD convention; if they caused demonstrable harm, they might be liable under customary international law. Neither is the case, says the Concert. But the countries making the veil are happy to admit that they are breaking the norms of international relations in a way that might inconvenience, discomfort, even shock. Civil disobedience does that. When there is a just cause to be fought for, the Concert argues, and when there is no forum in which the fight for that cause shows any sign of making progress, then something like civil disobedience is called for. To disobey the tenor of times, they say, is not a crime. It is a duty.

The practical aim of its action, the Concert explains, is straightforward and limited. It does not intend to stop or reverse warming; it intends only to slow it. It plans to thicken the veil at a pace that its climate modellers think will keep the rate of warming at or below 0.1°C a decade. A limit to the rate of change of the temperature, the historically minded are reminded, had been a widely canvassed objective when action on greenhouse gases was first mooted in the 1980s. The Concert's target, if it could deliver it, would mean that over the rest of the century the temperature would rise about as much as it did over the twentieth century. Cumulative change by the end of the century would remain below the 2°C limit. How thick the veil might have to be to achieve this would depend on future emissions; even if they went unchecked, the veil would not reach Pinatubo thicknesses for almost a century.

But the Concert makes it clear that it doesn't want emissions to go unchecked. It wants other nations to commit, as its members have, to quite steep cuts in emissions. And it is happy to welcome to its ranks nations that make such commitments, especially if they also commit to the development and deployment, over time, of technologies for carbon-dioxide removal. As new members of the Concert, those acceding nations get a say in decisions about revisions to the veilmaking plan in view of new monitoring data and new understanding of the earthsystem – the original Concert members are aware that they started with inadequate monitoring and sketchy knowledge, and are keen to reduce the risks that brought with it – as well as revisions that might be required by new trends in the politics of emission. Other nations do not.

The discussion moves with more than deliberate haste from the climate summit to the Security Council, as the Concert had known it would. Indeed, the announcement had been timed to coincide with a session of the Council when one of the Concert's members had a seat on it. The Concert's reckless attempt to seize power over the climate – to mount 'a coup against the planet', as the elderly Al Gore put it – is decried by various nations, including some of the Council's permanent members. A resolution that authorizes the use of military force to shut down the veilmaking facilities is put forward under

Chapter VII of the UN charter, which deals with threats to the peace, breaches of the peace and acts of aggression. It is vetoed by one of the permanent members. A separate resolution calling on the Secretary General to convene a conference with the aim of working towards the drafting of a UN Convention on Climate Engineering and Protection (UNCCEP) passes, as the leaders of the Concert had hoped it would.

Just as there is uproar in the UN, so there is uproar in the nations of the Concert. Their citizens had not been told what was going on, and many are shocked, both by what their leaders had done and by the clandestine way in which they had done it. Two governments fall, one of them in one of the nations which hosts the Espedair airfields. In the second key nation, though, a national referendum backs the plan. It is the first democratic vote ever taken on the Earth's radiative forcing. It is not a remotely representative vote – the electorate is 0.04 per cent of the population affected by the decision, which is to say the population of the whole world – but nor is it the last.

Within a couple of years a number of other countries, including some well outside the tropics, have started negotiations to join the Concert, and the number of facilities from which the veilmakers can fly has grown. The Concert makes it clear, though not explicit, that at this stage it does not want any of the world's large economies to join, and it fixes the mitigation 'price' required for accession in such a way as to dissuade them. The negotiations towards a convention, meanwhile, move slowly. It is not easy to craft an agreement that suits the Concert, its passionate detractors, the growing number of countries tacitly supporting the Espedair aircraft and the uncertain majority.

Even though not a member of the Concert, one of the world's largest economies makes direct reference to the changed situation when, a couple more years on, it says it will increase the speed of its emissions reductions beyond that which it had previously agreed to at UNFCCC negotiations; slowing the build-up of greenhouse gases has always been a good idea, its leadership says, and if doing so faster means that the Concert sprays less aerosol, then that is an added attraction. It urges others to follow its lead. Another large economy relaxes its previous plans; it no longer feels able to say to its fossil-fuel and heavy-industry lobbies that reducing emissions is as pressing as had previously been claimed. Would the same governments have tightened policy or loosened it under other circumstances? There is no real way to say.

Though the Chapter VII debate at the UN had felt genuinely tense, the possibility of the new arrangement provoking military conflict seems to recede over the years. If the Concert had been one of the great powers, its veilmaking would feel like an intolerable imposition of geopolitical will. But it isn't. Its founding members were countries of middling-to-more-or-less-no consequence. They laid no claim to a world agenda other than having decided that it would be good to limit climate harm and wanting a way to act towards that end, and it is hard to argue that they have hidden motives. They enjoy a version of what in the 1980s Václav Havel called 'the power of the powerless' – what can be done about them, when they are basically of little other account?

Public reactions are all over the place. Many people feel uneasy; some are deeply disturbed. Chemtrailers shout from the rooftops that they have been right all along (though they never reach agreement on what the Concert is really up to, or who is really behind it . . .). As people in geoengineering research had long feared, some in America and elsewhere perform the 'superfreak pivot', turning overnight from the position that global warming requires no emissions reduction because it isn't worrying to the argument that the Concert has it all covered. Green politicians and activists mostly condemn the Concert's climate vigilantism outright; those who do not call on it to put its power into the hands of the UN. A direct-action group called the Sky Shepherds blockades two of the more accessible Espedair airfields on and off for years with a succession of balloons and microlight aircraft flown over the runway approaches. In the long run, though, it cannot get new aircraft to the area as fast as the authorities can impound the ones already there.

Most people who take an interest are worried, or at least disconcerted; some are relieved; a few genuinely welcome the development. The lottery which offers people from anywhere on the planet a chance to ride as a passenger on one of the Espedair aircraft proves a hit; the God's-eye view and the sense of simultaneous participation in the earthsystem and world history is intoxicating, even if the conditions are a little claustrophobic. Down on the ground, people scrutinize sunsets with a new attention, comparing them in their imaginations with those they remember from their youth, or from just a few years ago. Though many convince themselves they are seeing a difference, at this stage they really aren't.

This stringing together of speculations is obviously intended to make solar geoengineering look like a somewhat attractive possibility. What, though, of the beads on this string? Considered in isolation, independent of the way that they are strung together, are they plausible? To a large extent, I think they are.

Would it be feasible for a few small states with airfields in the tropics and a tech-billionaire benefactor to try and put together a small aerosol geoengineering effort? Yes. The Aurora study discussed in Chapter Four argues that a much larger effort could be undertaken for about $2 billion a year. A first-generation system scaled to lift only tens of thousands of tonnes in its early years could be a lot cheaper. Designing an aircraft capable of flight at 22 kilometres is not that demanding if it doesn't have to do much else, and building such aircraft is within the capabilities of companies within well over a dozen countries, not all of them part of the developed world.

It is partly because of this technological feasibility that the billionaire-geoengineer is a well-worn trope of speculation: within the geoclique, it is called the 'Greenfinger' scenario, an allusion to James Bond introduced, I think, by David Victor, a political scientist now at the University of California, San Diego. After all, billionaires building spaceships – either very publicly, as Richard Branson does with Virgin Galactic and Elon Musk does with SpaceX, or more privately, as Jeff Bezos does with Blue Origin – is almost a commonplace. In its pure form – a billionaire who tries to take over the climate more or less by him or herself – I think the Greenfinger idea is highly implausible. Consider the case of Bill Gates (as many do, in this regard). He has funded some geoengineering research; what is more, he is part holder of the patent on the device Stephen Salter thought up to disrupt hurricanes by cooling the surface of the ocean, an idea that was fleshed out during a discussion in which Gates took part. But that does not mean that Gates wants to see such technology implemented, still less that he thinks it would be a good idea to implement it on his scofflaw own, still less that he thinks that the American government would let him: billionaires are not above all law.

But the fact that Bond villains don't exist does not mean that rich people do not seek out technological interventions that will leverage their wealth in world-changing ways. By funding vaccination programmes, Bill and Melinda Gates have played a part in saving millions of people; programmes they have paid for are saving lives on a scale similar to that on which Hitler, Stalin and Mao Zedong brought death, an accomplishment which strikes me as utterly remarkable. Elon Musk really does intend to make human space travel much cheaper and thus more routine – which I think he may well achieve – and sincerely believes that by doing so he will change the course of history. The idea of a Greenfinger motivated by a particular conception of the greater good rather than megalomania, and operating from and on behalf of a small group of sovereign states that is aiming only for a modest effect, is not something to dismiss out of hand.

The question of whether such a thing could be carried out in secret is harder to gauge. It would require a high level of security, a compartmentalization of information and some convincing cover stories on the part of the Concert. A good cover story makes it possible to do quite important things in secret even when they are visible to other people's satellites: witness the fact that the American government did not know that the Saudi government was fielding a force of non-nuclear ballistic missiles purchased from China in the 1980s until after the first squadrons were operational. It helps if the security services that might spot what you are up to aren't looking too hard. I have no idea whether that would be the case for a smallish covert geo-engineering effort or not. I tend to think intelligence agencies always have more call on their analysts than they have capacity, and that it would not be something people would be looking for hard. At the same time, I know that the two reports on climate geoengineering produced by America's National Research Council in 2015, were commissioned and paid for in large part by the CIA; that suggests that someone there has some interest in the possible development of the field, though not that there is any sort of intelligence-gathering effort aimed at spotting rogue programmes.

And being spotted by other governments would not necessarily impede the Concert in its work. The purpose of its activities might be inferred by analysts at an intelligence agency but not taken seriously by their overseers. The idea could be taken seriously but nothing done in response because no agreement on how to respond could be reached. There could be a response in the form of some sort of below-the-radar threat or promise that yields no results. Or there could be a non-response that is a tacit encouragement to continue. If some of the people shocked by the announcement at the climate summit of 202– were feigning their surprise, it wouldn't be the first time.

What of the Concert itself, a small group of small states trying to change the world? Climate negotiations, like trade negotiations, routinely throw up common interests among diverse parties, sometimes catalyzed by a small group of policy entrepreneurs. Thus, for example, the presence at the UNFCCC of the Alliance of Small Island States, which was put together in the early 1990s to give a collective voice to a set of countries that saw an existential threat in rising sea levels. No one who follows climate-change negotiations can easily forget the moment at a UNFCCC meeting in Bali when Kevin Conrad of Papua New Guinea, a part of that Alliance, told the United States: 'We ask for your leadership, we seek your leadership, but if for some reason you are not willing to lead . . . please get out of the way,' and America promptly caved. This scenario is not a *roman à clef*, and I don't have candidate members for the Concert in mind. But I think the dynamic necessary for its creation is possible. Climate risk falls inordinately on developing countries, and some of those countries may try ever more novel ways to get action out of the world's major emitters.

What is more, the system which the Concert would bring into being is one that might prove very well suited to climate negotiations in general. David Victor and others have argued that a 'club' approach to climate could yield significant benefits. The clubs in question would be groups of nations comparatively tightly bound together by agreements on mitigation, adaptation and other action in ways that suited their particular interest. Large emitters making multilateral or even bilateral agreements in such clubs – arrangements that might involve quid pro quos beyond the scope of UN climate diplomacy – might be able to offer each other inducements to deeper cuts than can be arranged through UN processes, which have to come up with agreements equally acceptable to everyone from Saudi Arabia and Bolivia.

The Concert's approach seeks to build on the general

benefits such clubs can provide by making it possible to link climate geoengineering to pledges on mitigation, and possibly to other things, too. The link between geoengineering and mitigation is normally taken to be an either/or – the existence of geoengineering is taken to mean a lowered likelihood of mitigation. But, as the legal scholar Edward Parson has argued, some forms of linkage with climate geoengineering could make mitigation easier to coordinate, and thus mean that the world sees greater reductions in emissions.

This is not true for all geoengineering scenarios. As Parson points out, if geoengineering is seen only as a response to an emergency at some unspecified time in the future, attempts to link it to mitigation actions in the present – for example, by saying that if people mitigate strongly today they will be allowed to geoengineer if it becomes really necessary, or that if they don't they won't – look entirely impractical. There is no imaginable way to make such a link binding enough for people to be sure that their successors several decades hence would honour it.

But if geoengineering is actually on the table, or already happening, Parson argues, linkage becomes much more feasible. It would require impressive diplomatic achievements. But every scenario that imagines strong climate action has to imagine international agreements put together through intelligent and subtle diplomacy. If you are willing to imagine such negotiations in the absence of climate geoengineering, it seems unfair to rule them out in its presence. If you are not willing to imagine such negotiations at all, you are ruling out any large diplomatic contribution to emissions reduction. This point may by now sound familiar – it has come up in other guises before -- but I think it worth repeating. To compare approaches to climate risk that use geoengineering only with the best possible non-geoengineering approaches is to distort your view of the world so much as to make it deceptive.

The possibility that there could be a linkage between

veilmaking and mitigation which does not trade one off against the other – as the moral-hazard worry says might be the case – but which instead encourages both applies well beyond the artifice of the Concert scenario. It is worth looking at a little more, because it has implications for how any movement towards geoengineering should be governed.

Climate geoengineering is often, and correctly, said to raise new challenges in international governance, and an absolute need for those challenges to be addressed prior to its deployment is frequently asserted. Many see the lack of a good model for that process as a reason for avoiding research that might get anywhere close to the technologies of deployment. Their worry stems from the intuitively obvious (though not provable) belief that society might have been much better served in the past if a wide range of technologies had received more anticipatory consideration. In the absence of such forethought it is more likely that the technology will be deployed in ways that predominantly serve the interests of already powerful groups, which will often mean that it does not serve the common good as well as it might. The damage done may be greater than the benefits achieved, and the two will both be unevenly distributed.

Against that, though, one should weigh the certainty that governance-in-advance will never be perfect. There are obvious problems in trying to develop governance structures after the genie is out of the bottle – ask Bernard Baruch – but there are also problems in trying too specifically to define all the governance systems in advance. Excessive precaution may lead to things which could in fact have been governed in safe, just, equitable ways not developing far enough for those possibilities to be realised. It may also make people feel more justified in rejecting the governance framework altogether, and pressing on regardless.

Better, as I have argued earlier, to let governance and technology co-evolve; to have researchers take governance into account when studying possible approaches, to have people concerned with governance take a closer look at where research is actually going. An example: some people are very worried that solar geoengineering requires an unending commitment if there is not to be a termination shock. That makes it a good idea to look at scenarios where the commitment is of limited duration more or less by definition. So David Keith and colleagues developed a scenario in which a stratospheric veil is used to slow the rate of warming to half of what it would otherwise be. This rule means that, as greenhouse gas levels start to stabilize, the veil starts to thin. The Concert scenario, in the form given above, is similarly self-limiting. The fact that such scenarios exist does not stop critics of veilmaking insisting that it means, of necessity, a commitment that would last for centuries or millennia. But it should.

There are risks in having engineering and its governance co-evolve, most notably the risk of capture; when a technology and its governance develop hand in hand there is often a tendency, over time, for the people thinking about governance to concentrate more and more on how the technology should develop, rather than on whether it should develop any further at all. But although societies do sometimes get 'locked in' to particular technological paths, they can also unlock themselves. There was a lot of work done to bring about an age of supersonic passenger transport in the 1960s, but it did not get any other form of SST than Concorde built. Germany had a large well-regulated nuclear industry; but that didn't stop it being closed down. Nations and individuals turn away from technologies despite the undeniable forces that try to make their development or use seem a foregone conclusion.

In general it seems to me that there is a good case for letting the development of ideas about how to do geoengineering and about what sort of geoengineering might be done, and under what circumstances, to develop hand in hand as part of a broad response to climate change. And it seems to me that the possibility of linkage between mitigation and veilmaking strengthens the case for such an approach. Geoengineering has attributes that inadvertent climate change lacks, and they make coming to agreements about who should do what easier.

International agreements on climate change such as those sought under the UNFCCC are very hard to coordinate because of the uncertainties and timescales involved. The places where the costs of mitigation fall hardest will not in general be the places where the benefits of mitigation are greatest, and the time at which the benefits will be felt will be decades to centuries later than the time at which the costs start to be paid. Solar geoengineering is much more straightforward. You don't have to coordinate the actions of many different players in advance. You don't have to wait for another generation to see the effects. There are clear responsibilities and prompt effects, and that would seem to make the problem inherently more tractable. Developing a governance approach to climate geoengineering that yokes these characteristics to the harder question of governing mitigation could give you a more workable system. A certainty? Of course not; as we shall see, minority action and prompt response could also make sunshine geoengineering very destabilizing. But a chance? Yes. That is one of the things that the Concert scenario is meant to demonstrate.

Small Effects, and Bad Ones

Those are my main reasons for thinking that the Concert scenario brings out useful truths about the politics and technology of geoengineering. What I haven't discussed is what the scenario means for the earthsystem. And there is a reason for that: in the near term it would probably mean very little, and its effects would be impossible to ascertain. As the different rates of warming in the 1990s and the 2000s show, climate varies a fair bit from decade to decade. Against this background, the effects of a modest intervention like that of the Concert would not be clearly visible for quite some time.

This problem of attribution would not make the effort pointless. A run of decades with less warming than would otherwise be expected would be a good thing to have in terms of lowered impacts and easier adaptation whether it were demonstrably the effect of the Espedair veilmakers or not. The same might hold for possible diplomatic benefits from linking mitigation efforts to geoengineering. An international policy process designed to define what limits should be put on such efforts would be a net plus whether they were achieving much or not, and so would operational experience that allowed the effects of aerosols on ozone, on other chemistry and on the water content and temperature structure of the stratosphere to be studied in detail.

This, to me, is the most important thing that the Concert scenario brings out: the knowledge that a solar geoengineering programme could be useful and at the same time small in its absolute effects. Solar geoengineering is normally talked about as a way to stop global warming, or indeed to reverse it, and there are three reasons why that is understandable. Stop-or-reverse is the sort of clear-cut effect that people expect when they start off thinking they are looking for a solution. Climate models are much better at showing the effects of large interventions than those of smaller ones, and so modellers have quite reasonably focused on big interventions. And stop-or-reverse is the technology's unique attribute; from what is known today, it seems pretty clear that a large-scale programme of stratospheric veiling (and possibly, though less certainly, of marine-cloud brightening) could stop the planet warming any further for quite a while, or indeed cool it for a bit; nothing else could do so on the same timescale. But none of that means that solar geoengineering has to be used for stop-or-reverse.

Slowing warming down could prevent a lot of harm. Using the slowed warming as a breathing space in which to deploy more and better zero-emission technologies would be a good strategy. Even if that wasn't the strategy followed, and mitigation proceeded at exactly the pace that it would have done otherwise, there would be benefits. Slower warming is a way of adapting globally that would make adapting locally easier.

But that doesn't mean that things would necessarily turn out for the best, or even well. Here are a couple of ways in which the Concert could fail.

The first is the moral-hazard nightmare. While some nations increase their emission-reduction efforts as a response to the Concert's initiative, more do the reverse. They stop caring, and burn fossil fuels wherever it suits them – after all, they argue, if things look bad, they can thicken the veil in response. The Concert nations disapprove, but they have few options by way of response. They try to win their use of the technology, and thus their strategy, a privileged status in the new UNCCEP negotiations in the same way that the nations which already had nuclear weapons got a special status in the Nuclear Non-Proliferation Treaty; they hope that international law could in some way be fixed so that their form of geoengineering, with some level of UN oversight, could be made the only acceptable form. They don't get what they want.

The unrepentant emitters of carbon dioxide develop their own veilmaking fleets and commit themselves to thickening the veil enough to stabilize world temperature at 1.5°C above the pre-industrial level. The world acquiesces to the new approach in part because the Concert has acclimatized it to the idea that veilmaking need not be catastrophic, in part because its adverse effects have, to date, been negligible, in part because one of the new stabilizers has a lot of military muscle and a permanent seat on the Security Council, and in part because there are, as ever, other more pressing matters to attend to. The technology for the removal of carbon dioxide that the Concert had tried to encourage languishes undeployed, rather as CCS technology does today. The response to climate change has gone from 'Make me chaste, Lord – but not yet' to 'Chill, Lord, we've got this.'

By the end of the century the carbon-dioxide level is close to 750ppm and still rising. Human emissions are falling from their peak, but they are doing so slowly and they are still substantial. As the oceans have warmed, the fraction of the emitted carbon dioxide they are able to soak up has decreased, so more of the emissions stay in the atmosphere. More than half of the warming that would have been expected from 750ppm is being held back by the aerosols, and the mismatch between aerosol cooling and greenhouse warming is becoming apparent – the everyone-a-winner sweet spot that low levels of solar engineering had seemed to promise has been left far behind. There are permanent changes in the hydrological cycle in much of the tropics and sub-tropics; the rhythms and intensities of El Niño and La Niña events have shifted. Ocean acidification has devastated many reefs and some other ecosystems, and its effects are worsening. And though the temperature has stabilized, various changes the engineering nations had wished to avert have come about regardless. Arctic sea ice is barely present in most summers. The Greenland ice sheet looks worryingly unstable.

This is not a good outcome. In many ways it is a better outcome than a 750ppm world in which there was no solar-geoengineering cap; high temperatures do real harm to people, to crops and to wildlife. In other ways it might well be a significantly worse outcome. Whether it would be worse overall is impossible to say – not just because today's models don't have the sort of detail that would enable you to make such a verdict, but because 'worse' always depends on how you value the prospects for different people. And the engineered planet now has the extra risk of a termination shock. No one can say how termination would come about, since it seems in no one's interests – but the fact that it might weighs heavy on some minds.

If you really believe that the world would largely abandon

mitigation once the solar-geoengineering option were made real and that, as long as geoengineering is firmly off the table, there will be a surge in the world's ambitions when it comes to emissions reduction which will persist for the next century, then this is the scenario you should use as your cautionary tale. It seems to me, though, that that pair of beliefs leaves you with what might be seen as a rather narrow 'window of wisdom' through which to escape from both the risks of calamitous climate change and the risks of climate geoengineering. If the world is insufficiently wise, then there is no strong mitigation and you don't make your escape: you get a lot of climate change. If the world is wise enough to understand that even with geo-engineering there is still a strong case for mitigation, you don't make your escape: you get geoengineering. It is only if the world is wise-enough-but-no-wiser that you avoid both massive climate change and geoengineering.

But these are not things that anyone can know for sure. It is clearly quite possible that in a world which already had some solar geoengineering some people might choose to increase the amount heedlessly and others might be unable or unwilling to stop them. That certainly needs to weigh on our thinking.

An interesting minor variation to this moral-hazard nightmare is a scenario you might call the naturalism schism. In this a small island nation, one of the founder members of the Concert, starts to develop a technology for liming the ocean in its immediate vicinity, increasing the alkalinity in order, it hopes, to preserve its treasured, beautiful and much-visited coral reefs. Other nations in the Concert strongly object to this; they see it as reducing the coral-island nation's commitment to encouraging global emission cuts.

From its inception, the Concert, like Paul Crutzen in 2006, always sought to have it both ways on climate geoengineering. Its members saw their scheme as offering a possible benefit in itself and as a way of frightening people into taking mitigation more seriously. Some of the Concert's members were more motivated by the first, some more by the second, but they all realized that the two were not mutually exclusive. But by trying to nullify the risks of ocean acidification, at least for itself, the coral-island nation pushes too far, and the middle ground breaks down. The Concert falls apart, other veilmakers step in, and the world goes down the moral-hazard-nightmare path while people try all sorts of new ocean-chemistry modifications, some of which go badly wrong.

To some old hands in geoengineering debates, it seems almost surreal that countries already engaged in a radical programme of solar geoengineering are falling out because of the moral hazard posed by small-scale action against ocean acidification. In retrospect, though, the Concert's idea that only one aspect of the earthsystem would be geoengineered was never likely to be stable. The dynamic in which geoengineering in one area would lead to geoengineering in others went back, said some, to well before the first flights of the Espedair squadrons. As soon as humans replumbed the nitrogen cycle they were more or less committed to dealing with the follow-on effects of moving from a 3.5-billion-person planet to one three times fuller. The idea that, once the earthsystem was being manipulated so blatantly on a global scale, it should not be tweaked more and more both globally and locally seemed, if not completely irrational, at least hard to defend. At the same time that it has a huge appeal, though, there is something about trying to opt out of ocean acidification that seems to be a step too far, and it upsets a lot of people. Not that there is anything they can do about it.

The second bad outcome to consider is one that turns up in almost all discussions of potential problems with stratospheric veils: conflict between great powers as the result of a dramatic failure, or sequence of failures, in the Indian monsoon. When people warn that solar geoengineering could affect the livelihoods of billions, mention of a climate emergency caused by a failed monsoon frequently follows. Such an emergency played a role in the set of scenarios examined at a Yale University meeting in 2013; in 2014 it featured in an imaginative future history called *The Collapse of Western Civilization* by the historians Naomi Oreskes and Erik Conway – indeed failure of the monsoon precipitated the collapse referred to in the title.

In the Concert scenario it would be hard to attribute such a climate emergency to the Espedair veilmakers with any certainty. The monsoon is variable to begin with. And a slow-building geoengineering strategy like that of the Concert would not have anything like as strong an effect on the land–sea temperature difference as large volcanic eruptions can.

What is more, it is unlikely that the Concert's actions would be the only, or even the biggest, human influence on India's climate over the coming decades. If India's economic development continues, it is very likely that its aerosol air pollution will decrease, which will undoubtedly affect the monsoon and might very well do so more than a thin, reasonably homogeneous layer of stratospheric sulphates would.

But none of this means that there could not be a very weak monsoon, or a series of very weak monsoons, during the early decades of the Concert's action. The economic consequences of such a failure would be grave, as would their human consequences; the lives of hundreds of millions would be made worse. The chances of a large death toll would be low – as the great economist Amartya Sen has shown, large famines are almost unheard of in democracies around which information spreads freely – but a nation does not need to undergo mass mortality to have a strong grievance. It seems entirely plausible that the Indian people and their government might full-throatedly blame the Concert for such an event even if the evidence for its culpability was weak.

So, to pick up the story: the failed monsoon of 204– leads

India to take a much more hostile line to the Concert than it has done previously, and this change of heart in the world's most populous country leads to the Espedair programme being shut down. Over the following years the warming the veil had suppressed makes itself felt quite unambiguously – the sudden removal of the aerosols produces a much clearer change in the behaviour of the earthsystem than their slow build–up did. This 'termination shock' is not particularly dramatic, but it produces a prompt warming of a few tenths of a degree that is attributable to human action. So when two very hot summers in a row cause calamity in another great power, that is attributable to human action, too. The people in that second power call for the veil to be reinstated; their leaders listen, and act.

The well-meaning Concert that I cooked up as a way of making things look not–too–threatening is now entirely superseded; the programme becomes a bone of contention fought over by at least two great powers with mutually incompatible goals, the first wanting no veil, the second insisting on one. The conflict escalates, and exacerbates another issue of longer standing between the same powers. War breaks out. And through a series of miscalculations, that war goes nuclear.

This counts as a very bad outcome even before the notional nukes start flying. A world in which two or more great powers believe that they have strong and opposed interests in the matter of solar geoengineering could be quite as scary as one in which geoengineering drained mitigation efforts of all ambition. Geoengineering might do more harm as a *casus belli* than through unintended and unwanted earthsystem side effects. And I think geoengineering lends itself more readily to the triggering of conflict than unintentional climate change does, for much the same reason that it is also lends itself better to negotiation; it has its effects quickly and you know who is doing it. That allows a disagreement about geoengineering to escalate quickly.

How big is this risk? Maybe not that large. Although

geoengineering might add to the stock of things over which great powers could fight, there is no reason to think that it would be a uniquely difficult problem for them to solve without fighting. The great powers of the world disagree over quite a lot and go to war very little. To bring about the coming of war in this scenario I first had to evoke a climate emergency attributable to geoengineering striking one major power, then an emergency attributable to the cessation of geoengineering striking another, and then have the subsequent conflict escalate to nuclear war. An unlikely sequence of events.

But certainly not an impossible one. Such possibilities have a clear place in the debit column of any risk ledger, where many might weigh them as the deciding factor.

Before leaving the world of monsoons and climate emergencies, though, another variation on the theme – this one a little merrier.

In 204– a volcano in Kamchatka explodes in the largest eruption since Tambora. It is far enough to the north that, as with El Chichón in 1982 and Katmai in 1912, its veil stays almost entirely on that side of the equator. As both history and models show quite clearly, a veil which cools the north of the planet but not the south would shift the intertropical convergence zone where the weather patterns of the two hemispheres meet in a way that leads to drought in the Sahel.

Veilmaking in the north stops before the eruption is over. But after an extremely tense set of discussions between its members, the Concert decides that veilmaking in the southern hemisphere should not only continue but should be quickly ramped up in a deliberate attempt to even out the volcano's lopsided cooling. Brazil is angry about this; India, already fearing the worst for the monsoon in the aftermath of the volcano, is outraged. This is no time, it says, for further tinkering with the climate. Combat air patrols from the carrier INS *Vishal* start to intercept the Espedair aircraft flying over the southern Indian Ocean.

The fighters do not shadow the Espedair aircraft; though faster, they cannot sustain flight at such altitudes. Instead, they flash past them on parabolic trajectories to demonstrate the slower planes' vulnerability and intimidate their operators. On a few occasions they pass close enough that the Espedair pilots almost lose control in the jetwash.

If the Espedair aircraft had been drones – many had long wondered why they weren't – the Indians might well have shot them down. The presence of pilots stays their hand. As the men and women flying the Espedair aircraft had known from the start, part of why they were up in the stratosphere was to increase the political cost of trying to interfere with the programme – to give the aerosol veil a human shield.

After a tense few weeks the Indians decide that they will not escalate their operations against the Concert's planes, or its bases. Within months much of the northern hemisphere is unseasonably cool. The Indian monsoon fails, as it was always going to do in the shadow of such an eruption. But the thickened veil in the south keeps the intertropical convergence zone more or less in place. For the first time in history a major eruption in the northern hemisphere does not lead to a drought in the Sahel.

You will have noticed that there is much that is speculative in this chapter's spinning of scenarios. This wrinkle, though, is quite robust. There will be major volcanic eruptions this century, as there are every century, and there is a reasonable chance of one taking place up in the North. The finding that such eruptions are followed by drought in the Sahel is about as well supported as any statement about a specific climate expectation in this book. The idea that if veilmaking technology were fielded in the southern hemisphere in such a way as to balance out that northern cooling a drought in the Sahel could be averted is highly plausible.

Does that mean the world should build some veilmakers just in case, to wait on the ground until the next big northern eruption? It seems an implausible plan. And as Jim Thomas of the ETC group pointed out in an online discussion he and I had about this, the people of the Sahel would probably prefer to see forms of development that would leave them at much less risk when the rains failed. That rings true. Nevertheless, if the emergency struck and that desirable change had not taken place, I am not sure that they would be averse to having the rains preserved.

And Straight on 'til Morning
Finally: what if nothing much goes wrong?

While writing this book I have tried to anticipate, and be fair to, arguments people might raise against what I have said. I am sure that some readers will think I have not done enough to this end. I suspect that others may have had their appetite for on-the-one-hand-on-the-other-hands more than sated. So it hardly needs saying, at this point, that there are risks. But there are not only risks.

So here is the radical end to the Concert scenario. It works.

By the end of the twenty-first century humans are emitting only about 2 gigatonnes of carbon a year, a lot less than today's 10 gigatonnes and a great deal less than the mid-century peak. Processes that could be decarbonized over a few generations have been decarbonized; that includes essentially all electricity production. Hydrocarbons are still used for some recalcitrant industrial processes and some forms of transport where alternatives are hard to find, like flight, and while some of them are specially made for the purpose, others are still sourced from the Earth's crust. But technologies for the removal of carbon dioxide are now nullifying some of the effects of these residual fossil-fuel emissions. While the world still has some oil and gas wells, it is putting quite a bit of the carbon it takes out of the ground back into the ground. There are no more industrial coal mines – only artisanal ones.

There is a persistent veil in the stratosphere. Following changes made in 205–, when the UNCCEP finally came into force, the sulphur-based system used in the first decades of the Concert was replaced by a set of aerosols which were more ozone friendly and had a better ratio of back scatter to forward scatter. The ozone layer's recovery has been slower than it might have been – but not that much slower, and no specific harm has ever been attributed to the delay. Less forward scattering means that new aerosols whiten the sky even less than the sulphates did. But they still redden the sunsets; evenings are more vivid than they once were.

They are also warmer, but not much. The average surface temperature has not reached the 2°C limit. Towards the end of the century the UNCCEP took the veilmaking protocols beyond those the Concert originally planned. The justification for this was threefold. First, emissions had started dropping quite quickly in all large economies. Second, it was now clear that cuts could go deep; some countries were emitting 90 per cent less than they had at their most fossil-fuel-dependent. Both these things lessened worries that more geoengineering would weaken mitigation efforts. The third factor was a dearth of dangerous side effects from the veilmaking to date. Rainfall patterns had changed a bit; the changes were sometimes for the better, though, and always within the limits of adaptation, so while there have been issues, they have not been huge.[*]

So in the last two decades of the twenty-first century the veil was thickened enough to slow warming to a crawl. Natural variation now means that scientists expect some decades in the early twenty-second century to be cooler than others – the first period in which this will have been true since the 1950s. That

---

[*] The huge and hard-to-resolve issues of this age are in a very different arena. Like many great issues, they look like a surprisingly old set of concerns arranged in a rather shocking new way. But which concerns, and in what way, I cannot say.

said, all decades are still roughly a degree warmer than they were a century back.

Because the world has warmed, sea levels have risen; people who live near shores have had little option but to adapt or move – some cities have built impressive sea walls, some low-lying rural areas exist now just in the memories of millions of refugees. Ocean acidification has been bad for some species; most marine ecosystems have undergone change as a result, and in some cases that change has been dramatic. But the oceans are not altered beyond recognition, or beyond the use of fish farmers, who produce most of the protein humans get from them. In some particularly sensitive areas liming programmes designed to reduce the effects of the decreased alkalinity – programmes related to some of the more successful efforts at carbon dioxide removal – are having some success, though at significant expense. In other places, due sometimes to poor implementation and sometimes to unforeseen circumstances, they have failed.

The higher carbon-dioxide level is having effects on land, too. Helped along by attempts to produce a more powerful West African monsoon using cloudships in the Gulf of Guinea, the vegetation of the Sahel is spreading north, and there are hopes that large parts of the Sahara may yet be reclaimed. There are desert lovers who hate this idea. While their arguments have not held the day in Africa, suggestions that monsoonal flow off California might be similarly strengthened have been seen off by those keen to keep the Mojave and Sonoran deserts as close as possible to how they have always been.

Regional programmes to modify other aspects of the earthsystem are getting under way. In parts of the Andes and the Alps a glacier unprotected from the sun is now a rare sight. In Greenland there are projects aimed at refreezing the base of the icecap to the island's bedrock at various strategic points by pumping liquefied air down arrays of boreholes. It takes a great deal of effort and energy – the effort is similar to that needed to exploit a large tar-sand play, not that anyone does that any more – but it does seem to be stopping the flow of some of the bigger glaciers towards the sea, and thus bottling up some of the avenues which would allow the ice sheet to collapse catastrophically. Whether it can stop such a collapse indefinitely is as yet unknown. Projects to divert the winds circling Antarctica using huge floating wind turbine arrays, thus increasing the precipitation on that driest of continents, thickening its ice cap and slowing sea-level rise, are still at a more speculative stage, and highly controversial.

Some see a beauty in these new forms of engineering – the grandeur of holding back great sheets of ice and re-routing planet-spanning currents of air thrills them. But they also see them as representations of a greater beauty, felt not seen – the beauty of an earthsystem newly cared for, newly loved. They draw analogies between the sunset-reddening veil and the golden lacquer used to bind together the fragments of broken pots in the Japanese practice of kintsugi – an art that treats its creations as all the more whole, and all the more beautiful, for having been broken and mended.

Others see such talk as dangerous poppycock.

Despite these arguments, the 'naturalism schism' which some had feared would split the Concert has not done so. More or less unavoidably, the world's view of what is natural has changed. Nature is appreciated in terms of processes and potentials more than as a thing in itself. There are still people who look for the unchanged and unspoiled, and they are increasingly frustrated; the planet as a whole is ever more subject to human empire. But at every scale from the window box to the rainforest, the willful and spontaneous processes of nature can still be seen and appreciated. The world and the earthsystem serve as settings for each other – settings as in jewellery, not settings as in thermostats. But they are not inseparably conjoined, nor indistinguishable; they still have their separations, and there is always room for the unintended.

Perhaps the all-encompassing nature that people once yearned for has gone — perhaps it was never really there. The idea that there are things, processes and places which can express nature has survived, as it has through all sorts of cultural shifts. The patchwork ecosystems created by the rewilding projects of a century ago — those that re-introduced once-common species like the wolf and beaver to landscapes from which they had been removed — now seem like places where nature is at work even though they are manifestly contrived. The sense of the natural as a process persists even when some genetic engineering has gone into the contrivance, as in the case of the mammoths now rooting up some Siberian and Canadian forests, thereby doing their bit to raise the planet's albedo as well as to delight its inhabitants. There is a nostalgia for other, older, purer natures, just as there has so often been, and recreations cannot entirely soothe it. But the world does not feel wholly artificial, and though there is loss, there is consolation.

If the Concert did not succumb to a naturalism schism, though, it has changed in other ways. After its early years its members decided to accept accession states only if a majority of the electorates in those states agreed to be part of the geoengineering process through a referendum. By 208– the states that had joined under those rules make up more than half of all humankind. Specific decisions about which aerosols to use in the veil, and indeed what level of radiative forcing the veil should provide, are increasingly being made by a panel of people's representatives with long terms of office. The members of this Council of the Atmosphere, which also has a growing voice in questions about emissions and their further reduction, are at the moment chosen by the Concert governments. In the future, though, they may be chosen by global plebiscite; the idea is gaining ground.

There are many people who feel an abiding unease at the fact that the world is out of equilibrium. They talk of living in a Potemkin Holocene — the planet's appearance is largely as it used to be, but the earthsystem maintaining that appearance has changed a lot. They worry about the idea that the Greenland ice sheet may need permanent restraints if it is not to raise sea levels far beyond the metre they have risen so far – and about the possibility that restraints that are meant to be permanent may prove not to be. They worry about half a degree or more of prompt warming being held at bay by the veil, like a potential flood of sunshine pent up behind a dam. They agitate for crash programmes to reduce the carbon-dioxide level – or, failing that, for the Council of the Atmosphere to decide that the world is rich enough and good enough at adaptation that the veil should slowly be thinned down to nothing and the temperature allowed to rise. Others argue that the veil should be thickened, and temperatures brought down to those of the twentieth century. Non-human life would benefit from temperatures closer to what it evolved for, and humans could adapt to any second-order problems in precipitation and the like. Humans have, after all, been getting some experience at adaptation.

The Council, aware of the arguments, is biding its time. And there are no serious challenges to its power or to its right to do so. Though there is nothing like a world government – there are, after all, huge issues unresolved in other fields of endeavour – there is a sort of limited government of the atmosphere. And this has come to seem reasonable. The thermostat has its hand, the lever its fulcrum.

At the beginning of the twenty-second century, just as at the beginning of the twentieth, all people who say anything about the size of the world agree that it is small. The fact that, to the handful of people living in the very expensively provisioned villages on Mars, the Earth is just a not-very-spectacular evening star is taken as underlining the case – though in truth it shows not that the Earth is small, but how much the human world has grown. Just as at all other times when the world has been seen as small, it is still full and complex, riven by opposing interests, impossible to reduce to rationality. Try and change it without the right levers and you soon find how large its resistance can be.

But some bits of the earthsystem really are small. Though the sky seems boundless, the atmosphere has always been small in planetary terms; it weighs only a little more than the Mediterranean Sea. It is because there is really not all that much of it that, along with the yet tinier biosphere, the atmosphere is the part of the system that humans have been most able to change.

Like any small sea, the atmosphere matters immensely to those on its shores, which means everyone except those Martians. But, also like a small sea, there are only so many problems it can pose. Build your houses in a way that respects its storms; fortify yourself against any enemies it might bring; keep it open for trade and its waters clean for fish; study what it contains and learn how it works; appreciate its beauty just a little more often than you would if you weren't paying attention: at that point you have done most of what there is to do. Those are not the hardest rules to agree to, or to hold people to. As technological options that produced less effluent became cheap and widely available, the Mediterranean got a lot cleaner.

Finding a set of principles for governing the small sea in the sky, a task which seemed almost impossible at the beginning of the century, has not been beyond human ingenuity. Putting those principles into action has not been politically impossible. Indeed, in retrospect, it seems like one of the evolving worldsystem's lesser challenges. The ones that have been met always do.

Envoi

End, as Louis Armstrong once said, of story. There is just one more thing to note.

There is a word mostly missing from this book. It is 'we'. In the places it crops up, it refers more or less explicitly to you, the readers, and me, the writer. It is mostly used to point either to what has been said before or to what will be said later. There has been a sort of community between us in the writing and reading, and within that space shared across time, 'we' has seemed a fair word to use.

What there is not is the unqualified 'we' that crops up in so much writing about the climate, and other topics – the 'we' of surreptitious and spurious suasion. The 'we' used to align the writer, the reader and an ill-defined group of people who, it is implied, naturally agree and which thus seems to include all right-thinking people. The 'we' that says 'we know what to do, we just have to do it'. The 'we' that supposes that my interests are your interests, and that the interests of people in different countries and with different views – say, the interests of the poor who want to be less poor and the interests of the rich who are not fully aware of all the benefits those riches provide – can be easily aligned. The 'we' that seeks to speak for the world – a world that 'they' are letting down. And you know what 'they' are like, don't you? They are a dodgy bunch. They are not like us.

The 'we' that I have left out is the 'we' who are told we cannot let this go on – blatantly ignoring the fact that we do. The 'we' that seeks to speak for all people, all history, all species – as long as they agree with the author.

It is almost impossible to speak on political subjects without invoking this we. It is quite hard to write without it. But I thought it was worth doing, because that we is not an innocent illusion. It is a harmful one. The we that matters is not one summoned as a rhetorical tool by an author. It is one built by people who do the difficult work of actually agreeing with each other about what they need to agree on, and agreeing on the realm of disagreement, too.

Making a we is hard, and it is essential. It is the essence of politics; it is the essence, too, of love.

And as far as geoengineering goes – as far as almost all international issues go – there is no we. There is no 'we must try it' or 'we must not try it'. There are arguments and places to have them, some of which exist and some of which must be invented. There are movements which will come to speak, not for all, but for themselves. There may be a 'we' at the end of the process. But here, at the beginning, before the beginning, there is not. The politics have yet to be done.

So, for now, this is personal. I have been aware, writing this book, that my hopes for geoengineering, and the meanings that I ascribe to it, reflect feelings deeper than those I have about radiative forcing and the goings-on above the tropopause. The feeling that things press in, that history and the world embroil the present in a way that confines choice ever tighter; that is part of my temperament. The constant sense that, if only there could be a bit more room, a bit less pressure, if only the envelope could be expanded, the boundaries pushed back, if only there were time, and space, to breathe: that is my confined, asphyxiated perception. It comes from experiences within my life, some of which I understand a bit, some of which I doubtless don't. It probably chimes with the experiences of some of you, too, in some way and to some degree. Who has room enough, time enough, choice enough?

I am less aware of where the hope comes from – the hope that things might indeed be different. But there is a hope. A hope that people might take more care, and arrange the world in a way that lessens harm. A hope that people might see in this care not a diminishment of the degree to which the world is natural, but a reimagining of how humans and nature can intermingle, a new consciousness of what can be done for the planet rather than a blind deference to what are claimed to be its limits. A hope that there is joy and solidarity to be found in exercising compassion on a planetary scale, and that, given the opportunity, people might not just create a temporary autonomous zone in between the constraints of climate and the momentum of growth, but go on to use that space to bring about a worldsystem that better suits the days of their lives and the days to come. To create a we that shares in this space, and that can set a better course.

Up above and far away, too far for any eye but the mind's, a future lifted on long, strong wings starts a graceful, cautious turn. It seems almost beyond the bonds of Earth, but it does not fly in freedom; there are things it cannot do and must not do – many ways for it to slip and fall. The future is hemmed in on one hand by its design, on the other by the unforgiving laws of nature. But its heading and height can, with skill, be changed.

Above it is emptiness, below it the bright and brightening world.

The load on the future's wings shifts subtly as it banks; their tilt and their line give a little in response. They need to be long-lived as well as long, these arcing wings; they must yield in this way again and again. At takeoffs and landings they flex almost alarmingly as the future's weight is passed from the Earth to the sky and back again.

History and habit may yet render them brittle. Trusted as they are, they may yet fail. But for now the long wings can bend. And as they do, they bend towards justice.

Greenwich, St David's Day 2015

Acknowledgements

In the long and sometimes tortuous course of working on this book I have incurred debts of gratitude too numerous, and in some cases deep, to be done justice to here. That said, I am very grateful indeed to Ken Caldeira, David Keith and John Latham for a level of support for the project that went beyond helpful. Other members of the geoclique, fellow travellers and Anthropocene thinkers – as well as critics of some or all of the above – to whom I owe particular thanks for stimulating conversations, good fellowship, helpful introductions and occasional invitations include Thomas Ackerman, Greg Benford, Jason Blackstock, Tom Brookes, Lilian Caldeira, Denise Caruso, Peter Cox, Paul Crutzen, Jim Fleming, François Gemenne, Timo Goeschl, Clive Hamilton, Hugh Hunt, Tim Kruger, Bruno Latour, Mark Lawrence, Margaret Leinen, Tim Lenton, Jane Long, Mike MacCracken, Andrew Mathews, Duncan McLaren, Granger Morgan, Ted Parson, Colin Prentice, Phil Rasch, Steve Rayner, Stan Robinson, Alan Robock, Stephen Salter, Steve Schneider (with whom I would dearly love to argue over the issue some more, alas), Dan Schrag, Jesse Scott, John Shepherd, Vaclav Smil, Robert Socolow, Pablo Suarez, Mark Sutton, Jim Thomas, David Victor, Kelly Wanser, Andy Watson and Matt Watson. Special thanks to sometime co-conspirators George Collins and Andy Parker.

Nothing I write on earthsystem subjects will ever, I suspect, escape the influence of Jim Lovelock, for which I will remain always grateful; the same is true, when it comes to history, of Simon Schaffer.

Further thanks go to Julian Allwood, Jim Anderson, Meinrat Andreae, Govindasamy Bala, Bidisha Bannerjee, Scott Barrett, Steve Barrett, Richard Betts, Linus Blomqvist, Olivier Boucher, Stewart Brand, Wally Broecker, Holly Buck, Rose Cairns, Ralph Cicerone, Olaf Corry, Steve Davis, Matthias Dörries, Simon Driscoll, George Dyson, Paul Edwards, Erle Ellis, Paul Falkowski, Chris Field, Piers Forster, David Fowler, Jean-Baptiste Fressoz, Alan Gadian, Jim Galloway, Stephen Gardiner, Arnulf Grübler, Patrick Halloran, Steve Hamburg, Jim Haywood, Tracy Hester, Clare Heyward, Thomas Homer-Dixon, Alf Hornborg, Jo House, Mike Hulme, Pete Irvine, Andrew Jarvis, Anna-Maria Jenkins Hubert, Andy Jones, Richard Klein, Ben Kravitz, Richard Lampitt, Lee Lane, Andrew Lockley, Sean Low, Adam Lowe, Simon Lewis, David MacKay, Doug MacMartin, Ashley Mercer, Pat Mooney, John Moore, Juan Moreno-Cruz, David Mitchell (neither of the more famous ones, though they are good too), David Morrison, David Morrow, Armand Neukermans, Simon Nicholson, John Nissen, Helena Paul, Arthur Petersen, Roger Pielke Jr, Ray Pierrehumbert, Rafe Pomerance, Julia Pongratz, Colin Prentice, Greg Rau, Jesse Reynolds, Kate Ricke, Andy Ridgwell, Johan Rockström, Danny Rosenfeld, Dan Sarewitz, Gavin Schmidt, Hauke Schmidt, Vivian Scott, Russell Seitz, Drew Shindell, Ian Simpson, Vaclav Smil, Joe Smith, Jordan Smith, Rachel Smolker, Karolina Sobecka, Andy Stirling, Bron Szerszynski, Michael Thompson, Simone Tilmes, David Titley, Kevin Trenberth, Slawek Tulaczyk, Paul Valdes, Naomi Vaughan, Jean-Pascal van Ypersele, Peter Wadhams, Gernot Wagner, Tom Wigley, Jennifer Wilcox, Lowell Wood, Rob Wood and Pete Worden.

A number of these people – Ken Caldeira, Jim Galloway, David Keith, Ben Kravitz, Tim Kruger, John Latham, Jim Lovelock, David Morrison, Phil Rasch, Matt Watson, Jim Galloway, David Victor – are due extra thanks for reading and commenting on part or all of the book; and particular thanks in this regard go to

Olivia Judson, my brother John Morton and Francis Spufford. They all made it better; all the reasons that it is not better still are my own.

On top of the opportunities to listen, talk and socialise at various geoengineering meetings and summer schools in Asilomar, Berlin, Big Sur, Calgary, both Cambridges, Edinburgh, Heidelberg, Lisbon, Oxford, Potsdam, Santa Cruz and Waterloo, I have enjoyed similar stimulation at the Breakthrough Dialogues convened by Ted Nordhaus and Michael Shellenberger. I am also very grateful to NCAR for a media fellowship in 2009 and to the Skoll Foundation and Sundance Institute for their 'Stories of Change' project. At a number of these venues it has been a pleasure to work alongside various other writers interested in this most fascinating topic, including Catherine Brahic, Jamais Cascio, Christopher Cokinos, Jeff Goodell, Eli Kintisch, Fred Pearce, Andy Revkin, Jon Vidal and Gaia Vince. My various magazine pieces on the subject benefited a great deal from being edited by Alexandra Witze, Rich Monastersky, James Crabtree and Geoffrey Carr. My colleagues, first at *Nature* and then at the *Economist*, have been both supportive and full of inspiring questions. I am hugely grateful to John Micklethwait, Zanny Minton Beddoes, Daniel Franklin and Andrew Palmer for expressing that support by allowing and facilitating all manner of absences.

Further gratitude is due as ever to Simon Schaffer, Anita Herle, Eva Herle Schaffer and, *in absentia*, Sheila Schaffer, for the retreat in Brighton and much more. Similar thanks for familial support go to three generations of Bacons and a wide range of Mortons and Hyneses. Life is always enriched by John Harrison; I am sorry not to be able to share the finished product with Dominic Prior – or for that matter Iain Banks. Many thanks to Laura Joanknecht. The baristas and other staff of many cafés have put up with me as I cluttered up their tables while writing; particular thanks to those of the Redwood Café, Brighton, the Southsea

Coffee Company, Southsea and, in Greenwich, the Plumtree Café and, most frequently of all, the Buenos Aires Café. Similar gratitude goes to Cambridge University Library, the Central Science Library (alas, no more), Corpus Christi College and the Royal Society of Arts, though none of them is really in the game when it comes to a flat white. Thanks also, in this regard, to Andrea Burgess. And thanks to Ali Shaw for the flat.

I am grateful for permission to quote from the works of Leonard Cohen and the Flobots, and to Faber and Faber and the Eliot estate for the quotation from 'The Lovesong of J. Alfred Prufrock'.

I am also grateful to Kevin Trenberth for permission to use the diagram I think of by his name, and to Wesley Fernandes for redrawing it for me.

Sarah Chalfant, my wonderful agent, had significantly more trust in this project at some points in its development than I did; many thanks to her for that and everything else, and to Alba Ziegler-Bailey too. Many thanks also to Philip Gwyn Jones and Ingrid Gnerlich for buying the book in the first place. Bella Lacey at Granta has been a champion among editors, and hugely supportive as the book morphed into something other than it was; many thanks to her, to Christine Lo and to the excellent copy editor Mandy Woods and proofreader Francine Brody.

As with my previous books, my greatest and most over-whelming debt of thanks is to Nancy Hynes, and so I shall now repair to the garden she has fashioned so beautifully to drink her health by the exquisitely managed sunlight of a full moon, slightly fogged by ice crystals high in the atmosphere.

References, Notes and Further Reading

General

There are fine accounts of recent geoengineering research, its practitioners and some of its implications in Goodell (2011) and Kintisch (2011). Arguments for the furtherance of this research with an eye to deployment can be found in Brand (2009) and Keith (2014). Critiques of these ideas are offered by ETC Group (2010), Hamilton (2014) and Hulme (2014). Fleming (2012) lays out highly relevant history. Caldeira, Bala and Cao (2014) provide an up-to-date review of scientific research in the field; Preston (2013) covers many issues in the social sciences and humanities. Reports from the Royal Society (2009) and the National Research Council (2015a,b) are very useful resources.

To understand geoengineering in its proper context requires an understanding of the earthsystem and its history: I recommend Langmuir and Broecker (2012) and, at a somewhat more advanced level, Lenton and Watson (2011). Lunine (2012) and Kump et al. (2013) are fine undergraduate textbooks. Cornell et al. (2012) and IPCC (2013) provide more detail.

Introduction: Two Questions

The meeting at which I heard Robert Socolow ask his two questions (which I have slightly reworded) was the 'Direct Air Capture Summit' held in Calgary in March 2012. Those who have read Wagner and Weitzman (2015) will recognise Socolow's questions, which they too used to open their book; as they mention in their notes, this followed my use of the questions to introduce a discussion of geoengineering at MIT in 2013 which Gernot Wagner attended.

For an excellent overall account of the science of climate change, see IPCC (2013); for an enlightening discussion of why people answer Socolow's questions in many ways, see Hulme (2009), and for an authoritative account of the lack of progress towards an international climate regime capable of curbing emissions, see Victor (2011). For Arnulf Grübler's thoughts on energy transitions as cited, see Grübler (2012). The estimates of decarbonisation rates come from Anderson and Bows (2009). The pre-1980 history of nuclear power in America is discussed in Walker (2006), and in France in Hecht (2009); see also Morton (2013). The case for new nuclear power is made in Stone (2013) and Lynas (2014). A very good general account of renewable energy can be found in MacKay (2008). The report chaired by Jane Long is in Long et al. (2011) and the APS report on direct-air-capture costs chaired by Robert Socolow is in Socolow et al. (2011).

Chapter One: The Top of the World
For a sketch of Teisserenc de Bort's life, see Greene (2000). A fine and encyclopaedic account of the U-2's history is Pocock (2005) and a sense of flight in the U-2 can be gained from May (2009); I am also very grateful to General Patrick Halloran (ret'd) for discussion of the experience. The notion of the overworld is introduced in Holton et al. (1995). The quotation from Burke's correspondence is from Taylor (1973), cited in Nye (1993). The fallout controversy is dealt with in Jessee (2014). The history of the ozone layer is treated in Parson (2003), and its science in some detail in Solomon (1999); the thoughts from Harold Johnston are adapted from Johnston (1984). Commoner's laws are in Commoner (1975).

Chapter Two: A Planet Called Weather
Of many discussions of the Earth seen from space, two stand out: Cosgrove (2001) and Poole (2010); the 'Spaceship Earth' aspects of the issue are also dealt with in Anker (2010), which introduced me to capsule ecology. My take on these things, expanded in Morton (2010), is influenced by Ingold (2000). The Trenberth diagram in its current form is first found in Trenberth, Fasullo and Kiehl (2009); the version in Figure 1 is derived from Trenberth and Fasullo (2011). The history of the Sankey diagram is to be found in Schmidt (2008); the concept of 'human empire' is illuminated brilliantly in Williams (2013). Tansley's definition of the ecosystem is in Tansley (1935), and his links to Freud are in Cameron and Forrester (1999). Sharon Kingsland on Lotka is from Kingsland (1994); for more, see Kingsland (1995). The history of Dave Keeling's research is given in Keeling (1998); see also Howe (2014). Sayre (2008) traces the history of the dubious concept

of carrying capacity. Crary (1999) is revelatory on Turner.

Chapter Three: Pinatubo

The Pinatubo eruption is fully documented in Newhall and Punongbayan (1996) and well narrated in Oppenheimer (2012). The effects of eruptions on the climate are reviewed in Robock (2000); Dorries (2006) elucidates the history of science's understanding of these effects, and D'Arcy Wood (2014) is excellent on the global effects of Tambora. Pollock et al. (1976) lay out the sulphate cooling hypothesis in full. The effects of Pinatubo on climate and ozone are detailed in McCormick et al. (1995). For the prediction of its climate effects see Hansen et al. (1992) and Hansen et al. (1996), and for the use of the data to demonstrate the water–vapour feedback in the climate, see Soden et al. (2002). The effects on the hydrological cycle are in Trenberth and Dai (2007) and the indirect light effects in Gu et al. (2003) and Mercado et al. (2009).

Chapter Four: Dimming the Noontime Sun

Matt Watson told me of the Soufrière Hills experience in conversation. For the technological sublime, see Nye (1994). The Aurora report is in McClellan et al. (2010). An academic account of the chemtrailers and their beliefs may be found in Cairns (2014); for the beliefs in action see Ian Simpson's Look Up! site, Simpson (2015 onwards). For more understanding of the problem of sulphate particles simply getting bigger, and how to get around it, see Pierce et al. (2010). Ozone effects are dealt with in Tilmes, Müller and Salawitch (2008); non-ozone-depleting aerosols are discussed in Weisenstein and Keith (2015), and the abandoned idea about geoengineering away the ozone hole is in Cicerone, Elliott and Turco (1991).

Simon Schaffer's definition of Promethean science is in Schaffer (2010). For an archive of GeoMIP papers, see GeoMIP (2015 onwards); an overview is provided in Kravitz et al. (2013). To see why polar or equatorial under- or over-cooling is not a fundamental issue, see MacMartin et al. (2013). Jones et al. (2013) deal with the termination shock; Ray Pierrehumbert gives strident voice to his concerns about commitment in Pierrehumbert (2015). The climateprediction.net paper is Ricke, Morgan and Allen (2010); its conclusions are broadly upheld in Kravitz et al. (2014) – though see also Crook, Jackson and Forster (2015) for an apparently different conclusion.

Chapter Five: Coming to Think This Way
The sketch of the history of ideas about human influence on the climate draws largely on Fleming (1998) and Locher and Fressoz (2012), as well as on Golinski (2007); Grove (1995) is excellent on islands, Blackbourn (2006) on Germany, Meyer (2000) on America, and Davis (2004) on North Africa. The actual effects of humans on Europe's climate before the twentieth century are modelled in Betts et al. (2007). The review of Etzler is in Thoreau (1974).

On Martian geoengineering, see Lowell (1896) and (1906), Marklay (2005) or, indeed, Morton (2002); the link to William James is in Dolan (1992). For more recent links between the ethics of geoengineering the Earth and rebuilding Mars, see Robinson (1993), (1994) and (1996). Hamblin (2013) is eye-opening on ideas about environmental warfare and their influence on the catastrophic imagination; McNeill and Unger (2010) provide helpful context. 'How to Wreck the Environment' is MacDonald (1967). Popular reporting of climate modification schemes of the late nineteenth and early twentieth centuries is discussed in Collins (2005). Roger Revelle's ping-pong balls are in Appendix Y4 of the President's Science Advisory Council (1965). Lamb (1972) gives an account of human influence on the climate and possible geoengineering approaches in Chapter 19; another account of similar vintage is Schneider (1977). Founding papers on geoengineering as a response to greenhouse warming are Marchetti (1977) and Dyson (1977).

A sense of where geoengineering stood in the early 1990s can be taken from the National Academies of Science (1992); see also MacCracken (1991), a paper presented at the Palm Coast conference. The interlinked growth of carbon-dioxide politics and climate-change science are traced in Weart (2008) and Howe (2014). The lack of decarbonisation in British consumption, as opposed to production, is discussed in Helm (2012). The best account of why international progress on climate change is hard is in Victor (2011).

Chapter Six: Moving the Goalposts
The Livermore paper is Teller, Wood and Hyde (1997). Responses to Lowell Wood's Aspen talk are in Keith (2000a) and Bala and Caldeira (2000). The political context of Teller's contribution is detailed in Oreskes and Conway (2012). Ken Caldeira's short but highly influential initial contribution to the ocean acidification debate is in Caldeira

and Wickett (2003); on coral reefs, see Hoegh-Guldberg et al. (2007). The pivotal discussion of geoengineering is in Crutzen (2006). The superfreak pivot takes its name from Levitt and Dubner (2011); for a withering critique, see Pierrehumbert (2011). The proceedings of the NASA meeting that gave the world 'solar radiation management' are in Lane et al. (2007).

'Moral hazard' arguments are dealt with in many places – I find the account in Gardiner (2012) particularly useful in the emphasis it puts on passing risk, but not capacity, down the generations; see also Hale (2012) and Morrow (2014). The never published Bush-era report is in Khan et al. (2001). The Novim report which played a role in establishing the 'climate emergency' framing is in Blackstock et al. (2009); see also Royal Society (2009) and Bipartisan Center (2011). For emergency thinking in action, see the Arctic Emergency Methane Group (2014); for well-founded criticism, see Horton (2015). The 'breathing space' paper is Wigley (2006). On the ideas David Keith and colleagues have for a stratospheric experiment, see Dykema et al. (2014), and for a portfolio of possible experiments, see Keith, Duren and MacMartin (2014). Principles for the regulation of such research are addressed in Rayner et al. (2013); see also Parker (2014). Agar (2013) provides various sobering examples of the role of 'control' in the ideology of modern science; Latour (2011) teaches the Promethean duty of care. For an understanding of the need to make the hand as well as the thermostat, I am indebted to Andrew Mathews.

Chapter Seven: Nitrogen
Chapter 6 of IPPC (2013) is a very helpful overview of biogeochemical cycles and human interference in them. For an account of the nitrogen cycle before and after human intervention, see Galloway et al. (2004), and for a magnificent account of how human chemists took it over, see Smil (2001).

The Crookes speech is in Crookes (1898). For more on Marx and guano, and thought-provoking analysis, see Foster, Clark and York (2010). The assessment of Crookes 30 years on is in Enfield (1931); the dust-bowl outcome of some of the changes Enfield chronicles is described in Egan (2006). On Vogt and Osborn, see Desrochers and Hoffbauer (2009); see also Ehrlich (1968) and Ehrlich and Ehrlich (2009). On the scientific and policy background of the 'green revolution', Cullather (2012) is excellent. The way that nitrogen fixation changed the basic energetics of farming was laid out in Leach (1976); see also Smil (2008).

A sense of the mess that nitrogen fixation has brought about is provided by Fowler et al. (2013), Erisman et al. (2013) and Sutton et al. (2011). Bala et al. (2013) estimates the effects of all this on global productivity. The point about making a mess not being engineering is made in Keith (2000b). The calculation of the area required to feed seven billion without modern agricultural yields is in Burney, Davis and Lobell (2010).

Chapter Eight: Carbon Past, Carbon Present
The primary source for atomic optimism is Soddy (1909); see also his biography, Merricks (1996). A particularly helpful reading of the twentieth century's carbon history is in Mitchell (2011). For fine contemporary accounts of the history of the Earth over post-Soddy depths of time, see Langmuir and Broecker (2012) and Lenton and Watson (2011). Ruddiman's ideas are laid out in Ruddiman (2008) and Ruddiman (2014); an overview of early human impacts on the Earth is provided in Ellis et al. (2013). Powerful critiques of Ruddiman's views include Bala et al. (2007) and Singarayer et al. (2011). The idea that 1610 might mark the beginning of the Anthropocene is developed in Lewis and Maslin (2015), and a sense of the great gulf in human history that was opened up at that time can be found in Pomeranz (2001) and Mann (2011); the idea that such an analysis makes Anthropocene change the result of a specific economic and political system, rather than being caused by the species at large, is developed in Malm and Hornborg (2014).

For an account of the discovery of climatic flip-flops in the ice ages, see Kunzrig and Broecker (2008); for an overview of tipping points and climate change, see Lenton et al. (2007). Rowan Sage's ideas about carbon dioxide and the origins of agriculture are in Sage (1995); agriculture being mandatory in the Holocene is an idea found in Richerson, Boyd and Bettinger (2001). Estimates of agricultural yields in a geoengineered world are from Pongratz et al. (2012). The Holocene Saharan tipping point is discussed in Liu et al. (2006) and Claussen (2009); a sense of life on the shores of Lake Megachad in the early-to-mid-Holocene can be found in Broodbank (2014). For the Sahara terraformed by brute-force desalination and irrigation, see Ornstein, Aleinov and Rind (2009).

Chapter Nine: Carbon Present, Carbon Future
For an overview of all proposed carbon-dioxide-removal technologies, see McClaren (2012). For a technical account of CCS, see Wilcox (2012). Sea-water approaches are described in Caldeira and Rau (2000) and Rau (2011). Ocean iron-fertilization experiments are reviewed in Boyd et al. (2007) and Williamson et al. (2012); see also Kunzrig and Broecker (2008), Kintisch (2010) and Goodell (2010). For a statement from oceanographers opposed to further geoengineering studies of this nature, see Strong et al. (2009). For the use of forests as sinks, see Dyson (1977); the estimates of the carbon dioxide that could be stored up by reversing all historic deforestation are from House, Prentice and Lequéré (2002). Woolf et al. (2010) provide an optimistic account of the potental of biochar; Read (2008) is a key exposition of BECCS. For the long-lasting effects of emissions in the absence of geoengineering, see Solomon et al. (2009) and Archer (2010).

Chapter Ten: Sulphur and Soggy Mirrors
For the story of cloud seeding and Project Cirrus, see Fleming (2010) and also Byers (1974) and Cotton and Pielke (1995), from which the funding numbers come; for broader context, see Meyer (2000). Kurt Vonnegut remembered his time with Langmuir in Vonnegut (1977). The long-term trial is in Breed (2014) and an update on the state of play around the world is in Bruintjes (2013). Latham's first geoengineering paper is Latham (1990). On global cooling in the 1960s, see Weart (2008). The ice-age paper is Schneider and Rasool (1971), and its genesis is recounted in Schneider (2009); see also Schneider (1977) and, on recent exaggeration of the ice-age worries, Peterson, Connolley and Fleck (2008). Fred Hoyle's plans for Earth-warming in the event of an ice age are in Hoyle (1981), and Lovelock's aside on anti-cooling geoengineering is in Lovelock (1966); his discoveries about planktonic sulphur are recounted in Lovelock (2000), and the culmination of that research is in Charlson et al. (1987).

The complex effects of aerosols on precipitation are reviewed in Rosenfeld et al. (2008) and on a regional basis on Shindell et al. (2012a); their effect on the south Asian monsoon is explored in Bollasina, Ming and Ramaswamy (2011). The possible link between Chinese sulphate emissions and slowed increases in mean global temperatures is explored in Streets et al. (2013). The case for reduction of short-lived forcers in both health and climate terms is made in Shindell (2012b).

For the effects on health and climate of moves like the IMO restriction on sulphur emissions in international waters, see Lauer et al. (2009) and Winebrake et al. (2009). For the cloudships, see Salter, Sortino and Graham (2008), and for an overview of marine-cloud brightening that includes Phil Rasch's calculations of its effectiveness, see Latham et al. (2008); the Met Office paper that found problems in Brazil is Jones, Haywood and Boucher (2009). For urban brightening, see Akbari, Menon and Rosenfeld (2009); crop brightening is looked at in Ridgwell et al. (2009) and Doughty, Fields and McMillan (2011); and for brightening water surfaces, see Seitz (2011). A very idealised model of veilmaking as a response to heat waves is in Bernstein et al. (2013), and regional applications of marine-cloud brightening are discussed in Latham et al. (2014). Enhancements in precipitation over land following marine-cloud brightening are modelled in Bala et al. (2011). An assessment of the potential of many of these techniques is in Lenton and Vaughan (2009). The effects of veilmaking in only one hemisphere are described in Haywood et al. (2013); see also Oman et al. (2005). Some legal issues surrounding regional geoengineering are addressed in Hester (2013).

Chapter Eleven: The Ends of the World
Badash (2009) provides a very full account of the origins of and arguments over nuclear winter; the key papers are Crutzen and Birks (1982) and Turco et al. (1983) – the TTAPS paper. Levenson (1990) provides excellent context. Weart (1988) and Brians (1987) are useful on the cultural ramifications of nuclear fear; Johan Rockström is quoted in Rayner and Heyward (2013). For more on Huxley's speech, see Deese (2010). Of many sources available on John von Neumann, Dyson (2012) stands out; his supernova fears are quoted in Smith (2007) and the proceedings of the 'Infinite Forecast' meeting are in Pfeffer (1956). The point about the different academic milieux of weather modification and climate modelling is made in Hart and Victor (1993), and Paul Edwards' dissection of climate-nuclear doublethink is in Edwards (2012). The symmetrical treatment of deliberate and inadvertent human modification of the climate is nicely seen in Schneider (1977). The trail of catastrophist papers that led to nuclear winter comprises Ruderman (1974), Alvarez et al. (1980) and Toon et al. (1982). On Schneider and Sagan's falling out, see Schneider (2009). Turco and Sagan (1990) lay out the case for nuclear winter having

helped end the cold war; Gaddis (2005) is one of various histories that is unconvinced (and may have unconsciously inspired my use of the word 'deathboat'). Sagan and Pollack (1993) assesses the prospects of geoengineering in the context of technological transformations of other planets.

Chandler (2008) provides a lively account of the early days of the asteroclique; Duncan Steel's Byron quotation is from Medwin (1824). Morrison et al. (2002) set out the science and policy nexus, and Harris (2008) shows how the risk of impact with an undetected asteroid was reduced so markedly; Mellor (2007) addresses links between asteroid detection and deflection and the military. The winners-and-losers aspect of asteroid deflection is laid out in Schweickart (2004); Carl Sagan's worries about deliberate misuse of asteroid deflection are in Sagan and Ostro (1994) and his work on nuking the moon is discussed in Davidson (1999). David Keith's self-levitating aerosols are in Keith (2010) and the idea of 'counter-geoengineering' is discussed in Parker and Keith (2015). A fictional use of geoengineering to bring about planetary glaciation can be found in Cooper and Niven (2001).

Chapter 12: The Deliberate Planet
Steve Rayner suggested the possibility that geoengineering might be initiated as a form of civil disobedience at the 2014 Berlin meeting, and I am very grateful to him for the suggestion, which I heard him make. The scenario on halving the rate of warming that is mentioned is in Keith and MacMartin (2015). The discussion on linkage is informed by Parson (2014), and the idea of geoengineering as a simplification dates back to the magnificent Schelling (1997). Some arguments that might be relevant to decision making in the Concert are in Weitzman (2012). Technological lock-in is examined in Cairns (2014a). The Yale scenarios are to be found in Banerjee et al. (2013). The role of the mammoths is discussed in Doughty, Wolf and Christopher (2010), and I first came across the idea of glacial restraint at a talk by Slawek Tulaczyk at UCSC; both have since seen fictional use in Robinson (2012). I am very grateful to Duncan McClaren for the insight into the relevance of kintsugi; my use of it should not be taken as representing his thinking on the matter.

Bibliography

Contains books, papers and other documents referred to directly in the text and notes.

Abbott, Charles Greeley (1938) *The Sun and the Welfare of Man* Smithsonian Press

Agar, Jon (2013) *Science in the 20th Century and Beyond* Polity Press

Akbari, Hashem, Menon, Surabi and Rosenfeld, Arthur (2008) 'Global Cooling: Increasing World-Wide Urban Albedos to Offset CO2' *Climatic Change* **94** 275–286

Alvarez et al. (1980) 'Extraterrestrial cause for the Cretaceous–Tertiary extinction' *Science* **208** 1095–1108

Anderson, Kevin and Bows, Alice (2008) 'Reframing the Climate Change Challenge in Light of Post-2000 Emission Trends' *Phil. Trans. Roy. Soc. A* **366** 3863–3882

Angel, Roger (2006) 'Feasibility of Cooling the Earth with a Cloud of Small Spacecraft Near the Inner Lagrange Point (L1)' *PNAS* **103** 17184–17189

Anker, Peder (2010) *From Bauhaus to Ecohouse: A History of Ecological Design* Louisiana State University Press

Archer, David (2010) *The Long Thaw: How Humans Are Changing the Next 100,000 Years of Earth's Climate* Princeton University Press

Arctic Methane Emergency Group (2014) *Strategic Plan* http://a-m-e-g.blogspot.co.uk/2012/12/ameg-strategic-plan.html

Bachelard, Gaston (1964) *The Poetics of Space* (translated by Maria Jolas) Orion Press

Badash, Lawrence (2009) *A Nuclear Winter's Tale: Science and Politics in the 1980s* MIT Press

Bala, Govindasamy and Caldeira, Ken (2000) 'Geoengineering Earth's Radiation Balance to Mitigate CO2–Induced Climate Change' *Geophysical Research Letters*, **27** 2141–2144

Bala, Govindasamy et al. (2007) 'Combined Climate and Carbon-Cycle Effects of Large-Scale Deforestation' PNAS 104 6550–55

Bala, Govindasamy et al. (2011) 'Albedo Enhancement of Marine Clouds to Counteract Global Warming: Impacts on the Hydrological Cycle' *Climate Dynamics* **37** 915–931

Bala, Govindasamy et al. (2013) 'Nitrogen Deposition: How Important is it for Global Terrestrial Carbon Uptake?' *Biogeosciences* **10** 7147–7160

Banerjee, Bidisha et al. (2013) *Scenario Planning for Solar Radiation Management* Yale Climate and Energy Institute

Bernstein, Diana N. et al. (2013) 'Could Aerosol Emissions Be Used for Regional Heat Wave Mitigation?' *Atmospheric Chemistry and Physics* **13** 6373–6390

Betts, Richard A. et al. (2007) 'Biogeophysical Effects of Land Use on Climate: Model Simulations of Radiative Forcing and Large-Scale Temperature Change' *Agricultural and Forest Meteorology* **142** 216–233

Bipartisan Policy Center (2011) *Geoengineering: A National Strategic Plan for Research on the Potential Effectiveness, Feasibility, and Consequences of Climate Remediation Technologies*

Blackbourn, David (2006) *The Conquest of Nature: Water, Landscape and the Making of Modern Germany*, Jonathan Cape

Blackstock, Jason et al. (2009) *Climate Engineering Responses to Climate Emergencies* The Novim Group

Bollasina, Massimo A., Ming, Yi and Ramaswamy, V. 'Anthropogenic Aerosols and the Weakening of the South Asian Summer Monsoon' *Science* **334** 502–505

Bonneuil, Christophe and Fressoz, Jean-Baptiste (2013) *L'Evenement Anthropocene: La Terre, l'Histoire et Nous* Seuil

Boyd, Philip et al. (2007) 'Mesoscale Iron Enrichment Experiments 1993–2005: Synthesis and Future Directions' *Science* **315** 612–617

Breed, Daniel et al. (2014) 'Evaluating Winter Orographic Cloud Seeding: Design of the Wyoming Weather Modification Pilot Project (WWMPP)' *Journal of Applied Meteorology and Climatology* **53** 282–299

Brand, Stewart (2009) *Whole Earth Discipline: An Ecopragmatist Manifesto* Viking

Brians, Paul (1987) *Nuclear Holocausts: Atomic War in Fiction, 1895–1984* Kent State University Press

Broecker, Wally (and indeed Wallace) – see Kunzrig, Robert and Langmuir, Charles

Broodbank, Cyprian (2014) *The Making of the Middle Sea: A History of the Mediterranean from the Beginning to the Emergence of the Classical World* Thames & Hudson

Brown, Harrison (1954) *The Challenge of Man's Future* Viking

Bruintjes, Roelof (2013) *Report on the Expert Team on Weather Modification Meeting 2012/2013* World Meteorological Organization

Byers, Horace R. (1974) 'History of Weather Modification' in *Weather and Climate Modification* ed W. N. Hess, John Wiley & Sons

Buffon, Comte de [George Louis Leclerc] (1778) 'Epoques de la Nature' in *Histoire Naturelle*

Burney, Jennifer A., Davis, Steven J., and Lobell, David B. (2010) 'Greenhouse Gas Mitigation by Agricultural Intensification' *PNAS* **107** 12052–12057

Cairns, Rose C. (2014a) 'Climate Geoengineering: Issues of Path-Dependence and Socio-Technical Lock-In' *WIREs Climate Change* doi: 10.1002/wcc.296

Cairns, Rose C. (2014b) 'Climates of Suspicion: "Chemtrail" Conspiracy Narratives and the International Politics of Geoengineering' Climate Geoengineering Governance Working Paper Series: 009

Caldeira, Ken, Bala, Govindasamy and Cao, Long (2013) 'The Science of Geoengineering' *Annual Review of Earth and Planetary Sciences*, **41** 231–256

Caldeira, Ken and Rau, Greg (2000) 'Accelerating Carbonate Dissolution to Sequester Carbon Dioxide in the Ocean: Geochemical implications' *Geophysical Research Letters* doi: 10.1029/1999GL002364

Caldeira, Ken and Wickett, Michael 'Oceanography: Anthropogenic Carbon and Ocean pH' *Nature* **425** 365–365

Cameron, Laura and Forrester, John (1999) '"A Nice Type of the English Scientist": Tansley and Freud' *History Workshop Journal Issue 48*

Carson, Rachel (1962) *Silent Spring* Houghton Mifflin

Chapman, Clark R. and Morrison, David (1989) *Cosmic Catastrophes* Plenum

Chandler, David (2008) 'The Burger Bar that Saved the World' *Nature* **453** 1164–1168

Charlson, Robert J. et al. (1987) 'Oceanic Phytoplankyton, Atmospheric Sulphur, Cloud Albedo and Climate' *Nature* **326** 655–661 – the CLAW paper

Cicerone, Ralph J., Elliott, Scott and Turco, Richard P. (1991) 'Reduced Antarctic Ozone Depletions in a Model with Hydrocarbon Injections' *Science* **254** 1191–1194

Clarke, Arthur C. (1979) *The Fountains of Paradise* Gollancz

Claussen, Martin (2008) 'Holocene Rapid Land-Cover Changes – Evidence and Theory' in *Natural Climate Variability and Global Warming* eds Richard Battarbee and Heather Binney, Wiley-Blackwell

Collins, Paul (2005) 'Polar Eclipse' *New York Magazine* May 10th

Commoner, Barry (1975) *The Closing Circle: Nature, Man and Technology* Knopf

Cooper, Brenda and Niven, Larry (2001) 'Ice and Mirrors' *Asimov's Science Fiction* February 2001

Cooper, Gary et al. (2014) 'Preliminary Results for Salt Aerosol Production Intended for Marine Cloud Brightening, Using Effervescent Spray Atomization' *Philosophical Transactions of the Royal Society A* **372** doi: 10.1098/rsta.2014.0055

Cornell, Sarah E. et al, eds (2012) *Understanding the Earth System: Global Change Science for Application* Cambridge University Press

Cosgrove, Denis (2001) *Apollo's Eye: A Cartographic Genealogy of the Earth in the Western Imagination* Johns Hopkins University Press

Cotton, William R. and Pielke, Roger A. (1995) *Human Impacts on Weather and Climate* Cambridge University Press

Crary, Jonathan (1999) *Techniques of the Observer: On Vision and Modernity in the Nineteenth Century* MIT Press

Crook, Julia, Jackson, Lawrence and Forster, Piers (2015) 'A Comparison of Temperature and Precipitation Responses to Different Earth Radiation Management Geoengineering Schemes' *Journal of Geophysical Research* (in press)

Crookes, William (1898) 'Presidential Address to the British Association for the Advancement of Science' *Chemical News* **78** 125–139

Crutzen, Paul J. and Stoermer, E. F. (2000) 'The "Anthropocene"' *Global Change Newsletter* **41** 17–18

Crutzen, Paul J. (2006) 'Albedo Enhancement by Stratospheric Sulfur Injections: A Contribution to Resolve a Policy Dilemma' *Climatic Change* **77** 211–219

Crutzen, Paul J. and Birks, John W. (1982) 'The Atmosphere after a Nuclear War: Twilight at Noon' *Ambio* **11** 114–125

Cullather, Nick (2012) *The Hungry World: America's Cold War Battle Against Poverty in Asia*, Harvard University Press

D'Arcy Wood, Gillen (2014) *Tambora: The Eruption that Changed the World* Princeton University Press

Davidson, Keay (1999) *Carl Sagan: A Life* John Wiley and Sons

Davis, Diana (2004) 'Desert "Wastes" of the Maghreb: Desertification Narratives in French Colonial Environmental History of North Africa' *Cultural Geographies* **11** 359–387

Deese, R. Samuel (2010) 'The New Ecology of Power: Julian and Aldhous Huxley in the Cold War Era' in *Environmental Histories of the Cold War* eds J. R. McNeill and Corinna R. Unger

Desrochers, Pierre and Hoffbauer, Christine (2009) 'The Post War Intellectual Roots of the Population Bomb: Fairfield Osborn's "Our Plundered Planet" and William Vogt's "Road to Survival" in Retrospect' *Electronic Journal of Sustainable Development* **1** 73–97

Dolan, David Sutton (1992) 'Percival Lowell: The Sage as Astronomer' PhD thesis, University of Wollongong

Dörries, Matthias (2006) 'In the Public Eye: Volcanology and Climate Change Studies in the 20th Century' *Historical Studies in the Physical and Biological Sciences* **37** 87–125

Doughty, Christopher, Wolf, Alexander and Field, Christopher (2010) 'Biophysical Feedbacks Between the Pleistocene Megafauna Extinction and Climate: The First Human-Induced Global Warming?' *Geophysical Research Letters* doi: 10.1029/2010GL043985

Doughty, Christopher, Field, Christopher and McMillan, Andrew (2011) 'Can Crop Albedo Be Increased Through the Modification of Leaf Trichomes, and Could this Cool Regional Climate?' *Climatic Change* **104** 379–387

Dykema, John A et al. (2014) 'Stratospheric Controlled Perturbation Experiment: A Small-Scale Experiment to Improve Understanding of the Risks of Solar Geoengineering' *Philosophical Transactions of the Royal Society A* **372** doi: 10.1098/rsta.2014.0059

Dyson, Freeman J. (1977) 'Can We Control the Carbon Dioxide in the Atmosphere?' *Energy* **2** 287–291

Dyson, George (2012) *Turing's Cathedral: The Origins of the Digital Universe* Allen Lane

Edwards, Paul N. (2010) *A Vast Machine: Computer Models, Climate Data, and the Politics of Global Warming* MIT Press

Edwards, Paul N. (2012) 'Entangled Histories: Climate Science and Nuclear Weapons Research' *Bulletin of the Atomic Scientists* **68** 28–40

Egan, Timothy (2006) *The Worst Hard Time: The Untold Story of Those Who Survived the Great American Dust Bowl* Mariner Books

Ehrlich, Paul (1968) *The Population Bomb* Ballantine Books

Ehrlich, Paul R. and Ehrlich, Anne H. (2009) 'The Population Bomb Revisited' *Electronic Journal of Sustainable Development* **1** 5–14

Ellis, Erle C. et al. (2013) 'Used Planet: A Global History' *PNAS* **110** 7978–7985

Enfield, R. R. (1931) 'The World's Wheat Situation' *Economic Journal* **41** 550–565

Erisman, Jan Willem et al. (2014) 'Consequences of Human Modification of the Global Nitrogen Cycle' *Philosophical Transactions of the Royal Society B* **368** doi:10.1098/rstb.2013.0116

ETC Group (2010) *Geopiracy: The Case Against Geoengineering*

Etzler, John Adolphus (1836) *The Paradise within the Reach of all Men, without Labor, by Powers of Nature and Machinery* John Brooks

Fisk, Dorothy (1934) *Exploring the Upper Atmosphere* Oxford University Press

Fleming, James Rodger (1998) *Historical Perspectives on Climate Change* Oxford University Press

Fleming, James Rodger (2010) *Fixing the Sky: The Checkered History of Weather and Climate Control*, Columbia University Press

Foster, John Bellamy, Clark, Brett and York, Richard (2010) *The Ecological Rift: Capitalism's War on the Earth* Monthly Review Press

Fowler, David et al. (2013) 'The Global Nitrogen Cycle in the Twenty-first Century' *Philosophical Transactions of the Royal Society B* **368** doi: 10.1098/rstb.2013.0164

Gaddis, John Lewis (2005) *The Cold War* Penguin

Galloway, James N, et al. (2004) 'Nitrogen Cycles: Past, Present, and Future' *Biogeochemistry* **70** 153–226

Gardiner, Stephen M. (2011) *A Perfect Moral Storm: The Ethical Tragedy of Climate Change*, Oxford University Press

Garrels, Robert M., Mackenzie, Fred T. and Hunt, Cynthia (1973) *Chemical Cycles and the Global Environment: Assessing Human Influences* William Kaufmann Inc

GeoMIP (2015 onwards) – an archive of all papers can be found at http://climate.envsci.rutgers.edu/GeoMIP/publications.html (accessed May 15th 2015)

Gernsback, Hugo (1917) *The Scientific Adventures of Baron Münchausen* (compiled, abridged and edited by Robert Godwin) Apogee Books

Glick, Thomas F. – see Thoreau, Henry David

Godwin, Robert – see Gernsback, Hugo

Golinski, Jan (2007) *British Weather and the Climate of Enlightenment* University of Chicago Press

Goodell, Jeff (2010) *How to Cool the Planet: Geoengineering and the Audacious Quest to Fix Earth's Climate* Houghton Mifflin Harcourt

Greene, Mott T. (2000) 'High Achiever' *Nature* **407** 947

Grove, Richard (1995) *Green Imperialism: Colonial Expansion, Tropical Island Edens and the Origins of Environmentalism, 1600–1860* Cambridge University Press

Grübler, Arnulf (2012) 'Energy Transitions Research: Insights and Cautionary Tales' *Energy Policy* **50** 8–16

Gu, Lianhong et al. (2003) 'Response of a Deciduous Forest to the Mount Pinatubo Eruption: Enhanced Photosynthesis' *Science* **28** 2035–2038

Hale, Benjamin (2012) 'The World that Would Have Been: Moral Hazard Arguments Against Geoengineering' in *Engineering the Climate: The Ethics of Solar Radiation Management* ed Christopher J. Preston, Lexington Books

Hamblin, Jacob Darwin (2012) *Arming Mother Nature: The Birth of Catastrophic Environmentalism* Oxford University Press

Hamilton, Clive (2013) *Earthmasters: The Dawn of the Age of Climate Engineering*, Yale University Press – see also a wide selection of essays at clivehamilton.com

Hansen, James et al. (1992) 'Potential Climate Impact of Mount Pinatubo Eruption' *Geophysical Research Letters* **19** 215–218

Hansen, James et al. (1996) 'A Pinatubo Climate Modelling Investigation' in *The Mount Pinatubo Eruption: Effects on the Atmosphere and Climate* eds Girgio Fiocco, Daniele Fuà and Guido Visconti, Springer

Hart, David and Victor, David (1993) 'Scientific Elites and the Making of US Policy for Climate Change Research, 1957–74' *Social Studies of Science* **23** 643–680

Harris, Alan (2008) 'What Spaceguard Did' *Nature* **453** 1178–1179

Haywood, Jim M. et al. (2013) 'Asymmetric Forcing from Stratospheric Aerosols Impacts Sahelian Rainfall' *Nature Climate Change* doi: 10.1038/NCLIMATE1857

Heard, Gerald (1936) *Exploring the Stratosphere* T. Nelson and Sons

Heard, Gerald (as Heard, H. F.) (1947) 'The President of the United States, Detective' *Ellery Queen's Mystery Magazine*, March 1947

Hecht, Gabriel (2009) *The Radiance of France*, MIT Press

Helm, Dieter (2012) *The Carbon Crunch: How We're Getting Climate Change Wrong – and How to Fix it* Yale University Press

Herzog, Arthur (1977) *Heat* Pan Books

Hester, Tracy D. (2013) 'A Matter of Scale: Regional Climate Engineering and the Shortfalls of Multinational Governance' *Carbon and Climate Law Review* **3** 168

Hoegh-Guldberg, Ove et al. (2007) 'Coral Reefs Under Rapid Climate Change and Ocean Acidification' *Science* **318** 1737–1742

Holton, James R. et al. (1995) 'Stratosphere–Troposphere Exchange' *Reviews of Geophysics* **33** 403–439

Hornborg, Alf (2014) 'Technology as Fetish: Marx, Latour, and the Cultural Foundations of Capitalism' *Theory, Culture and Society* doi: 10.1177/0263276413488960

Horton, Joshua (2015) 'The Emergency Framing of Solar Geoengineering: Time for a Different Approach' *Anthropocene Review* (in press)

House, Joanna I., Prentice, I. Colin and Lequéré, Corinne (2002) 'Maximum Impacts of Future Reforestation or Deforestation on Atmospheric CO2' *Global Change Biology* **8** 1047–1052

Howe, Joshua P. (2014) *Behind the Curve: Science and the Politics of Global Warming* University of Washington Press

Hoyle, Fred (1981) *Ice: The Ultimate Catastrophe* Continuum

Hulme, Mike (2009) *Why We Disagree About Climate Change: Understanding Controversy, Inaction and Opportunity* Cambridge University Press

Hulme, Mike (2014) *Can Science Fix Climate Change? A Case Against Climate Engineering* Polity Press

Ingold, Tim (2000) 'Globes and Spheres: The Topology of

Environmentalism' in *The Perception of the Environment: Essays in Livelihood, Dwelling and Skill,* Routledge

IPCC (2013) *Climate Change 2013: The Physical Science Basis* – see http://www.climatechange2013.org/

IPCC (2014a) *Climate Change 2014: Impacts, Adaptation and Vulnerability* – see http://ipcc-wg2.gov/AR5/

IPCC (2014b) *Climate Change 2014: Mitigation of Climate Change* – see http://mitigation2014.org/

Jessee, E. Jerry (2014) 'A Heightened Controversy: Nuclear Weapons Testing, Radioactive Tracers, and the Dynamic Atmosphere' in *Toxic Airs: Body, Place, and Planet in Historical Perspective* eds James R. Fleming and Ann Johnson, University of Pittsburgh Press

Johnston, Harold (1984) 'Human Effects on the Global Atmosphere' *Annual Review of Physical Chemistry* **35** 481–505

Jones, Andy, Haywood, Jim and Boucher, Olivier (2009) 'Climate Impacts of Geoengineering Maritime Stratocumulus Clouds' *Journal of Geophysical Research* doi: 10.1029/2008JD011450

Jones, Andy et al. (2013) 'The Impact of Abrupt Suspension of Solar Radiation Management (Termination Effect) in Experiment G2 of the Geoengineering Model Intercomparison Project (GeoMIP)' *Journal of Geophysical Research: Atmospheres* doi: 10.1002/jgrd.50762

Keeling, Charles D. (1998) 'Rewards and Penalties of Monitoring the Earth' *Annual Reviews of Energy and the Environment* **23** 25–82

Keith, David W. (2000a) 'Geoengineering the Climate: History and Prospect' *Annual Review of Earth and Planetary Sciences* **25** 245–284

Keith, David W. (2000b) 'The Earth is Not Yet an Artifact' *IEEE Technology and Society Magazine* **19** 25–28

Keith, David W. (2010) 'Photophoretic Levitation of Engineered Aerosols for Geoengineering' *PNAS* **107** 16428–16431

Keith, David W. (2013) *A Case for Climate Engineering* MIT Press

Keith, David W., Duren, Riley and MacMartin, Douglas G. (2014) 'Field Experiments on Solar Geoengineering: Report of a Workshop Exploring a Representative Research Portfolio' *Philosophical Transactions of the Royal Society A* **372** doi: 10.1098/rsta.2014.0175

Keith, David W. and MacMartin, Douglas G. (2015) 'A Temporary, Moderate and Responsive Scenario for Solar Geoengineering'

Nature Climate Change doi: 10.1038/nclimate2493

Khan, Ehsan et al. (2001) *Response Options to Limit Rapid or Severe Climate Change: Assessment of Research Needs* US Department of Energy

King, Paul – see May, James

Kingsland, Sharon E. (1994) 'Economics and Evolution: Alfred James Lotka and the Economy of Nature' in *Natural Images in Economic Thought: 'Markets Read in Tooth and Claw'* ed Philip Mirowski, Cambridge University Press

Kingsland, Sharon E. (1995) *Modeling Nature: Episodes in the History of Population Ecology* Chicago University Press

Kintisch, Eli (2010) *Hack the Planet* John Wiley and sons

Klein, Naomi (2014) *This Changes Everything: Capitalism vs. the Climate* Simon & Schuster (2014)

Kravitz, Ben et al. (2013) 'Climate Model Response from the Geoengineering Model Intercomparison Project (GeoMIP)' *Journal of Geophysical Research: Atmospheres* doi: 10.1002/jgrd.50646

Kravitz, Ben et al. (2014) 'A Multi-Model Assessment of Regional Climate Disparities Caused by Solar Geoengineering', *Environmental Research Letters* **9** doi: 10.1088/1748–9326/9/7/074013

Kump, Lee R. et al. (2013) *The Earth System* (3rd edition) Pearson

Kunzig, Robert and Broecker, Wallace S. (2008) *Fixing Climate: The Story of Climate Science – and How to Stop Global Warming* Profile

Lamb, Hubert Horace (1971) 'Climate-Engineering Schemes to Meet a Climatic Emergency' *Earth-Science Reviews* **7** 87–95

Lamb, Hubert Horace (1972) *Climate: Present, Past and Future* Methuen

Lane, Lee et al. (2007) *Workshop Report on Managing Solar Radiation* NASA/CP-2007-214558

Langmuir, Charles H. and Broecker, Wally (2012) *How to Build a Habitable Planet: The Story of Earth from the Big Bang to Humankind* Princeton University Press

Latham, John (1990) 'Control of Global Warming' *Nature* **347** 339–340

Latham, John et al. (2008) 'Global Temperature Stabilization via Controlled Albedo Enhancement of Low-Level Maritime Clouds' *Philosophical Transactions of the Royal Society A* **366** 3969–3987

Latham, John et al. (2014) 'Marine Cloud Brightening: Regional

Applications' *Philosophical Transactions of the Royal Society A* **372** doi: 10.1098/rsta.2014.0053

Latour, Bruno (2011) 'Love Your Monsters' *Breakthrough Journal* **2** 21–28

Lauer, Axel (2009) 'Assessment of Near-Future Policy Instruments for Oceangoing Shipping: Impact on Atmospheric Aerosol Burdens and the Earth's Radiation Budget' *Environmental Science and Technology* **43** 5592–5598

Leach, Gerald (1976) *Energy and Food Production* IPC Science and Technology Press

Lem, Stansilaw (1986) *One Human Minute* (translated by Catherine S. Leach) Harcourt, Brace

Lenton, Timothy M. et al. (2007) 'Tipping Elements in the Earth's Climate System' *PNAS* **105** 1786–1793

Lenton, Timothy M. and Vaughan, Naomi E. (2009) 'The Radiative Forcing Potential of Different Climate Geoengineering Options' *Atmospheric Chemistry and Physics* **9** 5539–5561

Lenton, Tim and Watson, Andrew (2011) *Revolutions That Made the Earth* Oxford University Press

Leopold, Aldo (1949) *A Sand County Almanac and Sketches Here and There* Oxford University Press

Levenson, Thomas (1990) *Ice Time: Climate, Science and Life on Earth* Harper Collins

Levitt, Steven D. and Dubner, Stephen J. (2011) *SuperFreakonomics: Global Cooling, Patriotic Prostitutes, and Why Suicide Bombers Should Buy Life Insurance* HarperCollins

Lewis, Simon L. and Malin, Mark A. (2015) 'Defining the Anthropocene' *Nature* **519** 171–180

Liu, Zhengyu et al. (2006) 'On the Cause of Abrupt Vegetation Collapse in North Africa During the Holocene: Climate Variability vs. Vegetation Feedback' *Geophysical Research Letters* **33** doi: 10.1029/2006GL028062

Locher, Fabien and Fressoz, Jean-Baptiste (2012) 'Modernity's Frail Climate: A Climate History of Environmental Reflexivity' *Critical Inquiry* **38** 579–598

Long, Jane et al. (2011) *California's Energy Future: The View to 2050* California Council on Science and Technology

Lotka, Alfred J. (1925) *Elements of Physical Biology*, Williams and Wilkins

Lovelock, James (1966) 'Some Thoughts on the Year 2000' Lovelock archives, Science Museum

Lovelock, James (1977) *Gaia: A New Look at Life on Earth*, Oxford University Press

Lovelock, James (1988) *The Ages of Gaia: A Biography of Our Living Earth*, Oxford University Press

Lovelock, James (2000) *Homage to Gaia: The Life of an Independent Scientist*, Oxford University Press

Lowell, Percival (1896) *Mars* Longmans, Green and Co

Lowell, Percival (1906) *Mars as the Abode of Life* Macmillan

Lunine, Jonathan (2013) *Earth: Evolution of a Habitable World* (2nd edition) Cambridge University Press

Lynas Mark (2011) *The God Species: Saving the Planet in the Age of Humans* National Geographic

Lynas, Mark (2014) *Nuclear 2.0: Why a Green Future Needs Nuclear Power* Kindle Single

MacCracken, Michael C. (1991) 'Geoengineering the Climate' Lawrence Livermore National Laboratory preprint UCRL-JC-108014

MacDonald, Gordon J. F. (1968) 'How to Wreck the Environment' in *Unless Peace Comes* ed Neil Calder, Penguin

MacKay, David J. C. (2008) *Sustainable Energy – Without the Hot Air* UIT – See also the excellent associated website http://www.withouthotair.com/

MacMartin, Douglas G. et al. (2012) 'Management of Trade-Offs in Geoengineering Through Optimal Choice of Non-Uniform Radiative Forcing' *Nature Climate Change* doi: 10.1038/NCLIMATE1722

Maddox, John (1972) *The Doomsday Syndrome: an Attack on Pessimism* McGraw Hill

Malm, Andreas and Hornborg, Alf (2014) 'The Geology of Mankind? A Critique of the Anthropocene Narrative' *Anthropocene Review* **1** 62–69

Mann, Charles C. (2011) *1493: Uncovering the New World Columbus Created* Knopf

Marchetti, Cesare (1977) 'On Geoengineering and the CO2 Problem' *Climatic Change* **1** 59–68

Markley, Robert (2005) *Dying Planet: Mars in Science and the Imagination* Duke University Press

Masco, Joseph P. (2012) 'The End of Ends' *Anthropological Quarterly* **85** 1109–1126

May, James (2009) *James May at the Edge of Space* (documentary film, directed by Paul King), BBC

Mazower, Mark (2012) *Governing the World: The History of an Idea* Allen Lane

McClellan, Justin et al. (2010) *Geoengineering Cost Analysis (AR10–182)* Aurora Flight Sciences

McCormick, M. Patrick, Thomason, Larry W. and Trepte, Charles R. (1995) 'Atmospheric Effects of the Mt Pinatubo Eruption' *Nature* **373** 399–404

McCray, W. Patrick (2013) *The Visioneers: How a Group of Elite Scientists Pursued Space Colonies, Nanotechnologies, and a Limitless Future* Princeton University Press

McKibben, Bill (2003) *The End of Nature: Humanity, Climate Change and the Natural World* (revised edition), Bloomsbury

McLaren, Duncan (2012) 'A Comparative Global Assessment of Potential Negative Emissions Technologies' *Process Safety and Environmental Protection* **90** 489–500

McMenamin, Mark – see Vernadsky, Vladimir

McNeill, J. R. and Unger, Corinna R. (2010) *Environmental Histories of the Cold War* Cambridge University Press

Medwin, Thomas (1826) *Journal of the Conversations of Lord Byron: Noted During a Residence with His Lordship at Pisa, in the Years 1821 and 1822* Henry Colburn

Mellor, Felicity (2007) 'Asteroid Research and the Legitimization of War in Space' *Social Studies of Science* **37** 499–531

Mercado, Lina M. et al. (2009) 'Impact of Changes in Diffuse Radiation on the Global Land Carbon Sink' *Nature* **458** 1014–1018

Merricks, Linda (1996) *The World Made New: Frederick Soddy, Science, Politics, and Environment* Oxford University Press

Meyer, William B. (2000) *Americans and their Weather* Oxford University Press

Mitchell, Timothy (2011) *Carbon Democracy: Political Power in the Age of Oil* Verso

Morrison, David et al. 'Dealing with the Impact Hazard' in *Asteroids III* eds William J. Bottke Jr. et al, University of Arizona Press

Morrow, David R. (2014) 'Ethical Aspects of the Mitigation Obstruction Argument Against Climate Engineering Research'

Philosophical Transactions of the Royal Society A **372** doi: 10.1098/rsta.2014.0062

Morton, Oliver (2002) *Mapping Mars: Science, Imagination and the Birth of a World* Fourth Estate

Morton, Oliver (2003) 'The Tarps of Kilimanjaro' *New York Times,* November 17th

Morton, Oliver (2007) *Eating the Sun: How Plants Power the Planet* HarperCollins

Morton, Oliver (2010) 'Globe And Sphere, Cycles and Flows: How to See The World' in *Seeing Further: The Story of Science and the Royal Society* ed Bill Bryson, HarperPress

Morton, Oliver (2012) 'The Dream that Failed' *Economist* March 11th 2012

Mossop, S. C. (1964) 'Volcanic Dust Collected at an Altitude of 20km' *Nature* **203** 824–827

National Academy of Sciences, National Academy of Engineering and Institute of Medicine (1992) *Policy Implications of Greenhouse Warming Mitigation, Adaptation and the Science Base* National Academies Press

National Research Council (2015a) *Climate Intervention: Carbon Dioxide Removal and Reliable Sequestration* National Academies Press

National Research Council (2015b) *Climate Intervention: Reflecting Sunlight to Cool the Earth* National Academies Press

New York Times (1985) 'Miscasting the Dinosaur's Horoscope' April 2nd

New York Times (2003) 'The Shrinking Snows of Kilimanjaro' November 26th

Newhall, Christopher G. and Punongbayan, Raymundo S. (1996) *Fire and Mud: Eruptions and Lahars of Mount Pinatubo* — http://pubs.usgs.gov/pinatubo

Niven, Larry and Pournelle, Jerry (1977) *Lucifer's Hammer* Ballantine Books

Nye, David E. (1994) *American Technological Sublime* MIT Press

Oman, Luke et al. (2005) 'Climatic Response to High-latitude Volcanic Eruptions' *Journal of Geophysical Research* **110** doi: 10.1029/2004JD005487

Oppenheimer, Clive (2011) *Eruptions that Shook the World* Cambridge University Press

Oreskes, Naomi and Conway, Erik M. (2012) *Merchants of Doubt: How a Handful of Scientists Obscured the Truth on Issues from Tobacco Smoke to Global Warming* Bloomsbury

Oreskes, Naomi and Conway, Erik M. (2014) *The Collapse of Western Civilization: A View from the Future* Columbia University Press

Ornstein, Leonard, Aleinov, Igor and Rind, David (2009) 'Irrigated Afforestation of the Sahara and Australian Outback to End Global Warming' *Climatic Change* **97** 409–437

Osborn, Fairfield (1948) *Our Plundered Planet* Little, Brown

Parker, Andy (2014) 'Governing Solar Geoengineering Research as it Leaves the Laboratory' *Philosophical Transactions of the Royal Society A* **372** doi: 10.1098/rsta.2014.0173

Parker, Andy and Keith, David (2015) 'What's the Right Temperature for the Earth' *Washington Post*, January 29th

Parson, Edward A. (2003) *Protecting the Ozone Layer: Science and Strategy* Cambridge University Press

Parson, Edward A. (2014) 'Climate Engineering in Global Climate Governance: Implications for Participation and Linkage' *Transnational Environmental Law*, **3** 89–110

Peterson, Thomas C., Connolley, William M. and Fleck, John (2008) 'The Myth of the 1970s Global Cooling Scientific Consensus' *Bulletin of the American Meteorological Society* **89** 1325–1337

Pfeffer, Richard (1956) *Dynamics of Climate: The Proceedings of a Conference on the Application of Numerical Integration Techniques to the Problem of the General Circulation Held October 26–28, 1955* Pergamon Press

Pielke, Roger Jr (2010) *The Climate Fix: What Scientists and Politicians Won't Tell You About Global Warming* Basic Books

Pierce, Jeffrey R. et al. (2010) 'Efficient Formation of Stratospheric Aerosol for Climate Engineering by Emission of Condensible Vapour from Aircraft' *Geophysical Research Letters* doi: 10.1029/2010GL043975

Pierrehumbert, Raymond T. (2009) 'An Open Letter to Steve Levitt' –http://www.realclimate.org/index.php/archives/2009/10/an-open-letter-to-steve-levitt/

Pierrehumbert, Raymond T. (2015) 'Climate Engineering is Barking Mad' *Slate* February 10th

Pocock, Chris (2005) *50 Years of the U-2: The Complete Illustrated History of the 'Dragon Lady'* Schiffer Publishing

Pollack, James B. et al. (1976) 'Volcanic Explosions and Climatic Change: A Theoretical Assessment' *Journal of Geophysical Research* **81** 1071–1083

Pollack, James B. and Sagan, Carl (1993) 'Planetary Engineering' in *Resources of Near Earth Space* eds John S. Lewis et al, University of Arizona Press

Pomeranz, Kenneth (2001) *The Great Divergence: China, Europe and the Making of the Modern World Economy* Princeton University Press

Pongratz, Julia et al. (2012) 'Crop Yields in a Geoengineered Climate' *Nature Climate Change*, doi: 10.1038/NCLIMATE1373

Poole, Robert (2010) *Earthrise: How Man First Saw the Earth* Yale University Press

President's Science Advisory Panel (1965) *Restoring the Quality of Our Environment*

Preston, Christopher J. (2012) *Engineering the Climate: The Ethics of Solar Radiation Management* Lexington Books

Rau, Greg H. (2011) 'CO2 Mitigation via Capture and Chemical Conversion in Seawater' *Environmental Science and Technology* **45** 1088–1092

Rasool, S. Ichtiaque and Schneider, Stephen (1971) 'Atmospheric Carbon Dioxide and Aerosols: Effects of Large Increases on Global Climate' *Science* **173** 138–141

Rau, Greg and Caldeira, Ken (1999) 'Enhanced Carbonate Dissolution: A Means of Sequestering Waste CO2 as Ocean Bicarbonate' *Energy Conservation and Management* **40** 1803–1813

Rayner, Steve et al. (2013) 'The Oxford Principles' *Climatic Change* **121** 499–512

Rayner, Steve and Heyward, Clare (2013) 'The Inevitability of Nature as a Rhetorical Resource' in *Anthropology and Nature* ed Kirsten Hastrup, Routledge

Read, Peter (2008) 'Biosphere Carbon Stock Management: Addressing the Threat of Abrupt Climate Change in the Next Few Decades: An Editorial Essay' *Climatic Change* **87** 305–320

Richerson, Peter J., Boyd, Robert and Bettinger, Robert L. (2001) 'Was Agriculture Impossible during the Pleistocene But Mandatory During The Holocene? A Climate Change Hypothesis' *American Antiquity* **66** 387–411

Ricke, Katharine L., Morgan, M. Granger and Allen, Myles R. (2010)

'Regional Response to Solar-Radiation Management' *Nature Geoscience* doi: 10.1038/NGEO0915

Ridgwell, Andy et al. (2009) 'Tackling Regional Climate Change by Leaf Albedo Bio-Geoengineering' *Current Biology* **19** 146–150

Robinson, Kim Stanley (1993) *Red Mars* Harper Voyager

Robinson, Kim Stanley (1994) *Green Mars* Harper Voyager

Robinson, Kim Stanley (1996) *Blue Mars* Harper Voyager

Robinson, Kim Stanley (2012) *2312* Orbit

Robock, Alan (2000) 'Volcanic Eruptions and Climate' *Reviews of Geophysics* **38** 191–219

Robock, Alan (2008) '20 Reasons Why Geoengineering May Be a Bad Idea' *Bulletin of the Atomic Scientists* **64** 14–18

Rosenfeld, Daniel (2008) 'Cooling Earth's Temperature by Seeding Marine Stratocumulus Clouds for Increasing Cloud Cover by Closing Open Cells' AGU Fall Meeting poster U43A-0043

Rosenfeld, Daniel et al. (2008) 'Flood or Drought: How Do Aerosols Affect Precipitation' *Science* **321** 1309–1313

Royal Society (2009) *Geoengineering the Climate: Science, Governance and Uncertainty*

Ruddiman, William (2005) *Plows, Plagues and Petroleum: How Humans Took Control of Climate* Princeton University Press

Ruddiman, William (2013) 'The Anthropocene' *Annual Review of Earth and Planetary Sciences* doi: 10.1146/annurev-earth-050212-123944

Ruderman, Mal E. (1974) 'Possible Consequences of Nearby Supernova Explosions for Atmospheric Ozone and Terrestrial Life' *Science* **184** 1079–1081

Sagan, Carl and Ostro, Steven J. (1994) 'Long-range Consequences of Interplanetary Collisions' *Issues in Science and Technology* **10** 67–72

Sagan, Carl and Turco, Richard (1990) *A Path Where No Man Thought: Nuclear Winter and the End of the Arms Race*, Century

Sage, Rowan F. (1995) 'Was Low Atmospheric CO2 During the Pleistocene a Limiting Factor in the Growth of Agriculture?' *Global Change Biology* **1** 93–106

Salter, Stephen, Sortino, Graham and Latham, John (2008) 'Sea-Going Hardware for the Cloud Albedo Method of Reversing Global Warming' *Philosophical Transactions of the Royal Society A* **366** 2989–3006

Sayre, Nathan F. (2008) 'The Genesis, History, and Limits of Carrying Capacity', *Annals of the Association of American Geographers* **98** 120–134

Schaffer, Simon (2010) 'Charged Atmospheres: Promethean Science and the Royal Society' in *Seeing Further: The Story of Science and the Royal Society* ed Bill Bryson, HarperPress

Schelling, Thomas C. (1997) 'The Economic Diplomacy of Geoengineering' *Climatic Change* **33** 303–307

Schmidt, Mario (2008) 'The Sankey Diagram in Energy and Material Flow Management – Part I: History' *Journal of Industrial Ecology* **12** 82–94

Schneider, Stephen H. (with Mesirow, Lynne E.) (1976) *The Genesis Strategy: Climate and Global Survival* Plenum

Schneider, Stephen H. (2009) *Science as a Contact Sport: Inside the Battle to Save Earth's Climate* National Geographic

Schweickart, Russell L. (2004) 'The Real Deflection Dilemma' in *2004 Planetary Defense Conference: Protecting Earth from Asteroids* AIAA

Scott, James C. (1998) *Seeing Like a State: How Certain Schemes to Improve the Human Condition Have Failed* Yale University Press

Seitz, Russell (2011) 'Bright Water: Hydrosols, Water Conservation and Climate Change' *Climatic Change* **105** 365–381

Sherlock, Robert Lionel (1922) *Man as a Geological Agent : An Account of His Action on Inanimate Nature* H. F. and G. Witherby

Shindell, Drew T. et al. (2012a) 'Precipitation Response to Regional Radiative Forcing' *Atmospheric Chemistry and Physics* **12** 6969–6982

Shindell, Drew T. et al. (2012b) 'Simultaneously Mitigating Near-Term Climate Change and Improving Human Health and Food Security' *Science* **335** 183–189

Simpson, Ian (2015 onwards) *Look-up!* – www.look-up.org.uk (accessed May 23rd 2015)

Singarayer, Joy et al. (2011) 'Late Holocene Methane Rise Caused by Orbitally Controlled Increase in Tropical Sources' *Nature* **470** 82–85

Smil, Vaclav (2001) *Enriching the Earth: Fritz Haber, Carl Bosch and the Transformation of World Food Production* MIT Press

Smil, Vaclav (2008) *Energy in Nature and Society: General Energetics of Complex Systems* MIT Press

Smith, P. D. (2007) *Doomsday Men: The Real Dr Strangelove and the Dream of the Superweapon* Allen Lane

Socolow et al. (2011) 'Direct Air Capture of CO2 with Chemicals: A Technology Assessment for the APS Panel on Public Affairs' *American Physical Society*

Soddy, Frederick (1909) *The Interpretation of Radium and the Structure of the Atom* John Murray

Soden, Brian J. et al. (2002) 'Global Cooling After the Eruption of Mount Pinatubo: A Test of Climate Feedback by Water Vapour' *Science* **296** 726–730

Solomon, Susan (1999) 'Stratospheric Ozone Depletion: A Review of Concepts and History' *Reviews of Geophysics* **37** 275–316

Solomon, Susan et al. (2009) 'Irreversible Climate Change Due to Carbon Dioxide Emissions' *PNAS* **106** 1704–1709

Stone, Robert (dir) (2013) *Pandora's Promise* (documentary film) Robert Stone Productions – see www.pandoraspromise.com

Streets, David G. et al. (2013) 'Radiative Forcing Due to Major Aerosol Emitting Sectors in China and India' *Geophysical Research Letters* **40** 4409–4414

Strong, Aaron et al. (2009) 'Ocean Fertilization: Time to Move On' *Nature* **461** 347–348

Sutton, Mark A. et al. (2011) *The European Nitrogen Assessment: Sources, Effects and Policy Perspectives* Cambridge University Press

Tansley, Arthur G. (1935) 'The Use and Abuse of Vegetational Concepts and Terms' *Ecology* **16** 284–307

Taylor, Nicholas (1973) 'The Awful Sublimity of the Victorian City: Its Aesthetic and Architectural Origins' in *The Victorian City: Images and Realities, Volume 2* ed Harold James Dyos and Michael Wolff, Taylor and Francis

Teller, Edward, Wood, Lowell and Hyde, Ron (1997) 'Global Warming and Ice Ages: 1. Prospects of Physics Based Modulation of Global Change' Lawrence Livermore National Laboratory preprint UCRL-JC-128715

Thomson, George (1955) *The Foreseeable Future*, Cambridge University Press

Thoreau, Henry David (1854) *Walden or Life in the Woods* Ticknor and Fields

Thoreau, Henry David (1974) @paradise (to be) Regained' in *The Writings of Henry David Thoreau: Reform Papers* ed Thomas F.

Glick, Princeton University Press

Tilmes, Simone, Müller, Rolf and Salawitch, Ross (2008) 'The Sensitivity of Polar Ozone Depletion to Proposed Geoengineering Schemes' *Science* **320** 1201–1204

Toon, O. Brian et al. (1982) 'Evolution of an Impact-Generated Dust Cloud and its Effects on the Atmosphere' *GSA Special Papers* **190** 187–200

Trenberth, Kevin and Dai, Aiguo (2007) 'Effects of Mount Pinatubo Volcanic Eruption on the Hydrological Cycle as an Analog of Geoengineering' *Geophysical Research Letters* **34** doi: 10.1029/2007GL030524

Trenberth, Kevin E., Fasullo, John T. and Kiehl, Jeffrey (2009) 'Earth's Global Energy Budget' *Bulletin of the American Meteorological Society* **90** 311–324

Trenberth, Kevin E. and Fasullo, John T. (2011) 'Tracking Earth's Energy: From El Niño to Global Warming' *Surveys in Geophysics, Special Issue*, doi: 10.1007/s10712–011–9150–2

Turco, Richard P. et al. (1990) 'Climate and Smoke: An Appraisal of Nuclear Winter' *Science* **247** 167–168 – The TTAPS paper

Vernadsky, Vladimir (1989) *The Biosphere* (revised and annotated by Mark McMenamin, translated by David Langmuir) Copernicus

Victor, David G. (2008) 'On the Regulation of Geoengineering' *Oxford Review of Economic Policy* **24** 322–326

Victor, David G. (2011) *Global Warming Gridlock: Creating More Effective Strategies for Protecting the Planet* Cambridge University Press

Vince, Gaia (2014) *Adventures in the Anthropocene* Chatto & Windus

Von Neumann, John (1955) 'Can We Survive Technology' *Fortune* June 1955

Vogt, William (1948) *Road to Survival* William Sloane Associates

Vonnegut, Kurt (1963) *Cat's Cradle* Gollancz

Vonnegut, Kurt (1977) 'The Art of Fiction No. 64' *Paris Review* Spring 1977

Wagner, Gernot and Weitzman, Martin L. (2015) *Climate Shock: The Economic Consequences of a Hotter Planet* Princeton University Press

Walker, J. Samuel (2006) *Three Mile Island: A Nuclear Crisis in Historical Perspective* (revised edition), University of California Press

Weart, Spencer R. (1988) *The Rise of Nuclear Fear* Harvard University Press

Weart, Spencer R. (2008) *The Discovery of Global Warming* (revised and expanded edition), Harvard University Press – further expanded online at http://www.aip.org/history/climate/index.htm

Weisenstein, Debra K. and Keith, David W. (2015) 'Solar Geoengineering Using Solid Aerosol in the Stratosphere' *Atmospheric Chemistry and Physics Discussions* **15** 1179–1185

Weitzman, Martin L. (2012) 'A Voting Architecture for the Governance of Free-Driver Externalities, with Application to Geoengineering' NBER Working Paper 18622

Wells, Herbert George (1898) *The War of the Worlds* Heinemann

Wells, Herbert George (1914) *The World Set Free: A Story of Mankind* Macmillan

Westbroek, Peter (1992) *Life as a Geological Force: Dynamics of the Earth* W. W. Norton

Wigley, Tom (2006) 'A Combined Mitigation/Geoengineering Approach to Climate Stabilization' *Science* **314** 452–454

Wilcox, Jennifer (2012) *Carbon Capture*, Springer

Williams, Rosalind (2013) *The Triumph of Human Empire: Verne, Morris and Stevenson at the End of the World* University of Chicago Press

Williamson, Phillip et al. (2012) 'Ocean Fertilization for Geoengineering: A Review of Effectiveness, Environmental Impacts and Emerging Governance' *Process Safety and Environmental Protection* **90** 475–488

Winebrake, James J. et al. (2009) 'Mitigating the Health Impacts of Pollution from Oceangoing Shipping: An Assessment of Low-Sulfur Fuel Mandates' *Environmental Science and Technology* **43** 5592–5598

Woolf, Dominic et al. (2010) 'Sustainable Biochar to Mitigate Global Climate Change' *Nature Communications* 2010 doi: 10.1038/ncomms1053

Many of the researchers mentioned in the text – for example, Ken Caldeira, David Keith and Alan Robock – maintain full online archives of their publications. Some other useful websites include:

The Oxford Geoengineering Governance Program – http://www.geoengineering-governance-research.org/cgg-working-papers.php

Climate Engineering, hosted by the Kiel Earth Institute – http://www.climate-engineering.eu/home-35.html

Geoengineering Our Climate: A Working Paper Series on the Ethics, Politics and Governance of Climate Engineering – http://geoengineeringourclimate.com/

The Berlin Climate Engineering Conference (CEC14) – http://www.ce-conference.org/

The Forum for Climate Engineering Assessment – http://dcgeoconsortium.org/

Geoengineering Monitor – http://www.geoengineeringmonitor.org/about/

A constant flow of information and discussion is available at https://groups.google.com/forum/#!forum/geoengineering

My own blog is at http://heliophage.wordpress.com

Index

Abbot, Charles, 49

Ackerman, Thomas, 307–8, 323

adaptation and mitigation: costs, 106; future scenarios, 355–7, 359, 360–1, 362–4, 366, 370; as only solution, 29; reasons for neglect, 146–7; relationship with geoengineering, 158, 159–60, 161–2, 163, 165, 347, 355–7, 359, 362–4; and two-degree limit, 165

AEC *see* Atomic Energy Commission

aerosols: cooling effects, 63, 73, 84–5, 89–99, 275–80; and fossil fuels, 12; future scenarios, 370; and health, 281; and local climate modification, 297; and ozone layer, 93–4; and rainfall, 95–6; from sulphur, 275; warming effects, 73; *see also* veilmaking

afforestation *see* trees

Africa: aerosols over, 297; and colonialism, 178; early humans in, 230; and global warming, 116; Kilimanjaro's glaciers, 344–5; potential geoengineering effects, 121–2, 371; prehistoric climate, 241–2; *see also* Sahara Desert; Sahel; South Africa

agreements *see* air pollution: agreements; climate negotiations and agreements; nuclear weapons: treaties and test bans; UNFCCC

agriculture and farming: and algal blooms, 196; and asteroid strikes, 329; and carbon dioxide, 236–40; and climate change, 72, 237–8; energy intensity, 193–4; EU subsidies, 208; genetic modification, 289–90; and greenhouse gases, 224–5, 227; increasing crop reflectivity, 289–90; machinery, 184; and nuclear winter, 307; organic,

199–200; origins, 230–1, 234–5; separation of livestock from arable, 205; unintended fertilization, 199–200; *see also* fertilizers, synthetic; plant growth

Agung, Mount, 90, 98

air conditioning, 284

air pollution: in Asia, 280, 297, 365; international agreements, 208, 282–3; and nitrogen, 199; and sulphur, 274–5, 281–3

aircraft: B-52 bombers, 42; F-15, 104; Lockheed U-2, 42–3, 44, 45, 53; and ozone layer, 51–2; supersonic, 358; and veilmaking, 101–6, 352, 368

airships, hybrid, 106

albedo: and averting ice ages, 278; definition, 71–2; future scenarios, 282; and ice, 73, 231; ice–albedo feedback, 223, 276, 278, 342–3; reduced by mammoths, 373; and trees, 130, 260; and volcanic eruptions, 85

algal blooms, 95, 195–6, 253–4, 256

Algeria, 241–2

Alliance of Small Island States, 355

Alps, 371

Alvarez, Luis, 321–2, 327–8

Alvarez, Walter, 321–2, 327–8

American Association for the Advancement of Science, 138

Americas, effect of European 'discovery', 227–9

Ames Research Center, 156, 307

ammonia, 176, 199; *see also* nitrogen

Anderson, James, 53, 169

Andes, 371

animals: biodiversity and global warming, 257; cattle, 224; and European

415

'discovery' of Americas, 227–8;
extinctions, 25, 321–2, 328; livestock
farming, 205; polar bears, 95; rewilding,
373
Antarctica: ice as record of earlier climates,
222, 227, 321; ozone layer over, 53, 93,
110, 111; protecting the icecap, 372; *see
also* Southern Ocean
Anthropocene: attitudes to, 219–21, 225–
9; definition, 25–6; and greenhouse
gases, 222–9; origins of term, 52; start
date, 44–5, 225–9
anthropology, 129
Apollo programme, 60, 62, 77, 212–13;
Apollo 8, 60, 213
Archer, David, 267
Archimedes, 81
Arctic: climate research projects, 138;
future scenarios, 342; ice, 313, 362;
methane release, 241; and next ice age,
278; ozone layer over, 110; temperature
records, 108; and weather in temperate
zones, 316
Argentina, 16
arms-control agreements, 144
Armstrong, Louis, 375
Arrhenius, Svante, 88, 130–1, 243
'asteroclique', 333–4, 335–6
asteroids and comets, 139, 321–2, 327–40,
341
Aswan Dam, 198
Atacama Desert, 180
atmosphere: effects and properties, 38–40,
65–73; etymology, 40; layers, 41; size,
375; *see also* stratosphere; troposphere;
veilmaking
Atomic Energy Commission (AEC), 45,
316
Aurora Flight Sciences, 101–5
Australia, 271

Bachelard, Gaston, 35
back scattering *see* scattering
Bacon, Francis, 24–5
Bala, Govindasamy, 151, 292
Bali, 90
balloons, 105–6
Barrett, Scott, 106–7
Baruch, Bernard, 185, 188, 314
BASF, 182, 190

BECCS *see* biomass energy with CCS
Bennett, Hugh, 185
Bezos, Jeff, 353
biochar, 264
biodiversity, 257–8
biofuel, 263
biomass, 18, 263; *see also* renewable energy
biomass energy with CCS (BECCS),
262–4, 265
biotechnology, 289
Birks, John, 306–7
Blair, Tony, 144
Blue Origin, 353
Bolivia, 180
Borgstrom, Georg, 228
Borlaug, Norman, 189, 190–1, 192, 197
Bort, Léon Teisserenc de, 39–40
Bosch, Carl, 182, 190, 193
Boserup, Ester, 203
Boulton, Matthew, 226
Box, George, 71
Branson, Richard, 28, 353
Brazil, 16, 294
Brians, Paul, 309–10
Bristol, University of, 261, 289
Britain *see* UK
British Association for the Advancement
of Science, 178, 181, 184
Broecker, Wally, 231–2
Bronfman, Edgar, 28
Brown, George, 330
Brown, Harrison, 239–40, 243, 246
Bryson, Reid, 275–6
Budyko, Mikhail, 138, 276, 277
Buffon, Georges-Louis Leclerc, Comte
de, 124
Burke, Edmund, 41
Bush, George W., 159
Byers, Horace, 269
Byron, George Gordon, Lord, 332

cabin ecology, 75–8
Caldeira, Ken: background, 150; funds
source, 156–7; and Gates, 102, 156–7;
and geoengineering, 149–52, 157, 160,
238, 240, 286
Calgary, University of, 28, 101–2
Calgary direct-air conference (2012), 1,
23, 27–9, 249
California: energy prospects of, 20; 1970s

drought, 319–20
Callendar, Guy, 243, 267
Caltech, 239–40
Canada, 184
cancer, 58–9
Cancún summit (2010), 165
capitalism, 225–6, 228, 310; see also
    industrialization
carbon: carbon cycle, 216–19; carbon
    cycle and humankind, 221–9; fixing/
    reduction, 216–17
carbon capture and sequestration/storage
    (CCS), 246–9
carbon dating, 215–16, 226
carbon dioxide: atmospheric
    concentration, 245; BECCS, 262–4,
    265; and carbon cycle, 216; carbon
    markets, 144; CCS, 246–9; centrality to
    climate change politics, 141–2, 143–7;
    and climate change, 65, 72–3; direct-
    air capture, 22–9, 245–6, 249, 265;
    early geoengineering schemes, 137–8;
    early research, 75–6, 88; emissions
    reduction, 1, 3–4, 8–22, 145, 264–5;
    historical atmospheric levels, 222–9;
    increase in emissions statistics, 2; and
    the oceans, 152–4, 249–59, 265, 362,
    363–4, 371; and plant growth, 96, 117,
    232–42, 259–61; reduction/removal
    technologies, 243–67, 369; relationship
    between emissions reduction/
    mitigation and veilmaking, 347–52,
    355–7, 359, 360–1, 363–4
Carbon Engineering, 28, 101–2
carbonates, 233
Carnegie Institute, 102
Carson, Johnny, 324
Carson, Rachel, 58, 277
Carter, Jimmy, 132
cattle, 224
CCS see carbon capture and
    sequestration/storage
CFCs see chlorofluorocarbons
Chapman, Clark, 328
Chapman, Sydney, 48, 50
Charlson, Robert, 282
Charney, Jule, 312
chemical warfare, 340–1
chemtrails and chemtrailers, 102–4, 137,
    351

Chernobyl, 15
Chicago, University of, 267
Chile, 180, 189–90
China: afforestation, 260; agriculture,
    230; air pollution, 280; carbon dioxide
    emissions, 21–2, 143; Chinese guano
    workers in South America, 180; effects
    of volcanic eruptions, 86; fertilizer
    industry, 193; future scenarios, 342;
    industrialization, 229; population issues,
    187; potential for cloud-brightening
    experiments, 296–7; rainmaking
    schemes, 271; US cold war attempts
    to minimize influence, 192; and
    veilmaking models, 121; weapons sales
    to Saudi Arabia, 354
China Syndrome, 14
chlorofluorocarbons (CFCs), 52–3, 72, 93,
    109–10, 143–4, 287
Christo, 344–5
CIA, 354
cities: age of, 230; and albedo, 72; cooling,
    289; growth, 177; low-lying, 371; and
    nuclear weapons, 306; and pollution,
    128
Clarke, Arthur C., 63, 266
Clementine mission, 334
climate, worldwide interconnectivity,
    293–4
climate change: and agriculture, 237–8;
    cooling effects of sulphates, 84–5,
    89–93, 94–5; difficulty of rectifying by
    switch to renewables and nuclear, 1,
    3–4, 8–22; first discovery, 43–4; harm
    caused by gases other than carbon
    dioxide, 146; lack of uniformity,
    94; politics, 139–47; rate, 68; and
    reflectivity of Earth's surface, 71–2;
    reliability of predictions, 68–71; risks
    and responsibilities, 1–3, 5–8; tipping
    points notion, 231–2, 240–1; two-
    degree limit, 165–6; and volcanoes,
    83–99; workings of, 65–80; see also ice
    ages; ozone layer
climate geoengineering: and achieving
    the two-degree limit, 166; argument
    from aesthetics, 337–8, 372; attitudes
    to, 124–5, 154–5, 261–2, 308, 311–12,
    344–6; breathing-space approach, 162–
    3; case for, 22–7, 29–31; challenges of,

81–2; as control system, 170–1; current
debate, 152–65; definition, 26 ; as
doomsday device, 342–3; and emissions
reduction/mitigation, 347–52, 355–7,
359, 360–1, 363–4; feasibility of secret
execution, 354; first scientific proposals,
137–9; future of, 166–72, 335; future
scenarios, 347–75; as gradual process,
359–61; historical methods and
attitudes, 126–47; as hope for future,
375–8; in-case-of-emergency approach,
160–2; international governance, 256,
357–9, 374; and international relations,
364–8; moral hazard issues, 158–62,
335, 361–4; nuclear risk's effect on
attitudes to, 308, 311–12; and nuclear
weapons, 312–18, 319–20, 340; origins,
308; parallels with asteroid impact
work, 332–8; reasons currently ignored,
141–7; weaponizing, 341–2; winners
and losers, 120–2, 164–5
climate negotiations and agreements,
21, 140–1, 143–4, 355–6, 359; *see also*
UNFCCC
climateprediction.net, 120
Climatic Change (journal), 137, 153
Clinton, Bill, 144
clouds: brightening, 273–4, 279–88,
292–8, 323, 360; calculating droplets
in, 298–9; as complicating factors
in climate change prediction, 72–3;
and cooling the Earth, 273–4, 279;
dominating view of Earth from space,
63; increased scientific interest, 268–72;
lightning generation, 299–300; and
ozone layer, 93; and rainmaking,
268–71; seeding, 268–73; and sunsets,
268; veilmaking's effect, 111
cloudships, 283–5
coal: current usage, 8; future scenarios,
369; and industrialization, 229;
sustainability as fuel source, 180–1;
twentieth-century abundance, 211,
212; *see also* fossil fuels
Cohen, Leonard, 148
cold war, 192, 326; *see also* nuclear
weapons
colonialism, 177–8, 228–9
Columbia University, 27–8, 106–7
Comer, Gary, 27–8

comets *see* asteroids and comets
Commoner, Barry, 47, 276, 277
communism, US fight against, 192
Competitive Enterprise Institute, 232
Conrad, Kevin, 355
consumerism, 23
continental drift, 47–8
Conway, Erik, 365
Copenhagen climate summit (2009),
10–11, 21, 146; accord, 165
coral reefs: creation, 233; and fertilizer
run-off, 198; future scenarios, 362,
363–4; and heat, 292; and ocean
acidification, 152; protecting, 251, 292,
294–5; and volcanic eruptions, 95
counter-geoengineering, 341–2
Cox, Peter, 240
Crookes, Sir William, 178–84, 194, 201–2,
210
Crutzen, Paul: and the Anthropocene, 52,
153, 219, 220; and effects of nuclear
war, 305–7; and geoengineering,
152–6, 280, 286; and nox, 52; ozone
layer studies, 152–3, 201
cryosphere, 40–1
Cullather, Nick, 191

Dai, Aiguo, 96
Dawson, Terry, 330
DDT, 47, 58
deforestation *see* trees
deserts, 241–2, 368–9, 371; *see also*
individual deserts by name
developing countries: and adaptation,
147; and carbon-dioxide fertilization,
198, 236; and carbon emissions, 5–6,
143, 145; and climate negotiations,
140–1, 143–4; energy use, 9–11; future
scenarios, 347–75; and geoengineering,
115–16; population, 10; and synthetic
fertilizers, 192, 193
difluoromethane, 341
dinosaurs, 321–2, 328
direct-air capture, 22–9, 245–6, 249, 265
Dolan, David Scott, 132
Dominic (author's friend), 300
Draper, General William, 188
drones, 368
drought, 95–6, 237, 293, 319–20, 368–9,
371; *see also* water supply

dry ice, 268–9
Dubner, Stephen, 154–5
Dyson, Freeman, 137–8, 259–60

Earth: age, 215–16; appearance from space,
    60, 63, 65; end of, 214–15; historical
    changes in perception of, 57–60;
    human relationship with, 24–6, 78–80,
    219–21; isolation, 62; non-fragility, 61;
    saving the planet, 61–2; size, 374–5;
    surface reflection of sunlight *see* albedo
    effect; *see also* Anthropocene
earthquakes, 83
earthsystem: definition, 24; human
    relationship with, 24–6, 78–80,
    219–21; workings of, 57–80; *see also*
    Anthropocene
economic issues: constant progress, 29;
    effect of synthetic fertilizers, 205
ecosystems: etymology, 74; geoengineering
    and, 257; workings of, 76–7, 81
Edwards, Paul, 80, 319
Egypt, 198
Ehrlich, Anne, 188–9
Ehrlich, Paul, 188–9, 307–8
Einstein, Albert, 185, 314
Eisenberger, Peter, 28
Eisenhower, Dwight D., 167
Ekholm, Nils, 243–4
El Chichón, 90–1, 92, 98
El Niño events, 70, 91, 140, 294, 362
electricity: and CCS, 246; and fertilizers,
    182; generating from renewables, 18;
    Massachusetts use, 263; twentieth-
    century usage and technology, 212; *see
    also* fossil fuels
electron-capture detector, 287
Eliot, T. S., 1
Ellis, Erle, 225
Emissions Trading System, 144
energy: from sunlight, 62–71; measures,
    12; rising needs, 9–10; sources of world
    supply, 211–12
energy transitions: drivers, 13; time taken,
    8–12
ENIAC, 312
ENMOD *see* UN Convention on
    the Prohibition of Military or Any
    Other Hostile Use of Environmental
    Modification

environmental warfare, 135–7, 334–5,
    339–41
environmentalism: changes in, 57–60;
    and carbon politics, 142; and nuclear
    energy, 14–16; and renewable energy,
    11, 17–19; and CCS, 247–48; opposing
    the Concert, 351, 'saving the planet',
    61–2
Erice, 327
Espy, James, 269
ETC Group, 23, 255, 257, 369
Etzler, John Adolphus, 134
eugenics, 129, 186, 188
evolution, 75, 78–9
exajoules, definition, 211
Exeter, University of, 240
exosphere, 41
experiment: earthsystem as subject of,
    42–44, 91, 137, 253–54, 340; case for in
    geoengineering research, 169, 288
explosives, 189–90, 193
extinctions, 25, 321–2, 328
extraterrestrials, 324

farming *see* agriculture and farming
fertilizers, synthetic: environmental
    problems caused by, 195–201;
    mitigating problems, 206–8; overview,
    120, 175–6, 178–84, 190–201;
    unintended fertilization, 199–200;
    world without, 201–6
First World War (1914–18), 190
fishing industry, 371
Fisk, Dorothy, 49–50
Fleming, James, 157–8
Flettner rotors, 285
The Flobots, 305
floods, 116
food: energy intensity of production,
    194; feeding the world, 185–6, 191–2,
    194–5; feeding the world without
    synthetic fertilizers, 201–6; gained
    from trade and colonialism, 228; *see
    also* agriculture and farming; fertilizers,
    synthetic
forests *see* trees
forward scattering *see* scattering
fossil fuels: absolute usage statistics, 8;
    and algal blooms, 196; costs and
    availability, 12, 18–19, 20–1; difficulty

of replacing infrastructure, 8–9; future
usage scenarios, 369; health risk, 12,
16; as percentage of world's energy
used, 3; power stations and CCS, 246;
production, 216–19; twentieth-century
usage and technology changes, 211–13;
see also coal; natural gas; oil
Fourier, Charles, 134
fracking, 247
France, 11, 15, 127–8
Frank, Pat, 311
Frankenstein (literary character), 171
Franklin, Benjamin, 127
Freiburg, University of, 182
Fressoz, Jean-Baptiste, 129
Freud, Sigmund, 74
Fukushima, 15–16, 17
Fuller, R. Buckminster, 76

Gagarin, Yuri, 57
Gaia hypothesis, 75, 290–1
Gardiner, Stephen, 161
gas, *see* natural gas
Gates, Bill, 28, 102, 155, 156–7, 353
Gates, Melinda, 353
Gaud, William, 192
GCMs, 314, 317
Gellner, Ernest, 211
General Electric research laboratory,
    268–70
genetic modification, 289–90
'geoclique', 157–8, 163, 286
geoengineering, climate *see* climate
    geoengineering
geology, 40, 321–2
GeoMIP, 113–20, 122, 158
George, Russ, 254–5, 256
George C. Marshall Institute, 154
germ theory of disease, 129
Germany: explosives industry, 190;
    geoengineering research, 159; nuclear
    industry, 17, 358; renewable energy
    (*Energiewende*), 19, 20, 106, 159 ;
    scientific research, 182
Gernsback, Hugo, 243
glaciers and ice: Arctic melting, 313, 362;
    and cloud brightening, 294–5, 336;
    and nuclear fallout, 44; protecting,
    344–5, 371–2, 374; as record of earlier
    climates, 222–3, 227, 321, 344; and

tracking climate change, 222–7; and
    volcanic eruptions, 86, 88
global cooling, 275–9
global warming: and counter-
    geoengineering, 341–2; 'pause' in, 3,
    70, 108, 280; sulphur's masking effect,
    279–80; *see also* climate change
Goddard Institute for Space Studies, 276
Goldsmith, Oliver, 83
Goodell, Jeff, 157
Gore, Al, 349
GPS, 118–19
Gran, Haaken Hasberg, 252
green movement: and carbon dioxide
    emissions, 141, 143; and CCS,
    247; future scenarios, 351; and
    geoengineering, 28, 159, 261–2;
    influence on environmental policies,
    19–20, 141; moderate green views
    on climate change, 135; and nuclear
    power, 16–17
'Greenfinger' scenario, 352–4
greenhouse gases: and climate change,
    65–71, 72–3; and farming, 224–5,
    227; harm caused by those other
    than carbon dioxide, 146; historical
    atmospheric levels, 222–8; and ice ages,
    231; *see also* carbon dioxide; methane;
    nitrous oxide; water vapour
Greenland, 222, 342, 362, 371, 374
Grübler, Arnulf, 11
guano, 180

Haber, Fritz, 182, 190, 193, 202
Hadley Centre, 273
hail, 271
Hamblin, Jacob Darwin, 136, 309
Hamilton, Clive, 157, 248
Hampson, John, 278
Hansen, James, 90–2, 140, 276
Hardin, Garrett, 77–8
Harvard Forest, 97–8
Harvard University, 28
Havel, Václav, 351
Haywood, Jim, 293
HCFCs, 72, 146
health: effect of European 'discovery' of
    Americas on Native Americans, 227;
    and fossil fuels, 12, 16; germ theory
    of disease, 129; nitrogen pollution

of water, 195–9; and nuclear power, 15–16, 45; and ozone layer, 49–50; vaccination programmes, 353; and veilmaking, 112, 281; *see also* air pollution

Heard, Gerald, 41, 342

Hebrew University of Jerusalem, 296

Helin, Glo, 328

helium, 178

Hiroshima, 148, 190

holism, 77

Holmes, Arthur, 216

Holocene, 222–4, 226, 231, 236, 241

Hoskins, Brian, 69

House, Jo, 261

Hoyle, Fred, 278

human empire, 24–5, 125, 177–78, 209–10, 372

human prehistory, 229–31, 241–2

Hungary, 314

hurricanes, 284, 294–5, 295–6, 353

Huxley, Aldous, 41

Huxley, Julian, 313–14

Hyde, Roderick, 149, 151

hydrological cycle: future scenarios, 242, 362; and veilmaking, 114–18; workings of, 64, 67

hydropower, 3, 182

hydrosphere, 40

ice *see* glaciers and ice

ice ages: 1960s and 1970s fear of human-generated, 275–8; artificially starting, 342–3; averting, 149, 278; and carbon dioxide in the oceans, 252–3, 254; and climate change, 231; as climate change phenomenon, 130; and greenhouse gases, 222–4; and human development, 230–1; next, 266–7, 277–8; enduring question of origins, 87–8, 98; and plant growth, 233–4; Younger Dryas, 226–7

ice–albedo feedback, 223, 276, 278, 342–3

IG Farben, 190

IMO *see* International Maritime Organization

India: agriculture, 192; air pollution, 365; future scenarios, 364–6, 367–8; monsoons, 86, 364–6; population issues, 187; rainmaking schemes, 271; and veilmaking models, 121

Indonesia, 86–7

industrialization, 128, 177, 225–6, 228–9

infrared radiation, 65–6

Ingold, Tim, 57

interglacials, 222–4

Intergovernmental Panel on Climate Change (IPCC), 7, 140

internal combustion engine, 212

international agreements *see* air pollution: agreements; climate negotiations and agreements; nuclear weapons: treaties and test bans; UNFCCC

International Energy Agency, 3

International Institute for Applied Systems Analysis, 246

International Maritime Organization (IMO), 282–3, 297–8

interstellar travel, 139, 150

Intertropical Convergence Zone, 293

IPCC *see* Intergovernmental Panel on Climate Change

Ireland, 127

iron fertilization, 252–9, 265

Israel, 16

James, William, 132

Jameson, Fredric, 310

JASON group, 136, 321

Jeanne-Claude, 344

Jefferson, Thomas, 127

Jesus Christ, 125

jet streams, 46–7

Jevons, Stanley, 180–1

Johnson, Lyndon B., 137, 139

Johnston, Harold, 51, 201

Jupiter, 37, 333

Kaempffert, Waldemar, 49, 314

Kármán, Theodore von, 136

Keeling, David, 75–6, 96, 98, 239–40

Keith, David: background, 150; death threats, 104; funds source, 28, 102, 156–7; and geoengineering, 101–2, 107, 149–50, 156–7, 160, 169, 286, 342, 358

Kennedy, John F., 59, 340

Kilimanjaro, Mount, 344–5

Kingsland, Sharon, 79

Kintisch, Eli, 157

Klein, Naomi, 225

Koch, Robert, 129
Krakatau, 86–7, 108
Kravitz, Ben, 113, 116–17
Kruger, Tim, 163
Kyoto conference (1997), 3
Kyoto protocol (2005), 140–1, 144, 145

Lackner, Klaus, 27–8
Langmuir, Irving, 269–70, 272, 295
Latham, John: career, 272–3, 283; cloud
    work, 268, 272–4, 283–4, 285–8,
    294–5, 298–301, 323; home, 298
Latham, Mike, 268, 300
Latour, Bruno, 171, 271
Lawrence Livermore National Laboratory,
    148–51, 317, 319, 334, 339
Le Châtelier, Henri Louis, 182
Leith, Chuck, 317
Lenton, Tim, 290
Lesseps, Ferdinand de, 128
Levenson, Tom, 324
LeVier, Tony, 57
Levitt, Stephen, 154–5
Lewis, Simon, 227
Libby, Willard, 45–6
Liebig, Justus von, 178–9, 237, 251–2
lightning, 272, 299–301
lightning conductors, 112, 127
lime and liming, 250–1, 363, 371
lithosphere, 40
Livermore *see* Lawrence Livermore
    National Laboratory
Lobell, David, 236, 237, 238, 240
Locher, Fabien, 129
Long, Jane, 20
Lotka, Alfred, 75, 78–9, 175, 217–19
Lovell, Jim, 60
Lovelock, James, 75, 275, 278, 282, 287,
    290
Lowell, Percival, 131, 132, 133, 139

McCarthy, Cormac, 309
MacCracken, Mike, 319, 327
MacCready, Paul, 299–300
MacDonald, Gordon, 136–7
McKibben, Bill, 125
Maddox, John, 204
Malthus, Thomas, 180, 185–6, 203
Manhattan Project, 42, 148, 312, 321
Marchetti, Cesare, 137, 246, 259

Mariner 9, 89
Mars: canals, 131–3; colonizing, 139, 140;
    craters, 322; expensive village on, 374;
    stratosphere, 37, 89
Martin, John, 252–4
Marx, Karl, 179–80, 205
Maryland, University of, 225
Masco, Joe, 310
Maslin, Mark, 227
Mauritius, 127
Mead, Margaret, 327
measles, 227
Mediterranean region, 116, 198, 230, 241,
    375
Medwin, Thomas, 332
mending; 359, 372
mesosphere, 41
Meteorological Office, 293, 294
methane: and climate change, 65, 72; and
    farming, 224–5; historical atmospheric
    levels, 223; human responsibility for
    emissions, 72, 146; positive feedback
    due to, 241; *see also* natural gas
Mexico, 90–1, 189, 190–1, 192
Mexico, Gulf of, 186, 195–6
Middle East, 284–5
the military, and asteroid impact work,
    334–5, 339–41; and cabin ecology, 75;
    and geoengineering, 315; and climate
    modification. 158, 270; and cloud
    seeding. 270, 272; and geophysical
    warfare, 135–7; and nuclear energy, 16;
    and nuclear weapons, 42, 306, 308–9
mirrors, space-based, 149, 150–1
Mitchell, Edgar, 77
mitigation *see* adaptation and mitigation
monsoons: future scenarios, 364–6,
    367–8, 371; and geoengineering,
    292; prehistoric, 241; and volcanic
    eruptions, 86, 115
Montreal protocol (1987), 53, 110, 143–4
moon: *Clementine* mission to, 334;
    craters, 322; Earth seen from, 60, 63,
    65; planned human moonbases, 75;
    planned nuclear explosion on, 338,
    339; appearance changed by volcanic
    eruptions on Earth, 86
Mooney, Pat, 23
More, Sir Thomas, 124, 127
Morrison, David, 328, 330–1

Mossop, S. C., 90
Munch, Edvard, 86
Musk, Elon, 353
Myhrvold, Nathan, 155

Nagasaki, 148, 190
Namibia, 294
NASA: Ames Research-Center, 156,
    307; asteroid-impact research, 328,
    330; climate studies, 90; ozone-layer
    research, 53; telescope lifting balloons,
    106
National Academies of Science, 138
National Center for Atmospheric
    Research (NCAR), 66, 283, 325–6
nationalism, rise of, 211
Native Americans, 188, 227
natural gas, 8, 11, 50, 211–12; see also fossil
    fuels
naturalism schism, 363, 372–3
nature: attitudes to the era of climate
    change, 61, 135–47, 372–3;
    geoengineering's relationship with,
    125, 139–47; as hierarchy, 77; historical
    attitudes to, 41, 101, 125–35; human
    relationship with, 77, 79, 80, 220, 311,
    337; see also Anthropocene; earthsystem
NCAR see National Center for
    Atmospheric Research
negotiations see climate negotiations and
    agreements; UNFCCC
Netherlands, 184
Neukermans, Armand, 287–8, 323
New York University, Department of
    Earth Sciences, 150
Niagara Falls, 64
Nisbet, Euan, 344
nitrogen: denitrification, 176, 196–7, 207;
    environmental problems caused by,
    195–201; and explosives, 189–90; and
    fertilizers, 175–6, 178–84, 190–201;
    fixing, 176, 197, 200; mitigating
    problems, 206–8; reactive, 197–201;
    science of, 176–7; world without
    nitrate fertilizers, 201–6
nitrogen oxides (NOx), 50–2, 199, 201,
    306, 321
nitrous oxide, 146, 199, 200–1, 208, 254
Niven, Larry, 322–3
Nobel, Alfred, 190

Noble, William, 181–2
Normalised Difference Vegetation Index,
    235
North Korea, 16
Novim, 160
NOx see nitrogen oxides
nuclear energy: China, 22; and climate
    control, 312–18, 319–20, 340;
    compared to CCS, 247; discovery of
    radioactivity, 209–11; effects on the
    imagination, 213, 214–15; German
    nuclear industry, 17, 358; link to
    nuclear weapons, 16; and the military,
    16; origins of fear of, 14, 311; as
    percentage of world energy supply, 212;
    pros and cons, 13–17; as replacement
    for fossil fuels, 9, 11; risks assessed,
    15–16
nuclear weapons: and asteroid-impact
    risk avoidance, 331–2, 336–40;
    effect on climate, 305–8, 318–27;
    and environmental warfare, 339–41;
    excavation using, 339; fallout
    from, 43–7, 58, 226; fear of, 187–8;
    Hiroshima and Nagasaki, 148, 190;
    IBMs, 316; impotence in the face of
    drought, 319; link to nuclear energy
    programmes, 16; Manhattan Project,
    42, 148, 312, 321; moon detonation
    project, 338, 339; effect on climate
    change thinking, 305–27; nuclear
    submarines, 16, 308–9; nuclear winter,
    305–8, 320–7; Star Wars programme,
    156, 334; Teller's work, 148–9; treaties
    and test bans, 59, 144, 361
Nye, David, 101, 123

Obama, Barack, 144, 326–7
oceans: acidification from carbon dioxide,
    152–4, 251, 362, 363–4, 371; carbon
    in, 221, 222; and CCS, 246; and
    climate change, 151; cooling methods,
    292; currents, 64; fertilization as
    atmospheric carbon-reduction solution,
    249–59, 265; future scenarios, 362,
    363–4, 371; and levels of atmospheric
    carbon dioxide, 224–5, 260, 264–5;
    ocean deserts, 252; sulphur emissions at
    sea, 282–3; and veilmaking, 116
oil, 8, 211, 229, 248; see also fossil fuels

Oppenheimer, Robert, 148, 314–15
Oreskes, Naomi, 365
Osborn, Fairfield, 185–9, 191, 195
Ostwald, Wilhelm, 182
overworld, 45, 101
oxygen, 216–17
ozone: ozone levels and plant growth, 237; in troposphere, 199–200
ozone layer: Crutzen's work, 152–3, 201; damage to, 50–3, 152–3, 201, 232; description, 47–50, 60; and nuclear war, 306; and veilmaking, 109–11, 153, 370; and volcanic eruptions, 93–4; weaponizing, 136–7

Paine, Thomas, 127
Papua New Guinea, 355
Parson, Edward, 356
Pasteur, Louis, 129
Pauling, Linus, 190
pedosphere, 41
Peru, 180
pesticides, 47, 58, 192
Philippines, 83–6
photosynthesis: and carbon dioxide, 96, 117, 221–2, 233–4; and iron fertilization, 252–3; and light from below, 289; and diffuse light, 97–8, 238; workings of, 65, 216–17
Pierrehumbert, Ray, 119–20
Pinatubo, Mount, 83–6, 91–9, 107, 233
plankton, 233, 252–4, 275, 282; see also algal blooms
plant growth: and carbon dioxide, 96, 117, 232–42, 259–61; and climate change, 237–8; and nitrogen, 177, 197–8; and veilmaking, 111, 117; and volcanic eruptions, 96–8; see also fertilizers, synthetic
plants: BECCS, 262–4, 265; biodiversity and global warming, 257; and European 'discovery' of Americas, 227–8; genetic modification, 289–90; as method of reducing atmospheric carbon dioxide, 259–61
Plato, 127
polar bears, 95
polar ice see glaciers and ice
Pollack, James, 89–90, 307–8, 323
pollution: and cities, 128; nitrogen

pollution of water, 195–9; and shipping, 282–3, 297–8; see also air pollution; health
Pongratz, Julia, 238, 240
population: control of, 77, 185–9, 191–2; current and expected future, 9–10; feeding the world, 185–6, 191–2, 194–5; feeding the world without synthetic fertilizers, 201–6; and fertilizers, 183
Pournelle, Jerry, 322–3
power, measures of, 12
precipitation see rain and rainfall; snowmaking
Project Cirrus, 270, 295–6
'Promethean science', 112, 123, 127, 210, 270

radiation, instruments for measuring entry to earthsystem, 150
radiative forcings, 72
radioactivity: and carbon dating, 215–16; discovery, 209–11; effects on the imagination, 214–15; see also nuclear energy
rain and rainfall: and carbon dioxide, 242; future scenarios, 370; patterns during ice ages, 231; rainmaking schemes, 268–71, 295; and veilmaking, 114–18; and volcanic eruptions, 95–6; workings of, 64, 67
rain forests, 204
Rasch, Phil, 286
Rasool, S. Ichtiaque, 276–7
Read, Peter, 262
Red Sea, 95
renewable energy: in China, 22; costs, 12; development, 11–12; inadequacy as replacement for fossil fuels, 9; as percentage of world's energy used, 3; pros and cons, 17–20; subsidies, 19; see also biomass; solar power; wind power
respiration, 217, 221
Revelle, Roger, 43–4, 75, 137, 139, 276
Rhine River, 126
rice paddies, 224
Ricke, Katharine, 120–2
rickets, 49–50
Ridgwell, Andy, 289–90
Rio Earth Summit (1992), 99, 140
Robock, Alan, 113, 157

Rockefeller, John D. III, 191
Rockefeller Foundation, 189, 191–2
Rockström, Johan, 308
Roebling, John, 134
Roosevelt, Eleanor, 185
Rosenfeld, Daniel, 296
Royal Society, 87, 157, 160
Ruddiman, William, 223–5, 227
Ruderman, Mal, 321
Ruskin, John, 80
Russia see USSR
Rutgers University, 113
Rutherford, Ernest, 209–10, 215–16

Sagan, Carl, 89–90, 307–8, 323–7, 338–9
Sage, Rowan, 234–5, 236
Sahara Desert, 128, 241–2
Sahel, 293, 368–9, 371
St Helen, Mount, 90
St Helena, 127
Salter, Stephen, 283–6, 288, 294, 353
Sankey, Matthew, 74, 283
Sankey diagrams, 74, 75
Santer, Ben, 154
satellite-navigation systems, 118–19
Saturn, 37
Saudi Arabia, 354
scattering: back and forward, 97, 111; and
    cloud brightening, 273; description,
    62–3, 73; effects, 39, 62–3, 73, 97;
    future scenarios, 370; on Mars, 89; and
    veilmaking, 111, 149–50
Schaefer, Vincent, 269, 272
Schaffer, Simon, 112
Schellnhuber, Hans Joachim, 21
Schneider, Stephen: on Carson show,
    324; on Crutzen's influence, 154; as
    editor of Climatic Change, 137, 153; and
    geoengineering, 138, 276–7, 325–6;
    and Sagan, 324, 325–6
science fiction, 63, 272, 309, 322–3, 342
Scripps Institution of Oceanography, 75–6
sea levels, 2, 222, 252, 371, 372, 374
Second World War (1939–45), 187, 190
seismology, 41
Seitz, Russell, 291
Sen, Amartya, 365
Serres, Michel, 80
SETI, 324
Shackleton, RRS, 275

Shakespeare, William, 315
Shalizi, Cosma, 228
Shell, 278
Shelley, Mary, 171
shellfish shells, 233
Sherlock, Robert, 133
shipping, and pollution, 282–3, 297–8
Shoemaker, Gene, 322, 328, 333
Shoemaker-Levy 9 (comet), 333
Simon, Herbert A., 100
sky: brightness, 97; colour, 39, 62–3;
    veilmaking's effect on appearance, 111
Slingo, Tony, 273
smallpox, 227
Smil, Vaclav, 195, 212
smoke, 306–7, 319, 323
Smuts, Jan, 77
snowmaking, 270–1
sociology, 129
Socolow, Robert, 1, 28, 249
Soddy, Frederick, 209–11, 212–13,
    214–15, 230, 313–14, 318
soil erosion and conservation, 185–6, 264;
    see also fertilizers, synthetic
solar power: development of industry,
    11–12; and direct-air capture, 28;
    inadequacy as replacement for fossil
    fuels, 9; as percentage of world's energy
    used, 3; pros and cons, 17–18; and
    veilmaking and volcanic eruptions, 98,
    111; see also renewable energy
solar-radiation management see clouds:
    brightening; veilmaking
soot: and climate, 297; and clouds, 279;
    emissions at sea, 282; and nuclear war,
    307, 320; production, 274–5; self-
    lofting, 320, 342; warming effects, 73,
    146, 279, 307
Soufrière Hills, 100, 108
South Africa, 16, 271
South America, 230, 252, 254
South China Sea, 296, 297
Southern Ocean, 252–3, 255, 257
Soviet Union see USSR
space colonies, 339
space missions, 57, 60, 62, 77, 212–13, 353;
    interstellar, 139, 150
Spaceship Earth concept, 75–8
SpaceX, 353
Spain, 180

Spufford, Francis, 344
Stakman, E. C., 191
Stanford University, 236, 237
Star Wars programme, 156, 334
stars, twinkling effect, 39
steam power, 236
steam turbine technology, 212
Steel, Duncan, 332
strategic missile defence *see* Star Wars
    programme
stratosphere: and nuclear fallout, 43–7;
    overview, 35–56; and volcanic
    eruptions, 84–6, 89–99; *see also* ozone
    layer; veilmaking
Strauss, Lewis, 316
the sublime, 41, 101, 123, 300, 337–8, 345
Suess, Eduard, 44
Suess, Hans, 44
sulphur and sulphates: cooling effects,
    84–5, 89–99, 107, 275–80; cutting
    emissions, 282–3; and global warming,
    279–80; sulphur cycle, 274–5; and
    veilmaking, 107–12, 280–1
the sun, 97, 216
sunlight: and climate change, 65;
    effects, 62–5; historical changes in
    pattern, 222–3; and plant growth, 97;
    reflectivity by clouds, 273–4, 283–8,
    292; reflectivity by Earth's surface,
    71–2, 289–93; regional methods of
    reflecting, 289–93; and sunsets, 268;
    see also clouds: brightening; scattering;
    veilmaking
sunsets, 86, 268
sunshine geoengineering *see* clouds:
    brightening; veilmaking
supernovae, 316–17, 321–2
Sutton, Mark, 207–8

Tambora, Mount, 86, 93, 108, 122, 127–8
Tansley, Arthur, 74, 76–7, 81
Taubo, 93
Teller, Edward: and asteroid impact
    avoidance, 331, 334–5; background,
    136, 148; and environmental
    warfare, 136, 340; and GCMs, 317;
    life at Livermore, 148–9, 317; and
    nuclear explosives, 319–20; and
    nuclear weapons, 136, 148, 312; and
    veilmaking, 149, 151, 154

Texas, 248
Thailand, 271
thallium, 178
thermodynamics, 65–6, 74–5, 80, 245–6;
    second law, 214, 215
thermosphere, 41
thermostats, 129, 151, 164
Thomas, Jim, 369
Thompson, George, 134
Thoreau, Henry David, 134, 268
Three Mile Island accident (1979), 15
Titan (moon), 37
Titley, David, 116
Toba, 93–4
Toon, Brian, 89–90, 307–8, 323
trees: and carbon dioxide, 224–5, 233, 234,
    236, 259–61; and Earth's reflectivity of
    sunlight, 72, 130, 260
Trenberth, Kevin, 66, 96, 283
Trenberth diagram, 66–8, 73–4, 79
tropopause, 46, 111
troposphere, 46, 85, 199–200
tsunamis, 340
TTAPS, 307–8, 323–7
Turco, Richard, 307–8, 323, 326
Turner, Frederick, 177–8
Turner, J. M. W., 79–80
Twomey, Sean, 273, 277, 279
Twomey effect, 273–4, 282
Tyndall, John, 88

UK: carbon emissions, 11, 145; energy use,
    18; gains from colonialism, 228–9
ultraviolet radiation, 47–53
UN Convention on Biological Diversity,
    257
UN Convention on the Prohibition
    of Military or Any Other Hostile
    Use of Environmental Modification
    (ENMOD), 270
United Nations Framework Convention
    on Climate Change (UNFCCC), 99,
    140, 143–4, 145, 165–6, 355
University College London, 227
US Agency for International
    Development, 192
US Bipartisan Policy Center, 160
US National Research Council, 354
USA: agriculture, 184–5; anti-
    communism, 192; Brooklyn Bridge,

134; California drought (1976–77), 319–20; carbon market, 144; chemical warfare, 340–1; Dust Bowl, 185–6; Dyson's proposed afforestation scheme, 259, 260; emissions reduction, 11, 20; and environmental warfare, 334–5, 339–41; fertilizer use, 190–1, 193; future scenarios, 342; and global warming, 355; historical attitudes to climate, 127; hurricanes, 295–6; and Kyoto, 141; Massachusetts electricity use, 263; Native Americans, 188, 227; nuclear energy, 13, 15; nuclear weapons, 43, 45, 324; and ozone layer, 52; population control, 188; rainmaking schemes, 268–71; and Saudi weapons purchase from China, 354; Texan oil industry, 248

USSR: climate research, 138; and cold war, 314; environmental management projects, 133; nuclear weapons, 43, 319; US cold war attempts to minimize influence, 192

vaccination programmes, 353
Van Allen belts, 340
veilmaking: aerosol choice, 107–12; attitudes to, 281–3; background science, 275–81; chances of working, 335; costs, 102, 105, 106–7; counterengineering, 341–2; current work, 169; delivery methods, 101–7; as doomsday device, 342–3; early schemes, 138; effects overview, 112–18; future scenarios, 347–75; health risks, 112, 281; and international relations, 364–8; overview, 100–23; and plant growth, 238; pros and cons, 112–23; regional, 291–2, 297; termination shock, 118–20, 358, 362, 366; weaponizing, 341; winners and losers, 120–2, 164–5; Wood and Teller's work, 148–51; see also clouds: brightening
Venus, 89–90
Vernadsky, Vladimir, 74
Verne, Jules, 128
Victor, David, 353, 355
Vietnam War (1954–75), 137, 270
Virgin Galactic, 353
Virginia, University of, 223

vitamins, 49–50
Vogt, William, 185–9, 191, 195, 202
volcanoes: and geological carbon cycle, 233; and climate, 100, 107–8; effects of eruptions, 83–99; future scenarios, 367–8; and sulphur, 273
von Neumann, John, 312, 314–17
von Neumann machines, 316
Vonnegut, Bernie, 269, 272, 301
Vonnegut, Kurt, 272, 298, 301

Wanser, Kelly, 286–7, 288
warfare: caused by climate engineering, 364–7; chemical, 340–1; environmental, 135–7, 334–5, 339–41; using climate engineering, 341–2
water: and cloud brightening, 283–8, 292; increasing reflectivity of standing water, 291; nitrogen pollution, 195–9; plant water-efficiency, 238–9, 242; see also hydrological cycle
water supply: 1930s America, 185; future scenarios, 362; and global warming, 116; increasing to dry areas, 284–5; and veilmaking, 114–18
water vapour: and aircraft engines, 102; and climate, 72, 88, 92, 95, 111; and climate change, 65; and clouds, 72–3, 111; and geoengineering, 111, 115; and hydrological cycle, 67; and plants, 242; in stratosphere, 85
Watson, Andrew, 290
Watson, Matt, 100, 109–10
Watt, James, 226
weather forecasts, 312
Weaver, Warren, 191
Wegener, Alfred, 47–8
Wells, H. G., 131–2, 210, 211, 214, 272, 313–14
Westbroek, Peter, 209
wheat, 178–84, 189, 190–3
Wigley, Tom, 162
Wilber, Charles Dana, 127
wind power: and cloudships, 284–5; development of industry, 11–12; as percentage of world's energy used, 3; in Southern Ocean, 372; pros and cons, 17–18; wind farms' potential effect on climate, 72; see also renewable energy
winds, 64, 292, 372